Nature in Miniature

ILLUSTRATED BY THE AUTHOR

"*To him who in the love of Nature holds
Communion with her visible forms, she speaks
A various language.*"

WILLIAM CULLEN BRYANT

"*Thanatopsis*"

Richard Headstrom

Nature
in
Miniature

Alfred·A·Knopf·New York·1968

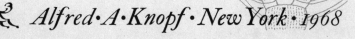

THIS IS A BORZOI BOOK
PUBLISHED BY ALFRED A. KNOPF, INC.

FIRST EDITION

© *Copyright 1968 by Richard Headstrom*

All rights reserved under International and Pan-American Copyright Conventions.
Distributed by Random House, Inc. Published simultaneously in Toronto,
Canada, by Random House of Canada Limited.

Library of Congress Catalog Card Number: 67-11122

Manufactured in the United States of America

To

MY WIFE

for her patience and understanding

ACKNOWLEDGMENTS

I want to take this opportunity to acknowledge my indebtedness to Mr. Angus Cameron of Alfred A. Knopf, Incorporated, who approached me with the idea of doing this book and whose suggestions and interest in the preparation of the manuscript have been most valuable. Also to my wife for her critical reading and typing of the manuscript.

Contents

Introduction

In recent years there has been a plethora of nature books—and more are being published. So why another one?

I do not claim that this book offers something unique, different, or new, though I would like to think so. But I do think it fills a need —the need to become better acquainted with the world in which we live. Not the world most people write about, but a world that is hidden from view; though this is not, perhaps, the way to put it. It is not exactly hidden from view, because it is revealed with the help of a piece of glass—not an ordinary piece of glass, but a piece of glass that has the property of magnifying whatever you may want to look at.

And so this book is designed to be the open sesame to a world of beauty and splendor, of mystery and the mysterious; an unbelievable, yet believable, world, one so fascinating as to stimulate one's imagination as never before.

Whatever your calling may be, whether a plumber, shoemaker, bookkeeper, secretary, telephone operator, or housewife, all you need is to be able to read, to have a curious nature, and to like being outdoors occasionally. And, of course, a magnifying glass which can be obtained for a few cents or a few dollars and, if you are interested enough, a microscope, which doesn't cost much these days.

You will be amazed at what you will see with one of these optical aids, and as you continue to examine one thing after another, you will gain an awareness of nature that should give a deeper meaning to life. You may also find a special sphere of interest as you continue to make further observations of the many wonders right at hand. Perhaps this book may even lead you to undertake investigations and explorations of your own, or it may provide you with a new hobby that offers an outlet for an increasing amount of leisure. Perhaps it will also serve as an escape, if only temporary, from the tensions and frustrations of modern living.

A Word About Equipment

Aside from a magnifying glass or hand lens, which is indispensable if the present book is to serve its purpose, and a microscope, which will be useful on a number of occasions, there are several other items of equipment that will prove helpful though not essential. One is an insect net for catching flying insects; another is a killing device that will render them immobile so they may be examined and studied at leisure. Both of these items may be purchased, or both may be made at home at a nominal cost. An insect net can be made from a piece of fine mesh or durable cloth and attached to a wire or metal ring, which in turn is fastened to a pole 2 or 3 feet long to serve as a handle.

A killing device is merely a jar containing a substance that will kill an insect placed in it. Professional entomologists use specially prepared killing jars containing cyanide, but one that is just as effective can be made from an ordinary jar and a substance far less dangerous than cyanide. A jar with a screw cover is preferable. A piece of absorbent cotton is attached to the lower surface of the cover and soaked with a fluid whose fumes will kill the insect. Several different fluids can be used, but carbon tetrachloride, which can be purchased at any drugstore, is the most serviceable, since it is not inflammable. One should not inhale too much of it, however.

For scooping up water insects and other forms of water life, a kitchen sieve or any other kind of scoop will serve, though professional entomologists use a water net. This can also be made at home. A water net is similar to an insect net, except that it is somewhat smaller and shallower. The wire ring supporting the net must be so rigid and so firmly attached to the wooden handle that it will not twist or bend when the net is pulled through the water, weeds, and muck.

A few assorted jars and bottles will come in handy for carrying animals home where they may be observed to better advantage. A few vials of alcohol might also be obtained for the small, fragile forms that must be preserved at once. A small aquarium will prove helpful in one or two instances. For further equipment a few razor blades, a pocket knife, some needles, a garden trowel, and a pair of tweezers or forceps for handling the animal forms and for the dissection of parts and structures will be useful. A pie plate can also be included in the list of equipment; it will serve as a background for field examination of certain forms whose beauty and luster quickly fade unless they are viewed when freshly taken from the water. And a carrying case or canvas knapsack is almost indispensable for carrying the equipment and for taking home specimens of both plant and animal life for more leisurely study.

A word in regard to the pocket lens. A pocket lens will vary in price according to its magnifying power. The least expensive one will serve in most cases, though obviously a more expensive one will give a higher magnification. A reading glass will prove useful in several cases, though it is not necessary. As for a microscope, the available funds will be the determining factor if a microscope is purchased at all. Space will not permit a discussion on the use of this instrument, but such information can be obtained from any number of books found in the local library. A microscope, however, is not essential except to those who would like to view plants and animals of microscopic dimensions; the present book will serve the purpose for which it was written without the instrument, though it must be added that for anyone who can afford one, it will be money well invested.

A list of supply houses is given in the appendix for anyone interested in purchasing equipment.

JANUARY

"And the streams which danced on the broken rocks,
Or sang to the leaning grass,
Shall bow again to their winter chain and
The mournful silence pass."

JOHN GREENLEAF WHITTIER

"The Frost Spirit"

JANUARY

NAMED AFTER JANUS, A GOD THE ROMANS held in the highest esteem, the month of January was the beginning of a new year on their calendar as it is on ours. But perhaps we are more fortunate in having the month come at midwinter, for in nature January is the nadir of the year—a period of repose or rest, akin to our night of sleep except that it is a longer night when nature rekindles her energies for a new year of activity. The leafless trees and shrubs have merely suspended operations and await a more favorable time to spring into renewed vigor and substance. The withered grasses and the faded flowers, although seemingly lifeless, are not altogether dead but live in countless seeds that have been scattered far and wide and need only the gentle rains of April and the warm rays of the spring sun to sprout and grow. The animals, too, except for those that brave the rigors of winter, have taken to retreats secure against low temperatures and swirling snow to lie dormant in one form or another until food again becomes abundant and they can resume a more active life. And the earth as well must have a time to recuperate and replenish its life-giving elements.

January may be wintry, or it may be mild and almost spring-like. I have known the month when the ground has been free of snow and the temperature so high that a light topcoat kept me warm; I have also known the month when there has been a deep cover of snow and the temperature has constantly flirted with the zero mark. I write of New England, where I have lived the greater part of my life; in the rest of the northern tier of states, the month follows fairly closely a pattern of blizzards and cold that is more like the traditional concept of winter. In the South, mild temperatures prevail for the most part, but here also nature can often be in a capricious mood: snow may fall and the temperature may plummet to well below freezing, with disastrous results to citrus groves and other crops and to trees and shrubs not hardened to such vagaries of the weather.

On the surface, the month does not seem to offer much of interest in the outdoors to those of us who live above the Mason-Dixon line, unless we belong to that breed of rugged individuals who are given to displaying their skill and prowess on frozen pond and snow-covered slope. But appearances are often quite deceptive, to use an overworked phrase; actually there is much for the nature lover to observe and study. There is much, too, for those who may not be aware of the satisfaction and inner peace to be gained from a contemplation of nature's handiwork.

January is not so entirely a month of desolation as it may appear to the unobservant and uninitiated, for along the roadside the purplish stems of the red osier give a bit of warmth to the wintry landscape, the scarlet pennants of the barberry gleam like lanterns in the gloom of tangled thickets, and along the woodland border the velvety red plumes of the sumac flash like flaming torches against the sky. In the rocky crevices of woodland hillsides the common polypody with its rich foliage softens the rugged outlines of the boulders among which it grows, and elsewhere the Christmas fern and the club mosses give a touch of summer cheer to the shadowy woodland floor. And as the beech's upright bole casts purple shadows on the snow and its polished brown stems describe an exquisite tracery against a backdrop of fluffy clouds, other trees form a varied array of silhouettes and assume a different character than when they are in full foliage.

We take our trees more or less for granted, and throughout the year we look upon them rather indifferently. Only if the trees beyond our doorstep were cut down and carted away during the night would we be conscious of something amiss. Trees are so commonplace—and for that reason, regrettably, unnoticed.

But pause a moment and select any tree, preferably one outlined against the sky and one that stands alone and is not hemmed in by others. Then select a second and a third, and perhaps scan the horizon for still others, and you will gradually discover that somehow they are all different and not the characterless objects we seem to think they are. One may be tall and slender and rise into the sky like a church steeple; another may be short and squat as if having been stunted by some blight; one may have ascending and gracefully outward-curving branches, another's limbs may be sharply diver-

FIGURE I

Sections of Buds

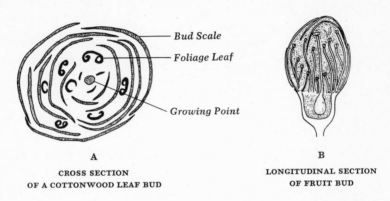

Bud Scale

Foliage Leaf

Growing Point

A
CROSS SECTION
OF A COTTONWOOD LEAF BUD

B
LONGITUDINAL SECTION
OF FRUIT BUD

gent; one may have horizontal branches, another lower pendant ones; one may taper broadly from the base to a point, and still another may have an egg-shaped head or crown.

Now, all these varied and distinct forms were not designed by man with his pruning shears, but by nature herself. It is only at this time of the year when, freed from its covering of leaves, a tree's architectural contours are revealed and its natural grace and charm are most clearly defined. Only at this time of the year does a tree acquire an identity of its own—something it is not quite able to do in complete foliage, when, as so often happens, it blends in with its neighbors.

But a tree, like everything else, is only the sum of its parts, and there are many other features of a tree that escape the attention of all but a few of us. Whenever we give or receive an unopened

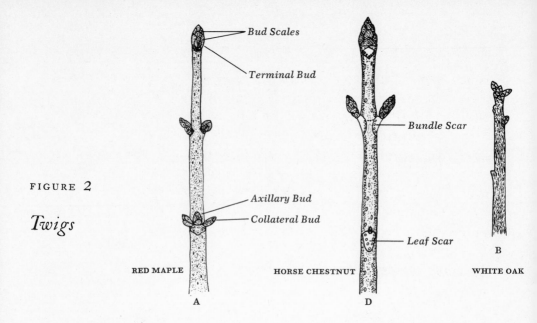

FIGURE *2*

Twigs

Bud Scales

Terminal Bud

Bundle Scar

Axillary Bud

Collateral Bud

Leaf Scar

B

RED MAPLE

HORSE CHESTNUT

WHITE OAK

A

D

flower as a token of esteem or affection, most of us glance only cursorily at the bud and, except for a fleeting thought, proceed to forget it until it has fulfilled its promise and become a floral delight. Indeed, when we think of buds at all, we generally regard them as unopened or unexpanded flowers; but this is only partially correct, for there are buds that produce stem growth and leaves. Many of the buds we find on the twigs and branches of trees and shrubs are rudimentary stems, consisting of a short length of partially developed stem with leaves in various stages of development. Remove a bud from any tree, cut it transversely with a knife or razor blade, then examine it with your hand lens, and you can see this for yourself. A bud of the cottonwood cut in this manner is shown in Figure 1A.

The buds of our woody plants are usually protected by several layers of overlapping scales called bud scales, which are really modified leaves. Bud scales are often covered with hair, as in the willow, and sometimes, as in the cottonwood, with a waxy secretion. Both are effective in protecting the enclosed tender structures from drying out and from mechanical injury. The buds of herbaceous plants and of tropical trees and shrubs do not need protection and hence lack scales. Such buds are said to be naked.

The buds on our trees and shrubs are formed during the summer and, because they persist on the twigs through the winter, are called winter buds. They may be leaf buds, flower buds, or mixed buds. Leaf buds contain a number of small or undeveloped leaves.

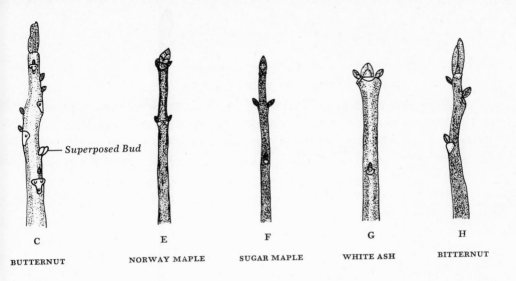

Superposed Bud

C	E	F	G	H
BUTTERNUT	NORWAY MAPLE	SUGAR MAPLE	WHITE ASH	BITTERNUT

Flower buds contain one or more miniature or undeveloped flowers, but no leaves. Such a bud may be found on fruit trees. For instance, if a bud from an apple or cherry tree is cut lengthwise, it will show miniature floral structures beneath the lens (Fig. 1B). You can even dissect out these structures with a needle if you are so disposed. Mixed buds (buckeye, oak), obviously, contain both undeveloped leaves and flowers. As a rule it is not possible to distinguish leaf buds from flower buds by merely looking at them (although in some cases, as in the elm, the flower buds are larger)—usually you have to dissect them.

A comparison of winter twigs from several different trees will show that the buds are not arranged in the same manner on all of them. Some twigs terminate in a bud called the terminal bud (Fig. 2A). Not all trees have a terminal bud; the catalpa, does not, for instance, and neither do oaks, which instead have a number of buds clustered at the ends of the twigs (Fig. 2B). The other buds found on a twig are termed lateral or axillary buds (Fig. 2A), since they develop in the leaf axils, which appear on a twig directly above the places where the leaves were attached. In some species, such as the horse chestnut, maples, and ashes, these buds are positioned opposite one another; in other species, as in poplars and hickories, they are arranged alternately on the twig. In addition to these lateral or axillary buds, others may be found. In maples such buds are formed alongside the lateral buds and are known as collateral buds (Fig.

2A); in ashes and in the butternut they are formed above the lateral buds and are known as superposed buds Fig. 2c). Then there are buds that may occur elsewhere on the twig, that is, any place except at the tip and leaf axils. They are the so-called adventitious buds.

A further comparison of the buds themselves will show considerable variation in size, shape, color, number of scales, and presence or absence of hairs. These variations may seem insignificant to the casual observer, but since they are constant for a species, they serve as identifying characters.

Consider, for instance, the buds of the horse chestnut, an Asiatic species which, since it is widely planted as an ornamental, especially in the Eastern States, should be available to most of us.[1] They are unusually large, particularly the terminal bud (Fig. 2D), a dark reddish-brown in color, and varnished with a gummy and resinous substance that is sticky to the touch. A closer inspection with the hand lens, especially of the terminal bud, shows that there are five pairs of scales arranged oppositely in four rows, and that they are rather thick with thin margins, the lower pair more or less keeled—that is, with a ridge—and frequently having abrupt sharp points.

The Norway maple is another exotic that has been extensively planted as a shade tree. The buds (Fig. 2E), rather large for a maple, are commonly red or yellowish-green toward the base; sometimes they are entirely tinged with green. The oval or ovate terminal bud is larger than the lateral buds, which are rather small and appressed, that is, they lie closely against the twig. All this is apparent to the naked eye. A closer look with the lens shows that the scales are more or less keeled, and if the outer ones are lifted up with a pin, the inner ones will be found covered with dark, rusty hairs.

From my window as I write, I can look out and see a venerable sugar maple of almost gigantic proportions. It stands by the roadside and, though somewhat disfigured by several jagged branches broken off in an ice storm a few years back, it is still a handsome tree by any standards. Its straight upright trunk and erect branches form a broad round-topped crown, and even at a distance the fine penciling of its twigs against the sky is quite discernible. However, it is

[1] This is the tree of Longfellow's poem "The Village Blacksmith."

not the tree but the buds that interest us at the moment. They are conical to ovate in form (Fig. 2F), sharply pointed, reddish in color, and rather downy, as revealed by the lens, especially toward the tip, with four pairs of overlapping scales (in some sugar maples there may be as many as eight pairs) whose margins are finely hairy.

At this time of the year the white ash can be recognized from almost any distance—depending, of course, on how good one's eyesight is—by the color of its buds. They are rusty to dark brown (a few are even black), and hence conspicuous on the wintry landscape (Fig. 2G). The buds are quite unlike those of other trees, as I think you will agree once you have looked at them. So, too, in their own way, are the buds of the bitternut (Fig. 2H). They are slender and strikingly yellow, and through the lens you can see that they are crowded with glandular dots and are slightly hairy beneath the scales which, strangely, do not overlap. In botanical terminology they are said to be valvate. But of all the buds to be found on our trees, none are so distinctive as those of the red and silver maples. Not only are they decidedly red, but there are also many collateral buds that, grouped in clusters, give the twigs a characteristic beaded appearance when viewed against the sky.

There are many other buds of equal interest; but buds are not the only clue to a tree's identity. There are other features that serve a like purpose, such as the light-colored scars apparent to anyone who has examined a twig. These were formed by the leaves when they fell off; hence they are known as leaf scars.

The shedding of leaves by many of our trees when cold weather approaches is nature's way of protecting the trees against excessive loss of water during the winter. At this season the supply is at a minimum, and it becomes necessary for every tree to conserve whatever water it can obtain from the soil. In addition to their food-making activities, leaves function as an outlet for any excess water that the trees may absorb from the soil at times when the water table is high. If this excess water were not eliminated, it would drown the trees. Since the leaves would continue to function in this manner were they permitted to remain on the twigs during the winter, it is quite possible that the outgo of water might exceed the intake, with resultant injury to the trees.

The leaf scars on most twigs are visible to the naked eye, but in some cases the glass is required to outline them more clearly. Like buds, leaf scars show considerable variation. In the horse chestnut, for instance, they are large and conspicuous and inversely triangular (Fig. 3A), whereas in the red maple they are much smaller and broadly U- to V-shaped (Fig. 3B). In the American elm they are semicircular (Fig. 3C). Leaf scars, as a matter of fact, may have almost any form. They may be circular (black ash, Fig. 3D), elliptical

FIGURE 3

Leaf Scars

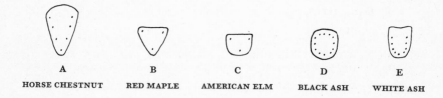

| A | B | C | D | E |
| HORSE CHESTNUT | RED MAPLE | AMERICAN ELM | BLACK ASH | WHITE ASH |

(catalpa), crescent-shaped (tupelo), and heart-shaped (Kentucky coffee tree). They may be narrow, as in the pear, and their upper margins may be flat, or convex, as in the black ash, or deeply notched, as in the white ash (Fig. 3E). Sometimes they form a band nearly surrounding the bud, as in the sycamore. In some instances they are dingy and inconspicuous; in others they are lighter than the twig and quite distinct, as in the elm. They may be level with the twig or more or less raised, and their surfaces may be parallel with the twig or may make various angles with it up to a right angle.

Much the same can be said of the bundle scars, in regard to their number, size, relation to the surface of the leaf scar (as sunken or projecting). and arrangement. The bundle scars are the little raised dots present in the leaf scars. They indicate the ends of the conducting vessels that carried water and its dissolved contents in and out of the leaves. In the horse chestnut there are seven bundle scars arranged in a simple curved line, whereas in the red maple there are only three. In the shagbark hickory they are numerous and generally scattered about, while in the black walnut they are arranged in three U-shaped clusters.

It is difficult to realize that a forest may have begun as a barren polished rock. Yet such may well have been the case, for if we could have kept that barren rock under continual observation, we would have seen appear on it one day a small grayish, crusty sort of material that would slowly spread and cover the rock with delicate traceries. As time went on, we would have observed how the rock slowly disintegrated and how, among the pulverized grains and meager amount of humus formed through the chemical activity of the peculiar grayish material that seems to possess the essence of life, small rock-loving mosses suddenly began to grow. And then as these mosses in turn made the soil still richer in humus, we would have seen grasses and ferns take root and flourish. But the reign of the mosses, ferns, and grasses would have been a short one. For as these plants in their turn made the soil even richer, the seeds of the higher plants, blown there by the wind or carried there by animals, would have found conditions most favorable for germination and growth, and before long the flowering plants would have dispossessed the mosses, ferns, and grasses. Thus from the time we first saw the crusty growth on a barren rock, enough soil would have formed to permit forest vegetation to thrive and to end in a climax forest of beeches and maples. This forest, able to reproduce itself, would endure indefinitely if nothing occurred to destroy it.

Such a succession of stages is known as ecological succession and is occurring today in countless places. Look about you, and you will surely find a rock or boulder covered with a crustlike formation, prostrate and lobed, looking like a leaf fragment cemented to the rock. Or you may find instead a yellow crusty rosette on a tree trunk, or a red coral growth on a decaying stump or log. Perhaps you will find all three. Doubtless you have seen them before and given them only a passing glance. But they are worthy of more than a fleeting thought, for they are among the most remarkable of plants. Able to gain a foothold and maintain an existence on a dry rocky surface, and capable of converting such an inhospitable substratum into one in which other plants can grow, these plants have been given a most important role in the vast scheme of things.

These astonishing plants are called lichens. As extremes of temperature have little effect on them, they can survive in places where other plants are unable to live—on a bare alpine peak, in the arctic wastes, in a tropical desert. They will, as a matter of fact, grow anywhere, given ample light, a firm substratum, and sufficient atmospheric moisture. In places where they have little competition, as in Iceland and Greenland, they are the only form of vegetation. But they cannot exist in polluted air and hence are not found in cities, since smoke, gases, and other impurities are lethal to them.

The mechanical contrivances that permit lichens to survive under what appear to be the most precarious of conditions have a strong appeal to the physicist, and the chemical processes that take place within them are of no less interest to the chemist, for among other things they secrete powerful acids that etch the rocks and break them up. And the biologist regards them with even greater respect, not only because of their ability to create soil in which other plants can grow but also because they are something of an oddity, being a closely knit relationship between two very dissimilar plants—a green algal plant and a colorless fungus plant—that live together for mutual benefit, a sort of partnership called mutualism.

The alga is a relative of the simple green plants that are found on damp stones and on the shady sides of houses and trees. The fungus is a relative of the mushrooms that appear magically on our lawns after a heavy rain, the molds that so often appear on our foodstuffs, and the mildews that disfigure our clothing during the hot humid days of summer. The fungus, in the form of filaments that form a tangled network of threads to which are attached the minute green algal cells, make up the bulk of the lichen (Fig. 4A). Lacking the green pigment chlorophyll, the fungus is unable to make its own food but instead has acquired the ability of absorbing and storing large quantities of water. The alga, on the other hand, is capable of absorbing water and carbon dioxide from the air and converting them into carbohydrates. Thus there is a give and take between the two plants; the alga provides carbohydrate food to the fungus, which in turn supplies the green algal cells with moisture that they need to keep from drying out.

The upper surface of a lichen is usually thickened and protective, while the lower surface is provided with anchoring hairlike structures called rhizoids. When dry, most lichens are a chalky-gray color, because the colorless fungal filaments so effectively cover the algal cells that their bright green is obscured. But when moist, the same lichens will appear green, because the moistened filaments transmit light quite well; hence the green algal cells can be seen through them. Lichens are occasionally other colors besides gray; some are black, especially when dry, and others are yellow or orange. These colors are due to certain acids produced by the fungus filaments. The acids may give the lichen a certain amount of protection in making it less desirable as food for such animals as snails and insects. It is also possible that the acids, formed on the surface of the filaments, prevent the spongy network from becoming saturated and waterlogged during the periods of submersal or heavy rains.

Lichens have quite a different appearance when viewed with a lens than with the naked eye, as you will discover when you examine them. And you do not have to go far from your doorstep to find them (unless, of course, you live in the city), for wherever we go—the marsh, the shore of a pond, the bank of a river, the woodland border, the woods themselves, the roadside, or a neglected field—we find lichens decorating tree trunks, stumps, fallen logs, fence rails, rocks and boulders, or almost any surface that will provide them with a foothold. We can find them at any time of the year, but they are at their best in the winter, for they like moisture. In the dry atmosphere of summer, they dry out, and many of them are hidden by foliage and other leafy vegetation and so are rather difficult to find.

The Star Lichen (Fig. 4B), one of our more common lichens, occurs on trees and rocks, where it forms small and inconspicuous silvery-gray or slate-gray rosettes. The lens shows small flat discs, dark brown, black, or frosted pale gray, with pale gray rims that may be smooth, toothed, or broken. They are the fruiting structures, in which spores are produced that perform the same function as the seeds of higher plants.

The Yellow Wall Lichen (Fig. 4C), a more or less northern species conspicuous for its brilliant orange color, usually forms small radiating growths closely attached to the rock or bark of trees on which it grows. Lungwort is a leaflike lichen common on tree trunks. The surface, leathery and tannish-gray when dry, but changing to a bright green when moist, is indented and marked with depressed areas which calls to mind the pitted surface of a lung. The Dog Lichen, a widely distributed species, grows in extensive patches on damp logs and earth. It is brown, almost black, with the reproductive structures projecting like the yellow teeth of a dog.

The coral-like lichens, such as the branching strap-shaped Iceland moss and the bushier reindeer moss (Fig. 4D), are commonly found in large patches on sand, earth, and level rocky ledges. Seldom do these plants exceed a height of 6 inches, but they make up in profusion what they lack in stature. The Scarlet-crested Cladonia (Fig. 4E), sometimes called the British Soldiers, is a common species throughout the Eastern states, occurring on the ground, on tree bark, on fallen and decaying logs, and sometimes on stones. When examined with a hand lens, the frosted green branches contrast delightfully with the little red tips.

When abundant, the gray-green threads of Old Man's Beard (Fig. 4F), hanging in clusters from the twigs and limbs of woodland trees, give the woods a hoary and venerable aspect and in our mind's eye provide a backdrop for the moonlight gambols of elves and other woodland sprites. This lichen superficially resembles Florida or Spanish moss, but the latter is a flowering plant of the Pineapple family found only in the Southern states.

Perhaps the most interesting of all our lichens is the Rock Tripe. There are a number of species, but unfortunately most of them occur at moderate and high elevations in the Northern states, although a few may be found at shore level. A typical species is smooth, dark green on the upper surface, sooty black on the lower. A central mass of rhizoids anchors the lichen to the rock on which it grows, allowing the curled-up edges to hang as they will. With every change of humidity the lichen curls and uncurls, writhes and twists, alternately covering the rocks with a green and black coating.

As you peer at lichens through your lens, you will become amazed at the unexpected beauty they reveal. Minute candelabra

FIGURE 4

Representative Lichens
and Cross Section
of a Thallus

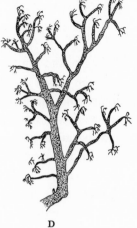

D

REINDEER MOSS

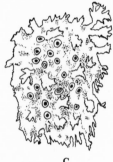

C

YELLOW WALL LICHEN

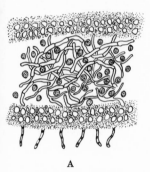

A

VERTICAL SECTION
OF A THALLUS OF A LICHEN

B

STAR LICHEN

E

SCARLET-CRESTED CLADONIA

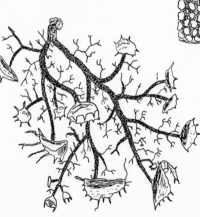

F

OLD MAN'S BEARD

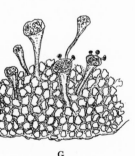

G

GOBLET LICHEN

appear as if by magic, and you will see tiny goblets (Fig. 4G), that may have served the potter as models for his art. And as you examine the unkempt and rather dirty incrustations on the surface of a rock, you may well wonder if some ancient sculptor did not find inspiration in their beautiful designs.

Across the road from where I live is a weedy, brushy field where many of our winter birds find good cheer in the brown stalks of weeds and grasses that stand in phalanxes against the sky. Only yesterday while I was outdoors attending to some chores, a flock of tree sparrows appeared as from nowhere and took temporary possession. As I watched them fly with cheerful industry from one brown patch to another, or hop about on the snow-covered ground in search of seeds that had been torn loose from their moorings, it seemed to me that their quest for food was not altogether a matter of life or death but also something of a frolic, if I interpreted correctly their gay notes that fell upon the wintry air like the tinkling of sleigh bells.

Sometimes goldfinches and juncos appear and look for seeds as meticulously as the sparrows. Goldfinches are at their best in late summer or early September, when they may be seen in flocks feeding on their favorite thistle seeds. But even in January, after their bright summer colors have faded to a more somber hue, they are always a welcome bird and a delight to watch. And not without good reason has the junco been called the snowbird, for not only does it seem to enjoy the snow and cold, but its color pattern also suggests winter: "leaden skies above, snow below." This color pattern is an excellent example of countershading—that is, the bird is darkest above, where it receives the most light, and lightest below, where it receives the most shadow. This sort of coloration tends to destroy the apparent solidity of the bird and to make it appear flat, so that it blends into the background and becomes part of it. Shore birds illustrate the same principle.

Beyond the field is a rather extensive woods, and here I often wander even in the winter, when I feel the need of an escape from the tensions of the day. At this time of the year I invariably find the tracks of a cottontail. Sometimes I follow his erratic course in the snow, curious to read the story it tells. The little animal is fond of rose hips and various berries, and his tracks often show where he has searched for them. But rose hips and berries are not always plentiful, and he must have recourse to the twigs and bark of small trees and shrubs. He is partial to sumac, and quite often his teeth marks show where he has been at work. When I follow his tracks, I often disturb a grouse in a tangle of bushes, and with a thunderous roar of beating pinions the bird will rise from the ground and take off among the trees. This usually sets a blue jay to screaming, and his raucous voice will echo through the woods and fall upon the ears like the sound from a discordant trumpet.

The prints of the cottontail are not the only ones I find; there are others, and all tell a tale if one can read them. Frequently they end abruptly—in tragedy. Pellets of bone and hair beneath the trees are also mute testimony that owls have been hunting in the vicinity. Most of their victims are meadow mice and deer mice, but owls also prey on other ground mammals—or at least those they can carry away and swallow, for owls swallow their victims whole and disgorge the fur and bones afterwards in the form of pellets.

Except for the woodchuck, the chipmunk, and a few others, most mammals are about during the winter; indeed, we often come across the prints of a mink in the snow along the banks of a stream or brook, and in almost any woods we can find the prints of a fox made when he prowled through the night. January is the month when raccoons mate, and in hollow trees bear cubs are born.

In the woods of which I write I often see the kinglet, feeding unconcernedly even though the north wind may howl and the branches may groan beneath their heavy burdens. The chickadees are always there and are always sure to enliven the winter scene with their amusing acrobatics and merry chatter. We always find a nuthatch or two in their company, since they hunt together all winter for beetles, caterpillars, and the pupae of insects. The brown creeper, too, spends most of its time searching for insects.

Except for a few species such as the snow bunting and Lapland longspur, which habitually seek open fields far from all cover, most of our winter birds prefer sheltered places; thickets and bush-grown roadsides, orchards, cedar and alder swamps, and stands of pines and other coniferous trees. Hairy woodpeckers are essentially birds of the forest, but with heavy snows they often appear in villages and orchards in search of food. The downies, of course, are always near at hand.

The woodpecker is admirably equipped for the kind of life it leads, but so, too, are other birds—the owl, heron, duck, vireo, and sparrow—by refinements in form, bill, feet, and wings. But though these refinements may be absorbing to contemplate, it is the presence of feathers that make birds unique and sets them off from other forms of animal life, for feathers are peculiar to birds. They are, moreover, highly specialized structures, and probably are better fitted to serve the various purposes to which they are put than any other structure we can name.

There are several kinds of feathers such as contour feathers, down feathers, semiplumes, filoplumes, and powder down; but the prevailing feathers are the contour feathers. Light in weight, they provide a durable protective covering, serve as implements of flight, and help to maintain the bird's body temperature. Air is an extremely effective insulating material, and feathers are replete with dead air spaces. Perhaps you have noticed how birds fluff out their feathers on a cold wintry day. This fluffing out, made possible by special muscles in the skin, creates more air spaces within the feather layers and thus increases the depth of the insulating material. Conversely, on warm days the birds hold their feathers close to the body, reducing the number of air spaces and permitting some of the body heat to escape.

To examine a feather it is best to obtain a contour feather, as it is generally more complete than the specialized flight and tail feathers. The flank feather of the ordinary fowl should be available to everyone. At first glance a feather (Fig. 5A) appears to resemble a leaf with its quill or calamus (petiole), shaft or rhachis (midrib), and vane or web (blade). A careful examination of the calamus with the lens shows that it is mostly hollow, with a minute opening at one end (the inferior umbilicus, through which nutrients and pigments were supplied to the developing feather) and a minute pit or depression (the superior umbilicus) at the opposite end. In some instances the feather may bear a replica of the main feather at the junction of the calamus and the rhachis. This replica is called the

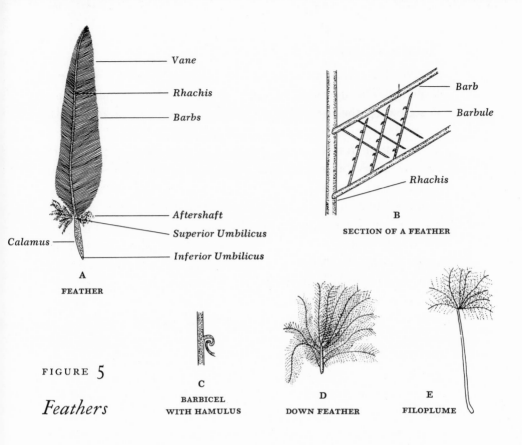

Vane

Rhachis

Barbs

Barb

Barbule

Rhachis

B

SECTION OF A FEATHER

Aftershaft

Superior Umbilicus

Calamus

Inferior Umbilicus

A

FEATHER

FIGURE 5

Feathers

C

BARBICEL
WITH HAMULUS

D

DOWN FEATHER

E

FILOPLUME

aftershaft or hyporhachis, and may be as long as the main feather, when the feather appears double. But in most birds it is considerably shorter, and in a great many species (as well as in the flight and tail feathers of all birds) it is entirely absent.

If a feather is held up to a bright light and a section of the vane is examined with a lens, what appears at first to be a continuous leaflike surface is found to be broken into a parallel series of closely spaced barbs (Fig. 5B). They are set at a slight angle and are provided with innumerable little branches, called barbules. These barbules overlap and interlock with the rolled edges of adjoining barbules by means of minute projections, called barbicels and hooklets (*hamuli*), so small that they can be seen only with a lens of high magnification or with a microscope (Fig. 5C). It is this arrangement of barbicels and hooklets that gives the feather its consistency and stiffness and makes it impervious to air and water. But since there is some amount of play between the hooklets and the edge of the adjoining barbule, the feather has also a degree of flexibility. The interlocking structures can readily be pulled apart by separating the barbs and can as easily be slipped back into place by drawing the feather through closed fingers.

The down feathers (Fig. 5D) we mentioned earlier provide the natal covering of newly hatched birds and completely invest the body of adult birds beneath the contour feathers. They lack the rhachis of the contour feathers, the barbs arising directly from the cala-mus, and as these in turn lack barbicels and hooklets, the feath-ers are without any stiffness and are entirely downy. Their pri-mary purpose is heat retention. Feathers intermediate between these two occur in some birds. In such intermediate feathers, too, the barbicels and hamuli are lacking. Called semiplumes, they are usually covered by the contour feathers and serve the same purpose as down feathers. Also, in some species specialized down feathers are found in large patches on the breast, over the hips, or in smaller groups scattered throughout the plumage. They grow all the time, and the tips of the barbs and barbicels are always breaking off in a fine powder that sifts over the rest of the plumage as a soft whitish bloom. Hence they are known as powder-down feathers.

Filoplumes (Fig. 5E) have an almost invisible rhachis with only a slight tuft of barbs at the tips and are scattered more or less throughout the plumage of some birds. Finally, the bristles that cover the nostrils of crows and jays and grow about the mouths of the whippoorwill and flycatcher are also modified feathers.

Except for a few flies that may be seen on a warm day, insects appear to be rare at this time of the year, at least when compared to the countless hordes we have with us during the warmer months. Yet they are just as abundant and may be found everywhere, if we know where to look. Many pass the winter as eggs which are quite unlike the more familiar eggs we have for breakfast, since insect eggs occur in all shapes, sizes, and colors, and some are even beautifully ornamented or exquisitely sculptured. Naturally they are more abundant in spring and summer, but even in the midst of winter we can find them attached to tree trunks and fences or hidden away in cracks and crevices. Do you want to see an egg that looks like a tiny flowerpot? Then, unless you live in the South, go outdoors and look on the branches of almost any tree, but particularly the apple and elm. You will find them in compact batches arranged side by side in regular rows. Through the lens, they appear as cut off at the top, with a central puncture and a brown circle near the border of

FIGURE 6

Winter Eggs of Some Insects

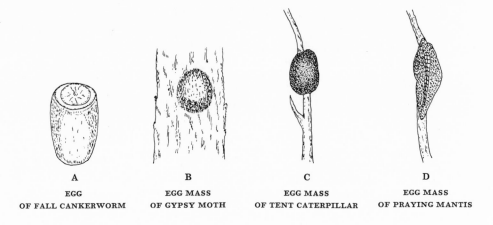

A	B	C	D
EGG	EGG MASS	EGG MASS	EGG MASS
OF FALL CANKERWORM	OF GYPSY MOTH	OF TENT CATERPILLAR	OF PRAYING MANTIS

the disk (Fig. 6A). They are the eggs of the fall cankerworm, an insect that is often a serious threat to fruit and shade trees.

Unlike the eggs of the fall cankerworm, most insect eggs found in the winter have some sort of protective covering. In New England and eastern New York State, the oval masses of the gypsy moth (Fig. 6B) are sometimes fairly conspicuous on the trunks of trees and on fences. Looking at them through the lens shows that they are covered with brownish hairs from the female's body.

Equally conspicuous on the twigs of various trees, chiefly wild black cherry and apple, are the eggs of the tent caterpillar, an insect widely distributed throughout the Eastern states and west to the Rocky Mountains. The eggs, numbering from three to four hundred, are laid in a brown varnished ring around the twigs and, though closely crowded together, are well worth looking at with a lens (Fig. 6C). Similarly, the eggs of the praying mantis are also deposited in a mass or cluster and are overlaid with a hard covering of silk. The egg mass is shaped like a short, broad cornucopia of dried foam (Fig. 6D) and is usually attached to the twigs or stems of plants, although I once found one fastened to a garden chair.

The egg sacs of the bagworm are most curious objects (Fig. 7A). The larvae[2] make them of silk, as revealed by the lens, to which is attached leaves or bits of sticks. When opened, some sacs will be empty; others will contain soft yellow eggs.[3]

Many insects spend the winter as larvae, nymphs (also an intermediate stage in the life history of certain insects, and one about which we shall have more to say), pupae, and even adults. Frayed cattail heads may seem a poor place to winter, but if we examine some of them we are likely to find that the larvae of the cattail moth find them quite serviceable (Fig. 7B). Spherical and elliptical swellings on goldenrod stems serve as winter quarters for the larvae of a fly and an adult moth, respectively. And if we look on the twigs of such trees as poplars and aspens, and the plum, cherry, and apple, we might come across the hibernaculum of the viceroy butterfly (Fig. 7C). This is simply a rolled-up leaf, but in it the larva finds a refuge against all the forces of the winter weather.

There are innumerable cases made principally of silk that serve as winter quarters for a variety of insects, but none are so odd-shaped as those that resemble a pistol (Fig. 7D) and a cigar (Fig. 7E). You may find them on various fruit trees, especially apple. When viewed with the lens, they are found to be a combination of silk, the pubescence from leaves, and excrement. They are inhabited by the larvae of two moths, called the pistol case-bearer and the cigar case-bearer.

The large silken cocoons of the cecropia and promethea moths (Fig. 7F) are as much a feature of the winter landscape as anything else we can name, for they are quite conspicuous on the naked branches of trees and shrubs. That of the cecropia moth is attached

---

[2] The larvae represent the second stage in the life history of such insects as butterflies, moths, beetles, flies, bees, and wasps. They generally are wormlike creatures.

[3] The empty ones were occupied by the male moths that deserted them after having transformed. The female moths also leave the egg sacs after having laid their eggs in them. Both male and female moths perish after having mated.

to a twig or branch, but not to leaves, as is the cocoon of the promethea moth. Examine the cecropia cocoon (Fig. 7G) closely. You will find it to be a tough waterproof structure with an exterior and an interior wall. Both of these are made of a strong paperlike silk, so that when cut open it appears as if one cocoon were hung within the other. Between the two walls many loose strands of silk provide innumerable air spaces that are an excellent insulation. It is

FIGURE 7

Winter Homes
of Various Insects

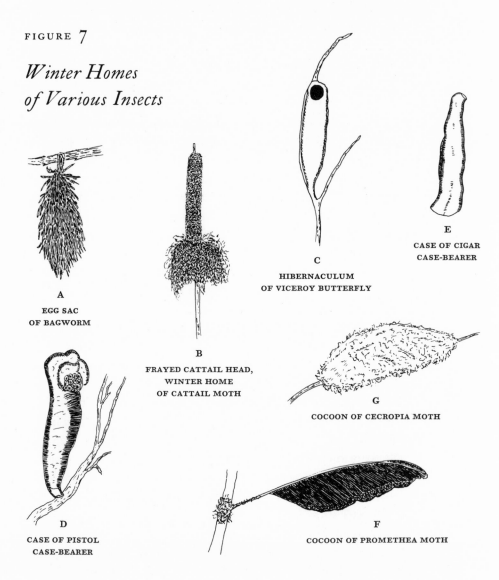

C
HIBERNACULUM
OF VICEROY BUTTERFLY

E
CASE OF CIGAR
CASE-BEARER

A
EGG SAC
OF BAGWORM

B
FRAYED CATTAIL HEAD,
WINTER HOME
OF CATTAIL MOTH

G
COCOON OF CECROPIA MOTH

D
CASE OF PISTOL
CASE-BEARER

F
COCOON OF PROMETHEA MOTH

an admirable winter shelter, protecting the helpless pupa from moisture, sudden changes in temperature, and most birds, although the downy woodpecker frequently punctures it.

Many insects that winter as adults may be found in all sorts of places: ground beetles, lady beetles, weevils, ants, leafhoppers, and aphis lions among the roots of grasses and other plants in open fields; bald-faced and yellow-faced hornets in the sphagnum moss of open lowlands and springy grasslands; thrips and other species among the woolly leaves of the mullein; wasps in stone walls and weather-beaten fences; and house flies, cluster flies, and mosquitoes in houses, barns, and other buildings.

Decaying logs often have a large winter population and may contain not only insects such as wireworms, click beetles, fireflies, and ground beetles, but also centipedes, millipedes, harvestmen, spiders, snails, and occasionally salamanders and wood frogs. The woodland floor is also well tenanted, for its covering of dead leaves, whose curved and twisted surfaces hold warm air, makes an excellent winter retreat for various insects such as the tarnished plant bug as well as other invertebrates: worms, snails, clubionid spiders, and occasionally small vertebrates like salamanders and spring peepers.

In the popular mind a bug and an insect are the same, and the two words are used indiscriminately and interchangeably. Moreover, the word bug is often applied to any small flying or crawling animal, although this is incorrect on two counts: in its entomological sense, a bug is a certain kind of insect, and an insect is an animal having certain structural characteristics not found in other animal forms.

If a butterfly, a lobster, and a spider were lined up side by side they would appear, at a casual glance, to be entirely dissimilar animals with little in common. But brief study would show that they all have segmented or jointed legs and a hard outer covering which is divided into sections or segments. Other animals, such as the crab, crayfish, centipede, millipede, scorpion, mite, and tick, have the same characteristics. Collectively they comprise a division of the animal kingdom called the Arthropoda.

Now a further comparison of all these animals would show that the butterfly, unlike the other anthropods, has three distinct body regions—a head, thorax, and abdomen—one pair of feelers or antennae, three pairs of legs, and two pairs of wings. Any animal, then, having the structural characteristics of the butterfly is an insect (with some slight exceptions, since some insects have only one pair of wings and some have none).

Again, a comparison of several different kinds of insects would show that they differ among themselves. Accordingly they have been classified into major and minor groups, as orders, families, etc., depending on how various structures have become modified. Butterflies and moths are grouped together because they have scaly wings; house flies, crane flies, and mosquitoes are placed in the same order because they have only one pair of wings; and ladybugs, fireflies, and weevils are all beetles because their first pair of wings is hard and horny. There are, of course, other differences that become quite apparent when a few actual specimens are examined.

Such specimens are easily obtained with a little effort. The tarnished plant bug, for instance, is a common inhabitant of the leaf mold in the woods and is readily collected by transferring some of the decaying leaves to a pail or other container with a garden trowel. It is a small insect (Fig. 8A) not much more than a quarter of an inch long, rather flattened, oval. Its color is generally brown, with irregularly mottled splotches of white, yellow, red, and black, which give it a somewhat tarnished appearance, and with a clear yellow triangle marked with a black dot on the lower third of each side. Examine the insect with your hand lens, and observe how it is divided into three regions and how these regions in turn are divided into segments. Then examine the wings, and note that the first pair, or forewings, are thickened or leathery at the base where they are attached to the thorax, and thin or membranous at the tips or extremities, which overlap and are folded over the second pair, or hind wings. Next look carefully at the head, and you will observe what appears to be a beaklike structure. Entomologically speaking, any insect with the kind of wings found in the tarnished plant bug and with a beak is a bug; any other kind of insect is a butterfly, beetle, bee, or grasshopper, as the case may be.

Insects are divided into two main groups: chewing insects and sucking insects. Obviously, insects which bite off and chew plant tissues must have a mouth structure quite unlike that of insects

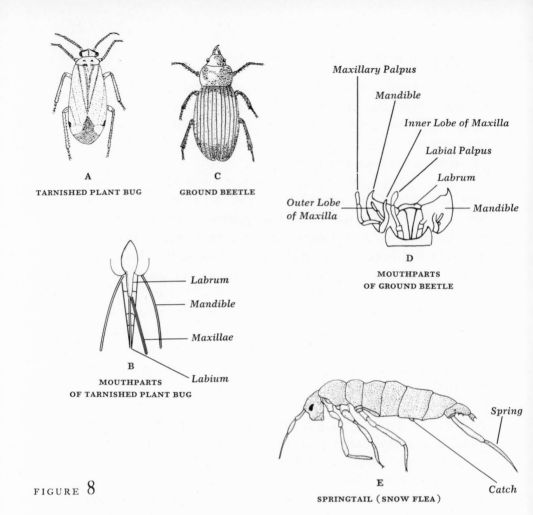

A

TARNISHED PLANT BUG

C

GROUND BEETLE

D

MOUTHPARTS
OF GROUND BEETLE

Maxillary Palpus

Mandible

Inner Lobe of Maxilla

Labial Palpus

Labrum

Outer Lobe of Maxilla

Mandible

B

MOUTHPARTS
OF TARNISHED PLANT BUG

Labrum

Mandible

Maxillae

Labium

E

SPRINGTAIL (SNOW FLEA)

Spring

Catch

FIGURE 8

*Spiders, Ground Beetle,
and a Few Winter Insects*

which pierce plant tissues and suck the fluids contained within them. The structures that serve insects for obtaining food are called mouthparts. They are most remarkable structures and vary considerably in form, as well as in function. Thus a predaceous insect must have mouthparts fitted for seizing and tearing its prey, whereas one that feeds on leaves must have mouthparts fitted for chewing this kind of food. Similarly, an insect that sucks the fluids contained within the tissues of a leaf must have mouthparts fitted

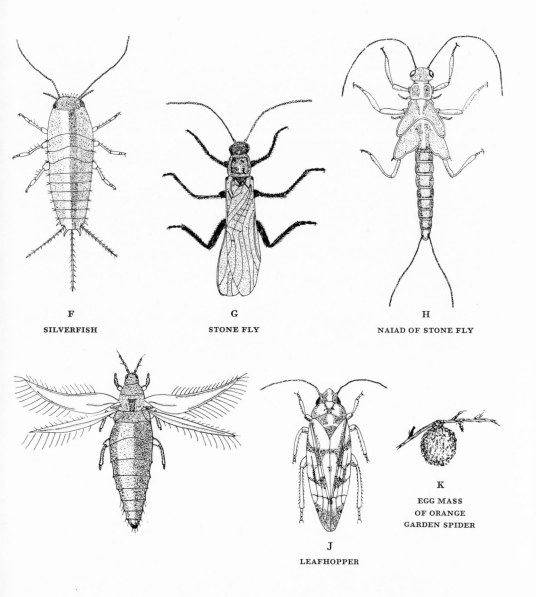

F
SILVERFISH

G
STONE FLY

H
NAIAD OF STONE FLY

K
EGG MASS
OF ORANGE
GARDEN SPIDER

J
LEAFHOPPER

for piercing the leaf surface, whereas an insect that feeds on nectar must have mouthparts which can be extended well into the nectaries.

Typically, mouthparts consist of an upper lip, the labrum; an under lip, the labium; and two pairs of jaws that act horizontally. The upper jaws are called mandibles, the lower ones maxillae. The maxillae and the labium are also furnished with feelers called the maxillary palpi and the labial palpi. In addition, there may be within the mouth one or two tonguelike organs, the epipharynx and the hypopharynx.

In the tarnished plant bug, the conspicuous beak is a well-developed four-segmented lower lip or labium. It serves as a sheath in which the upper jaws or mandibles and the lower jaws or maxillae, both of which have been modified into slender lancelike, sharply pointed piercing organs, are enclosed (Fig. 8B). The labium itself is not a piercing organ, though it may look like one; its function is rather to protect and direct the piercing organs and to determine, by means of tactile hairs at its tip, the place where the puncture should be made. The maxillae, which are fundamentally piercing organs, also lock together to form a sucking tube with two canals: an upper suction canal that takes up the fluid food, and a lower salivary canal that ejects the saliva. The labrum or upper lip is a short, much reduced structure and covers the upper part of the beak.

Another easily obtained insect specimen is a ground beetle. There are a number of ground beetles, but a typical species is shown in Figure 8c. It can be found by breaking open a decaying log. Examine the insect with your lens, and observe first of all that the forewings are greatly thickened and hardened and meet along the back in a straight line, and that the hind wings are membranous and are folded beneath the first pair. Then look at the mouthparts (Fig. 8D) and note that they are all well developed: a labrum distinct and more or less a flaplike organ; the mandibles thick and admirably adapted for chewing; the maxillae more complicated than the mandibles, with prominent palpi; and the labium equally complicated, also with prominent palpi.

Any time you are outdoors it is well to keep an observant eye out for a patch of snow covered with tiny dark specks that hop about like so many jumping jacks. Catch a few of them and look at them through your lens. You will find them the most grotesque and curious-looking creatures, with a distinct head, thorax, and abdomen. But they lack wings. And for this reason they were once believed to be degenerate descendants of winged creatures, but the present view is that they are primitive insects—in other words, that they originated before insects acquired wings.

These odd insects are a species of springtails and are known as the snow fleas (Fig. 8E). Look on the lower surface of the fourth abdominal segment and you will find a forked springlike process which is held in place when the insect is at rest by two small appendages on the third abdominal segment. When released, this mechanism straightens out and propels the insect into the air in the manner of a catapult.

Although the springtail lacks many of the specialized structures and the more highly developed parts of other insects, it is not the most primitive insect you can find. The silverfish (Fig. 8F) is even simpler in structure, because it lacks any kind of specialized organ such as the springing device of the springtail. A small, wingless, carrot-shaped insect with silvery scales, it is a very troublesome pest in laundries, libraries, museums, and households, for it gets into books, paper, starched clothing, and sometimes stored food. If you ever discover one (and it does find its way into the best-regulated households) and examine it with your lens, you will see a pretty little animal, covered with silvery scales. But you will probably have some trouble catching it, for it is most elusive and slips through the fingers by casting off its scales. The silverfish belongs to a group of insects called the bristletails because of their tail-like appendages. There are several species of these insects. Most are found under stones and other objects lying on the ground, but a few live in houses among the articles mentioned above. One species, called the firebrat, prefers the vicinity of fireplaces and similar situations. Once it was abundant in bakeshops, where it could often be seen running over the hot bricks in apparent disregard of the high temperature.

Since we usually associate insects with the warmer months of the year, we would hardly expect young ones to mature and transform into adults in winter. But surprisingly, certain species of stone flies (Fig. 8G) complete their nymphal lives in ice-rimmed streams and appear in the frosty air as adults that mate on the banks. You might visit almost any rapidly flowing stream and find them crawling about on the streamside trees and on the railings and walls of bridges. They are flattened, long-bodied and dull-colored insects, black, brown, or gray, with chewing mouthparts. Their membranous wings are held flat down their backs, the wide hind ones hidden beneath the front ones. Rather inconspicuous and with secretive habits, they take to wing awkwardly and fly slowly; indeed, their flight is so limited that they can often be caught by hand. They may be readily picked up when at rest, although if disturbed they are frequently able to make their escape by a sort of running gait.

The naiads, as the young are called, can usually be found throughout the year in swift water, swirling brooks, and waterfalls, and you should have little trouble finding them. They are elongate and flattened, in their body form already resembling the adults, with a pair of tails (cerci) at the end of the body (Fig. 8H). Most of them have filamentous gills, readily seen with the lens, that are always on the lower parts of the body, never on the back or sides, as in mayflies. Stone-fly naiads also have two claws on each foot, mayfly naiads but one. Stone-fly naiads are often more colorful than the adults, showing bright greens and yellows in ornate patterns.

Although stone-fly naiads are mainly carnivorous, feeding on mayfly naiads and midge larvae, the winter species are generally herbivorous. You can find them by lifting stones from the water and looking at the lower surfaces. However, as the naiads always seek the dark side of objects, the moment the stones are overturned and exposed to light they scatter and drop off the edges into the water. They are well able to climb among the rocks and to cling to them, because their claws are shaped like grappling hooks. When at rest, the naiads lie closely appressed to the rocks with their legs outspread so as to present the smallest surface area and thus offer the least resistance to the passing current.

The giant mullein, a common and picturesque wild flower of summer roadsides and pastures, is probably just as picturesque at this time of the year; even if the flowers have long since faded, the withered, tall, dried spike rising into the air from a rosette of gray-green woolly leaves certainly catches the eye with a sort of faded beauty. Next time you see the plant, look among the leaves, for they are a favorite winter retreat for thrips (Fig. 8I)—minute black insects that we so often see in summer threading their way in and out of daisy blossoms. As many as thirty or forty thrips may crowd themselves down among the leaves of one rosette, along with a mixed company of other small insects. Viewing a thrips with the lens will show that the wings are very short, perfectly transparent and practically without veins, but fringed with long delicate hairs. The mouthparts are a very curious shape and frequently differ on the two sides, that is, they are asymmetrical or lopsided. Another striking peculiarity of these odd insects is their feet; they are well developed, but terminate in a sort of cup into which is fitted a delicate protrusible membranous lobe or bladder. The bladder is withdrawn into the cup when not in use but protruded when the leg is brought in contact with an object. Since it is adhesive, it enables the insect to cling to smooth surfaces.

The roots of grasses and other plants are a favorite winter resort for leafhoppers (Fig. 8J). Small, slender insects, they have two pairs of membranous wings of uniform texture, and, as the lens will show, piercing and sucking mouthparts. When summer comes, we shall find many of these insects hopping around on all kinds of vegetation. But meanwhile, on a day when the ground is not too hard, you might dig up the roots of some accessible plant and become acquainted with them.

Spiders, like insects, spend the winter as eggs, young spiders, or full-grown spiders. They hide away in a variety of shelters. Many small spiders may be found in the egg sacs in which the eggs were laid. The pear-shaped egg sac of the orange garden spider (Fig. 8K) is a fairly common object and may be found attached to various plants such as dried thistle heads, securely fastened by many strands of silk so that winter storms will not tear it loose. Open it and a large number of young spiders (spiderlings) will crawl all over your hand. Viewed with the lens, they appear as small, glistening, animated beads. They are cannibalistic, the stronger feeding on their weaker brothers and sisters, so that from a sac which may contain a large number of spiderlings in early winter, a much smaller number of partly grown spiders emerge in the spring.

Equally common and also attached to dried stalks is the cup-shaped egg sac of the banded garden spider. This one, too, contains young spiders. Both of these spiders, incidentally, spin the large vertical webs so conspicuous in fields and gardens in late summer. And often attached to the undersides of stones you may find the brown, papery, disc-shaped egg sacs of the drassid spider.

January is one of the longer months of the year, but sometimes it seems longer than it actually is. Perhaps this is because we have difficulty getting back into the swing of things after the holiday festivities; perhaps it is because of the freezing temperature and heavy snowfalls, with the hazards of driving and walking on snow-encrusted and slippery streets; perhaps it is because we are anxious for spring to come. Perhaps it is none of these. Getting outdoors as often as we can, if only on a weekend—and for most of us, the weekend is the only time we can do so—losing ourselves for an hour or so in contemplation of the many things to look at, is to shorten one day at least, and before we know it January dissolves into February. And let February do its worst, for can we not paraphrase Mr. Shelley and say if February comes, can spring be far behind?

FEBRUARY

"An icy hand is on the land."

HENRY ABBEY

"Winter Days"

FEBRUARY

FEBRUARY IS USUALLY WINTER AT ITS WORST, at least in the Northern states, and often in many of our Southern states as well. With the landscape blanketed in a heavy cover of snow, with the temperature hovering for the most part about the zero mark, and with winds that sting our cheeks, it is not a month that tempts us outdoors, to wander leisurely in fields and woodlands, to stop for a moment and peer at something that may have caught our eye, or to poke among the snows for some hidden object of interest.

And yet there are days in February that are sunny and mild, when we feel the need to take a walk and shake loose the cobwebs that have festooned our brains these many winter days. Too, there is much to look at and inquire into, and we want to take the opportunity to do so whenever we can. For instance, there are the needles of pines, spruces, and firs: nature's plant-food factories, but smaller than those found on the oaks, maples, and poplars. We take considerable pride in our technological and engineering skill and our industrial know-how, but with all our scientific facility we have yet been unable to duplicate the alchemy of a green leaf, where the elements of air and soil are converted into food for both plant and animal. We may well examine a leaf with a hand lens or microscope and ponder the mystery that lies within its green tissues.

The pines and their relatives, so important to our welfare but generally neglected except by those whose business it is to know them well, have an ancient lineage. As we look at a pine, we little realize that it is a survival from a prehistoric age when its contemporaries were the Lycopods, the Sigillarids, and the Cycads whose remains form our coal deposits. Though eons have passed, it has retained the simplicity of floral structure characteristic of those plants of long ago. Look at a pine flower today, and you look back in time some millions of years.

I believe there are to be found, within the limits of the United States, thirty-nine species of pine. The most aristocratic of them all is the eastern white pine, a stately and beautiful tree and the tallest of the Eastern conifers. The trunk rises straight upward like an arrow, and in the forest towers branchless until it reaches the canopy, where it spreads out in a broadly conical crown. In the open, however, this pine assumes the form of all free-growing trees, with spreading horizontal branches set in whorls of five. The leaves, which grow in clusters of fives, are needlelike, slender, 3 to 5 inches long, flexible, and colored a soft bluish-green. They are also three-sided, which may surprise some of you. But if you remove a leaf, cut it transversely with a razor blade, and then examine the cross section with your lens, you will find that such is the case (Fig. 9A). Observe also the single conducting tube, called a fibrovascular bundle, and the two resin canals.

Unlike the white pine, the pitch pine is a tree of considerable diversity in form, habit, and development. The most virile and hardiest of the pines, it will grow under the most adverse conditions, clinging "like a limpet to a rock," as someone once remarked, to any substratum in which it can gain a foothold, even taking possession of the barren sands of the seashore. In the North it is commonly a small tree, and where exposed, it is often scraggly and grotesque; but in the southerly parts of its range it develops a fairly tall columnar bole and a small open crown. A more southern species, the yellow or shortleaf pine, is a medium-sized to large tree with a clear, well-formed bole that supports a small, narrowly pyramidal crown. The leaves of the pitch pine, rather short and stiff in a cluster of three, have two fibrovascular bundles with two to twelve resin ducts. Those of the yellow pine, which are as flexible and almost as long as the leaves of the white pine, usually grow in clusters of only two, and have two fibrovascular bundles with one to four resin canals.

The spruces and firs, pyramidal trees of arresting beauty, are largely restricted to the cooler regions of the Northern Hemisphere. But a few species—the red spruce, balsam fir, and eastern hemlock —occur in the Northeast and in the Appalachians as far south as northern Georgia, though the balsam fir in the southern part of its range is found only on the highest mountains of Virginia and West Virginia. Their leaves, unlike those of the pines, are linear in shape, flattened, laterally compressed (or more commonly four-angled in

FIGURE 9

Leaves, Twigs, and Seeds

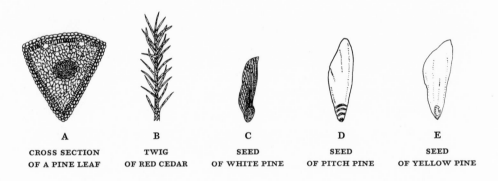

A	B	C	D	E
CROSS SECTION OF A PINE LEAF	TWIG OF RED CEDAR	SEED OF WHITE PINE	SEED OF PITCH PINE	SEED OF YELLOW PINE

the spruces), spirally arranged on the twig, and sessile (upon conspicuous peglike projections, sterigmata, in the spruces)—that is, without proper leaf stalks. When bruised, they usually give off an odor—ill-smelling in the spruces, aromatic in the firs. Frequently the spruces and firs are mistaken for the hemlock, or conversely the hemlock is mistaken for a spruce or fir, but the hemlock may easily be distinguished from either by its distinctly stalked leaves. In cross section, the leaves of the red spruce show one bundle and two ducts, those of the fir two bundles and two ducts; while those of the hemlock show one of each.

Possibly no evergreen tree of our flora has such a wide distribution as the red cedar, which is found from the Atlantic coastline west to the Rocky Mountains and from the Canadian border almost to the Gulf of Mexico. Tolerant of all kinds of soils, it grows easily on the gravelly slopes and rocky ridges of the North, as in the swamps and rich alluvial bottomlands of the South. Examine a branchlet of the tree and you will most likely find two kinds of leaves: awl-shaped and scale-shaped. There does not seem to be any set rule which determines their occurrence, except that the awl-shaped leaves always appear on young plants, but both may occur upon the same branchlet of a mature plant. The awl-shaped leaves are rigid, long, sharply pointed, ¼ to ¾ inch long, with a white bloom on the upper surface (Fig. 9B); and the scale-shaped leaves are minute, closely appressed, and acute or obtuse, and they usually bear a glandular disk on the back.

Sometimes the red cedar is confused with the white cedar and the arbor vitae, since all have similar scalelike leaves. But neither the white cedar nor the arbor vitae have the awl-shaped leaves of the red cedar. It is not quite so easy to distinguish between the white cedar and the arbor vitae when both are found within the same range. There are differences, of course: the leaves of the white cedar are bluish, whereas those of the arbor vitae are a yellowish-green, white cedar leaves are smaller than those of the arbor vitae, and the twigs are less distinctly fan-shaped. These are nice distinctions, really only apparent when the two are compared.

In the South many trees and shrubs remain green throughout the year, but by February in the North most of the woody plants have lost their leaves, or if they still persist on the branches, they are yellow or brown or some other drab color. Of course, the pines and other cone-bearing trees are green at this season, and so, too, are the ubiquitous mosses and the lycopods. And there are others—more than we suspect: wild cranberry, snowberry, bearberry, checkerberry, partridgeberry, mountain laurel, sheep laurel, pipsissewa, bog rosemary, leatherleaf, inkberry, shinleaf and trailing arbutus, the famous mayflower of Pilgrim devotion. Many of them still bear berries that provide food for wintering birds and the nonhibernating mammals. This is a difficult season for both birds and mammals, and as winter advances inexorably toward its climax and heavy snows cover the ground, they are faced with a diminishing food supply and must use every skill with which nature endowed them in order to survive.

How the shrews, for instance, manage to get through our northern winters has always been something of a mystery, since these diminutive animals require an enormous amount of food and literally eat all the time; if deprived of food for several hours, they starve to death. I frequently find their elfin tracks in the snow but seldom see the animals, since they usually work under cover. Occasionally I do come upon one of them poking his snout into a crevice in the bark of a tree trunk, or ferreting about in the leaf mold or among decayed pieces of wood. And at such times I cannot help comparing this tiny and hardy animal that has to hunt constantly for food with the clumsy, lumbering porcupine, who finds plenty to eat in the bark and leaves of evergreen trees and need bestir himself ever so slightly for his meals. Yet with food so near at hand, the quill pig will curl up in a dense treetop during a cold spell and take a prolonged nap, no longer interested in food. The opossum behaves in much the same manner, spending much of his time in a hollow tree and usually venturing out only in mild weather to search for something to eat. Occasionally, however, he will go abroad when zero temperatures prevail, only to get his naked ears and tail frostbitten. In the South he fares much better.

Deer wander as much as the snow cover will permit to feed on twigs and the foliage of evergreens, but when the snow gets too deep and travel becomes difficult, many perish from exhaustion and starvation or are pulled down and killed by wild dogs and other predators. The howl of the wolf is no longer heard in the woods, but the fox still barks on a glistening hilltop, and the scream of the wildcat may be heard almost anywhere. More disturbing to me is the loud "hoo, hoo hoo, hoo, hoo" of the great horned owl as his cry speeds through the frosty air, breaking the silence of a winter's night with unexpected shrillness. It has an eerie quality and a suggestion of nameless terror that doubtless strikes fear into the hearts of woodland creatures, for no living thing above ground, except the larger mammals and man, escapes his talons. Even the skunk is not exempt, for this implacable enemy, flitting through the woods as silently as a shadow, cares little for the disagreeable consequences of attacking such a pungent animal.

This is the time of the year when the weasel is not likely to kill for the mere pleasure of killing, for food is too scarce. Sometimes I wonder if his change of coat from brown to white is designed to help him capture his prey, or if it is of more value as a means of protection—for surprisingly, this fierce marauder is subject in his turn to the law of fang and claw. He often falls victim to foxes, wildcats, and birds of prey, although how they succeed in catching him is a mystery.

In his search for food the weasel will often cover several miles in a night, and if he is unable to find his evening meal he may invade a poultry yard. His relative, the mink, also roams far and wide; the streams that he usually haunts are generally frozen over, and he takes to the woods, searching for open rapids and warm springs or hunting rabbits and other animals he can capture. When he finds an opening in the ice of a frozen brook, he will run along the stream bed if the water has fallen away from the ice and left an air space and perhaps a narrow strip of turf uncoverd along the edge of the water. For it is in such places that the meadow mice often spend the winter, their burrows opening out into the banks.

Meadow mice, of course, are active all winter, scurrying about in their runways beneath the snow looking for the unblanched shoots of grasses, seeds, and hardy rootstocks. Many openings lead to the upper air, and at night the mice scamper back and forth across the snow. If we could read their tracery of footprints on the white surface as they lead from tree to tree and from stump to stump, what tales might they tell, tales of adventure and daring. For what other reason would they leave the comparative safety of their tunnels to venture forth and tempt fate in the form of a fox or owl?

Despite his cache of seeds, nuts, and other edibles, the white-footed mouse, too, is abroad. Even on the clear bitter nights of winter, when countless stars form a canopy over the treetops, biting winds hiss through the leafless branches, and snow is piled high over tangled brush, the little animal emerges from his home in a half-rotted stump or in a cavity of some venerable beech to skip over the snow.

Like the mammals, our wintering birds find it necessary to search assiduously for enough food to keep them alive. Since their favorite food may not be plentiful, they often eat berries and other fruits which they normally ignore. The scarlet pennants of the barberry, made sour presumably by a provident nature so that summer residents and fall migrants leave them untouched, now may become the difference between life and death. And the bitter, velvety-red plumes of the sumac are eaten with avidity by such birds as the grouse and bobwhite. But even such food sometimes becomes inaccessible when winter goes on the rampage. Only the crossbills seem unaffected by heavy snows, for the cones of the pines and other cone-bearing trees on which they feed are usually above even the deepest snow cover. Actually they do not feed on the cones but on the seeds, which they scoop out with their tongues after prying the scales apart with their curiously crossed bills.

With our lens we might examine the seeds of some of the conifers, as the pines and their relatives are known. On doing so we find that they come in all shapes, sizes, and colors, with expansions or wings so that they may be carried and distributed by the wind. The seeds of the white pine (Fig. 9c), which are ovoid, reddish or grayish-brown, and mottled with dark spots, have a straw-colored wing streaked with brown. The seeds of the pitch pine (Fig. 9d) are triangular to oval, dull black, sometimes mottled with gray, rough or pebbly to the touch, with a long, light-brown wing commonly showing darker longitudinal streaks. And the seeds of the yellow pine (Fig. 9e) are brown with black markings and have a straw-colored wing, broadest near the middle.

In spruces and firs, the seeds show similar differences. Those of the red cedar, however, are wingless, since they are not scattered by the wind but by the many birds which eat the dark-blue berrylike fruits.

In many ways February is a month of contrasts. Normally a month of cold and snow, there are days when the mercury climbs high and howling winds give way to gentle zephyrs, when a benign sun warms a frozen earth and melting snows cascade along rock-ribbed gullies, when butterflies flit about in sunny glades. Butterflies are certainly not a part of the winter scene—they belong to summer, flitting about in the sunshine,

> *Seeing only what is fair,*
> *Sipping only what is sweet. . . .*

FIGURE 10

Some More Insects Found in Winter

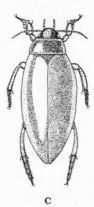

D

CLASPING ORGANS
OF MALE DIVING BEETL

C

DIVING BEETLE

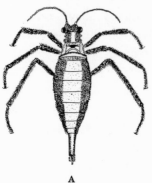

A
SNOW-BORN BOREUS

B
WINTER SNOW FLY

E
LACE BUG

—and yet a mourning cloak may often be seen on a bright, warm, sunny February day flying about in the snow-clad woods.

We usually do not associate insects with winter, but on warm days we may see gnats fly about, snow flies walking or hopping over the snow, and, in woodland pools where the ice has melted, diving beetles swimming in the water. The gnats and snow flies are more active at this time of the year than during the warmer months, and hence have received such names as the winter gnat, the snow-born Boreus (Fig. 10A), and the winter snow fly (Fig. 10B).

The gnats are very small insects and, except for the fact that all members of a swarm face the wind when flying, are of no particular interest. Far more rewarding to study are the curious snow flies. They belong to a group of flies called the crane flies—long-legged, two-winged insects that we see on our windows or flying about streams and meadows in the summer. But unlike the typical crane flies, the snow flies have lost their wings and instead have small knobs which you may be able to see with your lens; if not with your lens, certainly with a microscope. When seen for the first time on the snow, snow flies may easily be mistaken for spiders, since the shape of their bodies and their long hairy legs make them look more like arachnids than insects.

As a rule, insects do not mate in winter, but these snow flies are the exception. During other seasons of the year they may be found among fallen leaves and under moss and stones, but in winter they hide at the bases of tree trunks. When the temperature rises during a snow storm, they creep up the trunks; and when the sun shines again, they appear on the snow, where they usually mate. After having mated, the females again creep down beneath the snow close to the tree trunks and lay their eggs.

The snow-born Boreus is a scorpion fly, but unlike most other scorpion flies, it is wingless. In place of wings the female has a pair of tiny scalelike lobes and the male a pair of finely toothed spines. Boreus is a small black or brown insect, about ¼ inch long, which has a conspicuous beak formed by the trunklike prolongation of the entire front part of the head, a characteristic that makes it easily recognizable with the lens. When hundreds of these little scorpion flies gather on the snow and start to hop around on their long slender legs, they look like wind-blown cinders.

Although diving beetles are common inhabitants of our ponds and streams and may be seen at all times during the warmer months of the year, I particularly like to visit a woodland pool during a February thaw to watch them swim about in the water. They can be seen to best advantage at this time, for the water is fairly clear and not filled with submerged vegetation and debris.

Diving beetles (Fig. 10c) spend the winter in the bottom mud or under the banks in a dormant or semidormant state, except when they are attracted to the surface by a rise in temperature. They are well adapted for an aquatic existence, for their oval bodies offer the least resistance to passage through the water, and their long, flattened hind legs function admirably as propelling organs.

There are a number of species, some being an inch and a half long while others are minute, and they are usually black or brownish marked with yellow. But big or small, in either case quite innocent-appearing, they are fierce and voracious and a terror to various small inhabitants of ponds and streams. They frequently hang head downward from the surface of quiet waters, and if you observe one carefully you will see that just before it dives, it lifts its wing covers and takes a supply of air into the space beneath them for use while submerged.

Insects breathe by means of an elaborate system of branching air tubes, called trachea, that penetrate to all parts of the insect's body, including the appendages. They end in smaller, more delicate passages called tracheoles which divide and reunite constantly with one another, forming a capillary network of confluent tubes filled with a fluid rather than air. The trachea, which open to the exterior through small round openings called spiracles, function as conduits of air, actual respiration taking place in the tracheoles. In many species the spiracles are provided with hairs to exclude dust, and in a few insects they are furnished with a lid. By using a pan or some kind of scoop, you should be able to capture a diving beetle and see the spiracles with your hand lens. They open on the dorsal side of the abdomen beneath the wing covers (elytra).

In the males of certain species of diving beetles, the forelegs have a circular disk and little cuplike suckers. They serve as clasping organs to hold the female during coition and hence are secondary sex organs. They are quite evident with the lens (Fig. 10D).

The appearance of the diving beetles may lead one to conclude that they are the first manifestation of renewed activity in our ponds and streams. But this is not so; many animals living in our fresh waters are active throughout the winter season. Poke about in a swiftly flowing brook and you may suddenly see a two-lined salamander slither from beneath a flat stone. A small, slender amphibian, about 3 to 4 inches long, with a ground color that varies from brown to yellowish and with two dark lines on the back that extend from the eyes backward to the tail, it is a very active little animal and can move with amazing speed when disturbed, as you will discover if you try to catch one.

In almost any swift rivulet or spring-fed brook you should find various mayfly naiads. A common species is Ephemera, shown in Figure 11A. It is a burrowing insect, and to get a specimen it will be necessary to scoop up some of the bottom mud. Note the seven pairs of gills located on the abdomen, the three long, slender tail filaments, and the single claw on each leg. (Stone-fly naiads, as you will remember, have two claws on each foot, and the gills are situ-

FIGURE 11

Pertaining to the Mayfly

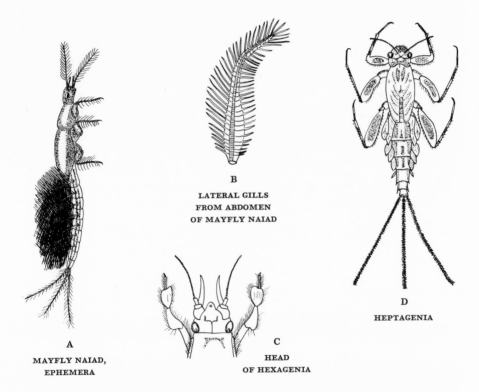

B
LATERAL GILLS
FROM ABDOMEN
OF MAYFLY NAIAD

D
HEPTAGENIA

A
MAYFLY NAIAD,
EPHEMERA

C
HEAD
OF HEXAGENIA

ated on the thorax). When viewed with the lens, the gills appear as in Figure 11B. They are hairlike expansions of the body wall, are abundantly supplied with trachea and tracheoles, and are extremely thin-walled to facilitate the transfer of gases between the tracheoles and the water.

If you can obtain a specimen of Hexagenia, another mayfly naiad, observe the bladelike feet, which are admirably adapted for burrowing, and the pair of enormous mandibular tusks that project forward from beneath the head (Fig. 11c). Sometimes a sloping bank will be mined with hundreds of Hexagenia naiads: their tails may often be seen in the openings to their burrows. If one of them is pulled out and placed on the mud, it will burrow in again with astonishing speed.

There are several other mayfly naiads that will well repay a little effort spent in looking for them, such as the chestnut-colored Blasturus with its double, leaflike gills and the flat-bodied Heptagenia (Fig. 11D), the latter usually being found clinging to the undersides of stones. All mayfly naiads are semiactive in the water, at least until the freezing point is reached, and all are vegetarian, feeding on the green algae that cover the rocks, or sifting and swallowing the soft silt. Desmids and diatoms[1] also serve as food, and so do the tender tissues of larger plants.

Because we have used the word naiad on several occasions, it might be well if we pause for a moment and consider how naiads differ from nymphs and larvae (caterpillars, maggots, and grubs), since all three represent the immature stage of various insects. Insects have a hard outer covering called the epidermis or cuticula, which is composed of a hard material known as chitin. The cuticula, which is inelastic, serves both as a protective covering or armor and as an external skeleton to which the internal organs are attached.

As a young insect eats and grows, the cuticula eventually becomes too small to permit further growth. At this point it splits open, and the insect, aided by a fluid that dissolves the cuticula, works its way out of it. Meanwhile a new cuticula has been formed from a liquid secreted by certain glands called hypodermal glands. The new cuticula is at first soft and elastic and can easily accommodate the increase in size; but it soon hardens or becomes chitinized, and after a while it, too, has to be discarded. A change of cuticula may occur a number of times—it varies with different insects—until the young has become full-grown, when it becomes an adult or else passes into a resting stage before it becomes an adult. The process of casting the old cuticula is called moulting, the period between successive moults is known as a stadium, and the form of the insect between moults is an instar.

[1] See the section on December for an account of these one-celled plants.

Except for primitive insects, the young of all insects have a number of active feeding instars, and in a large number of species the instars differ little from one another except in the gradual development of wings, genital organs, etc. In other words, the young, when they emerge from the egg, resemble the adults. In grasshoppers and their relatives, the termites, stink bugs and their relatives, the aphids, scale insects, and a few others, the young are terrestrial, as are the adults, eat the same kind of food as the adults, breathe

FIGURE 12

Hellgramite

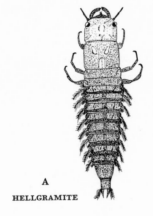

A

HELLGRAMITE

B

PART OF A TUFT
OF TRACHEAL GILLS
OF HELLGRAMITE

atmospheric oxygen as do the adults, and do not pass into a distinct resting stage before becoming adults. They are known as nymphs.

In mayflies, dragonflies, and stone flies, the young, although they resemble the adults, differ in habits. They are aquatic, breathe by means of gills, eat a different kind of food, and pass into a more or less distinct resting period before reaching the adult stage. They are known as naiads. Finally, in butterflies, moths, beetles, flies, bees, wasps, and a few others, the young are strikingly different in form from the adults, being more or less wormlike. They eat a completely different kind of food and pass into a distinct resting period known as the pupa, during which they undergo some remarkable transformations and change into adults. They are known as larvae (caterpillars, maggots, and grubs).

While exploring a swiftly flowing brook for mayfly naiads, you may unearth the hellgramite (Fig. 12A). I have found it occasionally, but more by accident than by intent. The hellgramite is also known by various other names as hell-diver, hell-devil, conniption bug, water grampus, flip-flap, snake doctor, dragon, goggle goy. Names usually mean little but in the present instance are quite descriptive, for the hellgramite is a forbidding-looking creature, and you will probably be loath to pick it up. If you do you will probably drop it in a hurry, for when touched or irritated in any way the insect arches its body, stretches its jaws savagely, and presents a terrifying appearance. But it is quite harmless, except to the naiads of mayflies and stone flies upon which it feeds.

The hellgramite is a flattened, sprawling creature with large hairy legs and a tough, thick skin. It may be dark-brown, slate-gray, or greenish-black in color, and when full grown it measures 2 or 3 inches in length. A tuft of white hairlike gills, which appears at the base of each of the lateral appendages on the first seven abdominal segments, looks like Figure 12B when viewed with the lens. At the end of the abdomen there is also a pair of stout, fleshy prolegs, each armed with twin grappling hooks. The animal shuns light and is seldom seen unless stones are overturned or pulled from the riffles of the stream. When thus disturbed, it either clings to the surface of the stones or hitches itself rapidly backward by means of the hooks on its prolegs. The hellgramite is the larval or immature stage of the dobson fly and looks so unlike the adult—a large winged insect with wings that measure 4 or 5 inches from tip to tip when fully expanded—that the two seem unrelated. But in the animal world there are countless similar instances of the young being unlike their parents.

Swift rivulets and spring-fed brooks have, at this time of year, a surprisingly large and diverse population, which may include (in addition to the mayfly naiads, hellgramites, and two-lined salamanders, as well as other insects we haven't mentioned) brook leeches (Fig. 13A) and scuds. We usually think of leeches as blood-sucking animals, but these leeches feed entirely on aquatic insects. They are oval, flat, and olive-green in color and are usually found clinging to the undersides of rocks and stones or resting in the sand or mud beneath them. If you find them, you may observe that some have young attached to the undersides of the body.

Unlike the sedentary leeches, scuds, which are distant cousins of shrimps, crabs, and lobsters, are very active little animals and may be seen swimming jerkily about searching for food and in turn being eaten by brook trout. Shaped like fleas, with arched backs and narrow bodies, and with legs adapted for climbing and ab-

FIGURE I3

Leeches and Other Water Creatures

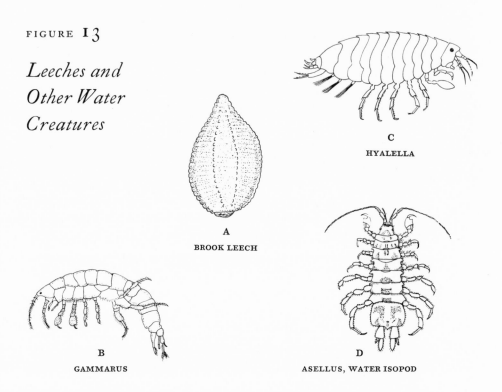

C
HYALELLA

A
BROOK LEECH

B
GAMMARUS

D
ASELLUS, WATER ISOPOD

dominal appendages for swimming and jumping, they can perform all these acrobatics with equal facility. Apparently nature was in an expansive mood when she created them. Although they can be seen with the naked eye, scuds are too small to watch in their native habitat and can be better observed in an aquarium planted with a few water plants. Here, at close range and with the aid of a reading glass, you can watch them perform all their acrobatic stunts at your leisure. They can be collected with a pan, or by holding a wide-mouthed bottle in the water and letting it flow in.

We have two species of scuds that are commonly active through the winter: Gammarus (Fig. 13B) and Hyalella (Fig. 13C). Both are primarily herbivorous, feeding on a great variety of living and dead plant life, though they have been seen to feed on snails, tadpoles, and other small animals. You can observe their feeding and mating habits in the aquarium. When mating, the male clasps the female with his clasping legs and swims about holding her beneath him for a period of several days. The female carries the fertilized eggs in a brood pouch beneath her body, and here the eggs hatch, in anywhere from nine to thirty days, as miniature adults quite capable of fending for themselves. Unlike other crustaceans, they do not have a larval life. Scuds are not as prolific as some other aquatic plant-feeding forms, but one pair of Hyalellas produced a clutch of a dozen and a half eggs fifteen times in 152 days. Frequently a female may carry a previous brood of young in her pouch during the mating swim.

Unless we live in some of the Southern states, most of us have never seen an armadillo except in a zoo. But we can find what looks like a miniature armadillo, at least when seen from above, if we examine the muddy bottom of a small woodland pool—especially one that looks rather unpromising for any sort of animal life. The animal I have in mind is a small, broad-backed, flattened, grayish-brown creature that can usually be seen crawling sluggishly about on the bottom mud or bottom litter of decayed leaves. It is a small crustacean known as a water isopod, with the scientific name of *Asellus communis* (Fig. 13D). Asellus is not entirely restricted to trashy pools but may also be found in ponds and streams; under certain favorable conditions, thousands may be seen in clusters swimming together just beneath the ice or resting on the bottom of a slow-moving stream.

Asellus's body is hairy and usually covered with silt, and it has seven pairs of legs. Look at them closely and you will observe that two pairs are adapted for grasping, the others for walking. Also look on the abdomen and you will find appendages called pleopods. They are provided with gills which extract the oxygen dissolved in the water. Asellus is of more than passing interest because of its long breeding period, which begins in February and lasts into the summer. A new brood of young is produced every five or six weeks,

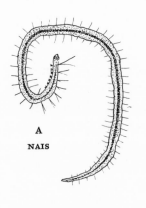

A
NAIS

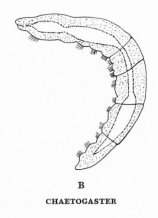

B
CHAETOGASTER

C
DERO

FIGURE 14

Bristleworms

and the females always seem to be carrying a brood pouch full of eggs or of young ones.

The water isopod is not the only animal that has the urge to mate in February. During this month yellow perch begin to migrate to their spawning places in shallows along pond and lake shores; skunks and squirrels begin to look for mates; and the great horned owl may be heard courting with loud hoots. If the owl's mating occurs early in February, the eggs may be laid before the month is over. The female sits closely on them during the cold days and long nights, and it is not uncommon to find her stolidly incubating under a thick blanket of snow.

With the ground frozen and covered with snow, earthworms would hardly be in evidence. Actually they are huddled below the frost line in closely packed balls to conserve the moisture of their bodies. But certain relatives of theirs are quite active by the millions in ponds and streams, where they overturn the ooze of the bottoms as effectively as the earthworms overturn the topsoil of the land. These are the bristleworms. Like the earthworms, they have segmented or ringed bodies, the segments clearly defined on the outside by constrictions easily seen with the lens, and set off internally by muscular partitions. Certain species, known as naids (family Naidae), are quite small, measuring about a quarter of an inch in length, with conspicuous tufts of chitinous rods or bristles. These occur on the segments in varying numbers and differ in length and form.

Worms in general seem rather drab, colorless, and uninteresting, though they play a significant role in the economy of nature; but bristleworms are worth viewing with the lens. They may be collected by scooping up some of the mud along the shoreline where the ice has melted and screening it through a fine-meshed net or sieve. One of the more common species is the graceful, transparent Nais (Fig. 14A). It is whitish or yellowish in color and is free-swimming, propelling itself along through the water by means of the bristles, which are usually longer toward the anterior end. Its internal organs are easily discernible with the lens. The species Chaetogaster (Fig. 14B) is somewhat larger than most and creeps about on its dense bristle clusters. The lively little Dero (Fig. 14c), with a broad funnel-like rear end from which extend two ciliated branchial appendages, often builds a small mud tube and slips in and out of it with astonishing speed. It even changes ends within the tube. All bristleworms reproduce principally by budding—that is, by automatic divisions of the body—which often results in chains of incompletely formed worms.

Should we have a prolonged winter thaw or at least a few warm sunny days, so that the snow cover disappears enough to reveal patches of spreading strawberry leaves in the fields and meadows, I suggest you look among the leaves for lace bugs (Fig. 10E). They are small, delicate, and, under the lens, beautiful insects with fore wings that are like fine lace.

The worst part of a February thaw is that we begin to think of spring, while on the morrow we may have a blizzard. But the sun is climbing higher in the sky, the days are getting appreciably longer, snow buntings are moving northward, and the geese may be seen winging their way to their breeding grounds. Black ducks, too, are moving toward the interior, a few woodcocks are putting in an appearance, starlings are beginning to whistle, and the cave bats, awakening from their long slumber, are making short flights in their winter quarters. And we may even hear the faint call of the peeper from a woodland pool.

MARCH

"*The stormy March is come at last,*
With wind, and cloud, and changing skies."

WILLIAM CULLEN BRYANT

"*March*"

MARCH

ARCH IS A CAPRICIOUS MONTH; WE NEVER know what the weather will be. There are years when March can be wintry, even worse than February, and there are years when it can be mild and springlike. There are years, too, when March cannot seem to make up its mind from day to day. There may be several days of sunny warm weather which lead us to believe that spring is on its way, and then a snowstorm will blanket the landscape and we are reminded that winter is lingering on.

Whatever the weather may be, spring has come for me, at least, when I see the first bluebird or the first robin, or hear the sweet thin "pe-ep, pe-ep, pe-ep" of the peeper. Look for the peeper and you will probably not find him, for as long as the air is still chilly he remains well hidden. No matter how you may poke around among the dead leaves and mosses at the edge of a woodland pool or carefully scrutinize every stick and bit of grass, he will still be a mysterious piping voice.

March is not the type of month to tempt us into the fields and woods, for the sodden ground makes walking difficult, and our muscles, unused to such strenuous exercise after a winter of inactivity, tire quickly. But your true nature lover doesn't mind. As March arrives with its promise of spring, he is eager to be afield and to note with never-flagging interest the many signs of reawakening nature.

For though spring may sometimes seem provokingly slow in coming, it is inexorably on its way. We may not be able to see the roots at work, but we know they have begun to function: sap has started to run and buds to swell, since there is a noticeable brightening of color in the contour twigs of groves and swamps. On wooded hillsides the maples, gray since November, are becoming suffused with a rosy hue; the tops of willows are turning a golden yellow; bramble and other shrubs are developing deep red and purple tints. And where the sun has melted the snow, or where we scrape away the frozen crust, we find tiny shoots sprouting into tender greenness.

Some of the male purple finches that have wintered in the vicinity and have been regular visitors to our feeding stations are beginning to show a little red. In the field across the road, where goldfinches and tree sparrows found winter cheer a month or so ago, the starlings already show touches of yellow in their bills. And on a warm day, bluebottle flies will appear and dart about in sunny places. If we can catch a glimpse of the weasel in a woodpile, we find that his winter coat of white is beginning to change to one of brown.

By the roadside pussy willows are commencing to show their furry coats, and in swampy ground the skunk cabbage is pushing its twisted horns above the mucky soil. The hepatica, too, adds its note of cheerfulness to the still-bleak woods and thrills a strolling wanderer with its unexpected beauty.

It is difficult to describe one's feelings upon finding this exquisite little flower among the winter snows, for we normally associate flowers with the warmth of summer. Yet looking at it a bit more closely, we find that it does not seem so delicate after all, for its leathery green leaves wrapped in fuzzy furs appear ample protection against the rigors of winter. Delicate or not, it is a capricious plant, for its flowers vary in color—blue, lavender, purple, lilac, pink, or white —and in fragrance as well. Sometimes it is the purple flowers that are sweet-scented, sometimes the white, sometimes the pink. Only by trying them can we find those that are fragrant. The odor is faint and reminds us of violets.

Unless you know where to look, you will find the hepatica only by chance, blooming beneath the decaying leaves of the woodland floor, in some hidden nook, or beneath the lingering snows that have withstood the warm rays of the spring sun. When you find it, examine the flowers with your lens (Fig. 15A) and you will see that what you think are petals are instead colored sepals, leaflike structures that enclose and protect the other floral organs within the bud before they are fully developed. You will also find three small,

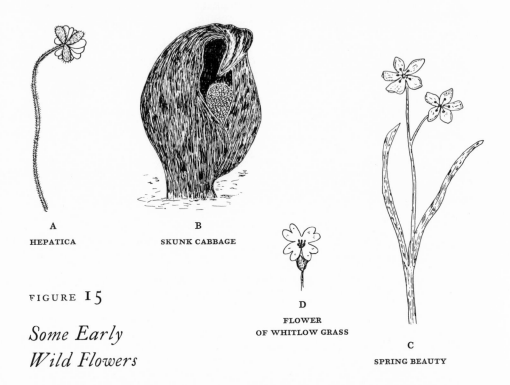

A
HEPATICA

B
SKUNK CABBAGE

FIGURE 15

Some Early
Wild Flowers

D
FLOWER
OF WHITLOW GRASS

C
SPRING BEAUTY

sessile leaves forming an involucre directly beneath the flower. They simulate a calyx and might easily be mistaken for one.

The skunk cabbage (Fig. 15B) may not evoke the same response as the hepatica, but I have always had a special feeling for this perennial harbinger of spring, though its name has little to commend it to most of us who look for a more sophisticated kind of beauty. If you can ignore its somewhat fetid odor, examine with your lens the minute flesh-colored flowers profusely scattered over the stout and fleshy spadix, and you will find they are not unpleasing to the eye. Note that the flowers have both stamens and pistils and that the grayish straw-colored anthers are fairly conspicuous.

Observe, too, that the hive bees are among the first of the insects to visit it, but being ungrateful creatures soon leave it alone when other and more attractive flowers appear. Perhaps it is just as well, for the hive bees never entered into the skunk cabbage's scheme of things, and many a bee which manages to gain entrance into the horn, that was manifestly designed for smaller insects, finds the going too slippery on its way out, falls back, and perishes miserably. But small flies and gnats that have lived under the fallen leaves during the winter and have gradually warmed into active life will soon swarm about the spathes.

We would not expect this malodorous plant to bear any relationship to the stately calla lily so widely used for floral decoration, but they are distant cousins. Compare the two and you will find many points of similarity, notably the presence in both of a spadix and spathe. It is the showy, solitary, beautifully colored spathe for which the calla lily is cultivated, but I think you will agree that the spathe of the skunk cabbage is not entirely without an esthetic interest and that there is a certain charm in the way the colors are blended.

The ponds and streams, no less than the earth, respond to the warm influence of the spring sun, and everywhere, even in the smallest spring pool, there are signs of renewed activity. As the ice of the ponds break up, muskrats leave their homes and feed on the shore, woodfrogs and spotted salamanders slip from the cover of the forest floor and enter the water to mate and lay their eggs, and water striders and whirligig beetles come to the surface and play about the water's edge.

Water striders (Fig. 16A) have always fascinated me, ever since I found them in a bubbling brook that ran through a woodland glen. It is surprising how memories linger through the years; I still vividly recall the day when I first chanced upon these curious, slender, long-legged insects and was completely lost watching them skate about on the surface of the water. Even today I often stop to watch them play, darting here and there, drifting with the current, or jumping up and alighting on the water without breaking the surface film. Sometimes they gather in schools in a quiet sheltered spot, scattering for shelter if alarmed, but quickly congregating again.

If you look closely you will see that in skating, water striders push themselves along with the middle pair of legs and steer with the last. They use the first pair for capturing other insects, dead or living,

FIGURE 16

Water Strider,
Whirligig Beetle,
and Wood Louse

C

HEAD OF WHIRLIGIG BEETLE
SHOWING DIVIDED EYE

A

WATER STRIDER

B

WHIRLIGIG BEETLE

D

WOOD LOUSE
OR SOW BUG

such as backswimmers, emerging midges that come up from the water below them, and leaf hoppers that fall on the water from plants that line the banks. Capture a water strider with a kitchen sieve, a water net, or any other kind of scoop; it will take a little doing, for they are extremely fast-moving. Examine it with your lens and you will find that both its feet and body are covered with soft, velvetlike hairs. These hairs prevent the insect from getting wet and so becoming heavy enough to sink. Water striders also envelop a silvery film of air, enabling them to submerge occasionally and remain under water for a short time.

Whirligig beetles (Fig. 16B) are the small steel-blue or black beetles we so often find swimming around in circles on the surface of brooks and quiet ponds. They may be seen at almost anytime of the year except when the ponds and streams are frozen. But it is in early spring when rising temperatures lure them from the mud that they appear in the largest numbers, gyrating on the surface and breaking the water into ripples, or basking like turtles on logs and stones. If alarmed or disturbed, they dash about in interlacing circles; so quickly do they move that the eye cannot follow any one beetle in its mad turnings and twistings. Occasionally they fly, if they can climb out of the water to take off. When captured, they squeak by rubbing the tip of the abdomen against the wings and emit a disagreeable milky fluid.

No matter how many there are or how fast they swim, they never collide as they dart about and around each other, nor do they bump into objects like floating sticks or protruding rocks or logs. Even if a number of them are transferred to the confines of an aquarium there are still no collisions. Darken a room and illuminate the aquarium with the red light of a photographer's darkroom lamp, to which the beetles do not respond, and they will still not collide. However, remove all dust from the surface of the water by skimming it several times with the sharp edge of a glass plate, or coat the inside walls of the aquarium with paraffin so that the water meniscus disappears, and they will not only run into one another but also bump the glass walls.

The answer to this apparent mystery is a simple one. As the beetles swim on the surface of the water, they set in motion floating particles of various kinds that in turn give rise to vibrations. The beetles are able to detect these vibrations, as well as those reflected from objects in the water and from the shore, and thus avoid collisions in somewhat the same manner that bats avoid colliding with obstacles by echolocation. The organ that appears sensitive to these vibrations is called Johnston's organ, a minute and complex structure composed mainly of tactile hairs and located on the second antennal segment. If this segment is removed, collisions occur.

Examine a whirligig carefully and you will readily see why it is such a capable swimmer. Note how the body is flattened, oval, and smooth, which helps the insect to overcome the resistance of the water. In some species (there are about thirty species that differ from one another mainly in length and color pattern) the ventral or lower side is shaped like a canoe, allowing the insect to move through the water with even greater facility. Observe, too, how the hind legs, fringed with hairs and shaped like paddles, are used like oars to drive the insect through the water in a rapid sculling movement. Also look closely at the eyes with your hand lens; you will find that they are divided by the sharp margins of the head, so that the insect can look up from the water with one half of the eye and down into it with the other half (Fig. 16c). Whirligigs frequently dive, and when they do they carry down with them a bubble of air under the tips of the wing covers. It glistens like a ball of silver and is easily discernible.[1]

No matter what kind of weather March may bring, I get outdoors as often as time permits, to stroll leisurely down the road, to wander in the woods, or to visit a not too distant pond to note the progress of the alder catkins. At such times I may see a cottontail, thin and ragged, scurry about in the rustling brush or find the footprints of a raccoon in the soggy woodland floor. Usually I hear the "kong-quer-ree" of redwings in the distance. If the day is warm and sunny, I invariably walk into a swarm of midges dancing in the air, and I am always sure to see the spring azure flying about like a "violet afloat."

Butterflies are such ethereal creatures that they seem out of place in our blizzard-swept winters, yet as we have seen, the mourning cloaks are often abroad during midwinter thaws. It is somewhat surprising to find that one of the first butterflies to appear in the spring is the small, dainty spring azure; it would seem more fit-

[1] The insect is not entirely dependent on the supply of oxygen that it takes down, but can replenish the supply from the water, up to a point. As the whirligig uses the oxygen in the bubble, the oxygen pressure is decreased and the pressure of the nitrogen increased. (Air is composed roughly of about one-fifth oxygen and four-fifths nitrogen, with very small amounts of carbon dioxide and the rare gases. The carbon dioxide which the insect gives off as a result of respiration does not remain within the bubble, but diffuses rapidly into the surrounding water). To replace the oxygen used, more oxygen from the water passes into the bubble and the nitrogen is forced out. Since the oxygen diffuses into the bubble three times as fast as the nitrogen is expelled, the net result is that the available supply of oxygen in the bubble becomes thirteen times that originally taken down in it.

ting if some larger and more robust butterfly were the first to venture forth after the snows have melted.[2]

To be sure, the spring azure is not the only butterfly we may see during the month. If we are lucky we shall also find the tortoise shell or the violet tip, perhaps on the branch of a sugar maple, sipping the sweet sap from a wound made, possibly, by a red squirrel. The little rodent is fond of the sugary liquid and taps the trees

FIGURE 17

Butterfly Anatomy

 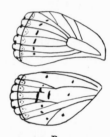

A
LOWER SURFACE
OF WING OF SPRING AZURE,
SPRING BROOD

B
LOWER SURFACE
OF WING OF SPRING AZURE,
SUMMER BROOD

by gnawing through the bark on the upper side of a branch. The cut forms a cavity in which the sap collects and serves as a "drinking fountain" which the squirrel may visit several times a day.

The red squirrel, always a noisy animal, seems unusually so at this time of the year. Perhaps it is because he and his cousins, now emerging from their vermin-infested retreats, are busy building clean, cool nests for their young, which will appear before the

[2] The spring azure is a creature of many fashions, for over a territory ranging from Alaska to the Gulf of Mexico we find one form in one locality, a different one in another. It is a source of endless labor to those of us who would study its protean forms, for even seasonally it differs to a marked degree. Thus we have, in the vicinity of Boston, Massachusetts, an early spring form which is small with large black markings on the undersurface of the wings (see Fig. 17A), a later variety which is larger with smaller black spots, and finally in summer a third form, still larger and with considerably fainter spots (see Fig. 17B).

month is out. Porcupines may also give birth during March, though frequently their young are not born until May.

Beneath stones and logs that have lain long on damp ground, wood lice (Fig. 16D) are now producing their first brood. The number of young wood lice may vary from a dozen to as many as two hundred. They are like the adults—slate-gray, spotted white, oval in form, and rather flattened. Wood lice are also known as sow bugs. They are not insects but terrestrial isopods, related to the aquatic forms such as Asellus with which we have already become acquainted. They are extremely common; you will find them under almost any stone or log you may overturn, provided it is in a damp place. Unlike Asellus, they breathe atmospheric oxygen, and instead of gills they have respiratory plates, which you will find on the abdominal appendages or pleopods.

We never know what we will discover beneath a stone or log, especially at this time of the year. Turn over almost any one that appears to have been in place for some time, and you will probably find millipedes and centipedes, various kinds of insects, snails, and spiders—and perhaps a red-backed salamander, so named from the reddish[3] stripe on its back. A very secretive little animal, hiding under all manner of objects—stones, logs, bark, tar paper, and other trash—it rarely seeks to escape when discovered, but remains quietly curled in the position in which it was found. Prod or even touch it, however, and it will usually run off, though "run" may not seem quite the word: its legs are relatively weak and must be aided by looping movements of the trunk and tail, so that the animal seems to progress over the ground in a series of rapidly executed leaps.

Salamanders are often confused with lizards, but the two have little in common and belong to two entirely distinct groups of animals. Salamanders have a smooth skin and are akin to frogs and toads; lizards are covered with scales like the snakes to which they are in a way related.

Speaking of snakes, you may also find the little northern brown snake (formerly called DeKay's snake) beneath a stone or log; indeed, beneath almost anything lying on the ground large enough

[3] The stripe may vary in color and be orange, yellow, or even light gray.

to cover it. It is widely distributed and quite common; look for it in bogs, swamps, freshwater marshes, woods, hillsides, fields, waste places, city parks, cemeteries, and vacant lots. Its general color is brown or deep reddish-brown. There are two parallel rows of blackish spots down the back, and the belly is pale yellow, brown, or pink, and unmarked except for one or more black dots at the side of each ventral scale.

You might examine the scales of the little snake with your lens. It is an inoffensive reptile and there is no danger in picking it up. It may give off a fluid from the anal scent glands, but the fluid is harmless and the odor inoffensive. The scales, unlike those of the fish that may be scraped off separately, hang together as a continuous armor and cannot be removed singly.

You will find that the dorsal or upper scales are arranged in both straight longitudinal rows and diagonal rows across the body. In most snakes the longitudinal rows are an odd number. The number is not necessarily constant but often varies on different parts of the body according to the amount of tapering. Should you count the rows of scales on the brown snake, you will find that there are seventeen oblique rows of keeled scales from one side of the abdomen to the other (Fig. 18A). A keeled scale (Fig. 18B), unlike a smooth scale (Fig. 18C), has a distinct linelike ridge running horizontally from the base to the tip. You will observe that the scales of the head are enlarged and are formed differently from those on the rest of the body. They are referred to as head shields or plates, and since there is a uniform normal arrangement of these shields among snakes, they are important in classification. These head shields, which may vary on the two sides of the head, are shown in Figure 18D. Compare the head of the brown snake with this figure, and you will find that the loreal scale is lacking.

With the exception of blind snakes, all snakes in this country have what are called ventral plates (or simply "ventrals") on the underside of the body. They extend entirely across the belly in most species, and in many they extend onto the sides of the body, with a more or less abrupt angle where they turn upward. The plates on the lower surface of the tail are called caudals, and if transverse like the ventrals, are said to be single, but usually they are in two rows.

With some variation, the ventrals and caudals correspond to the number of vertebrae. The scale or plate that covers the anal cleft is called the anal plate and may be single or divided (Fig. 18E): in

FIGURE 18

Scale Characters
of Snakes

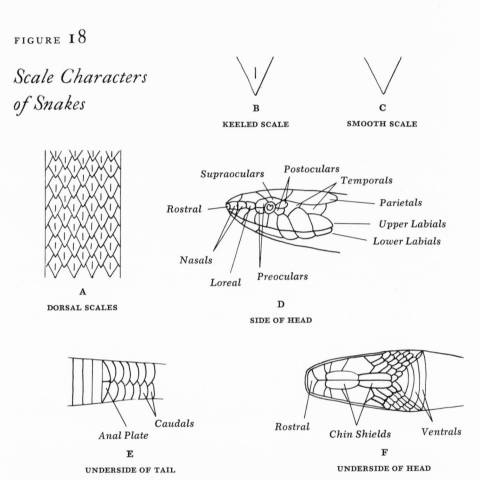

B
KEELED SCALE

C
SMOOTH SCALE

A
DORSAL SCALES

Supraoculars
Postoculars
Temporals
Rostral
Parietals
Upper Labials
Lower Labials
Nasals
Loreal
Preoculars

D
SIDE OF HEAD

Caudals
Anal Plate
E
UNDERSIDE OF TAIL

Rostral
Chin Shields
Ventrals
F
UNDERSIDE OF HEAD

the brown snake it is divided. On the underside of the head the scales that border the mental groove are called the chin shields. They are elongate scales and are in contact anteriorly with the lower labials (Fig. 18F).

Most of us are familiar with millipedes (Fig. 19A), popularly known as "thousand legs" though they do not have anywhere near this number of legs. They live in dark, moist places, such as beneath logs, where they feed principally on vegetable matter. Pick one up with the tweezers or forceps and look at it with your lens; you will observe that the body is nearly cylindrical and is divided into any-where from twenty-five to more than a hundred segments. Most of the segments have two pairs of leglike appendages that carry the

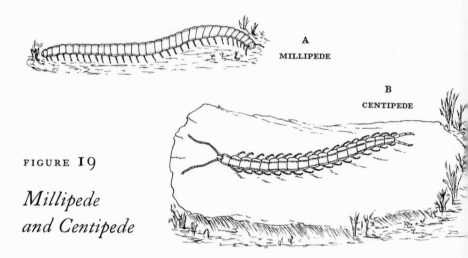

A
MILLIPEDE

B
CENTIPEDE

FIGURE 19

*Millipede
and Centipede*

animal over the ground rather slowly, which is not surprising. With so many legs, it is a wonder that the animal can move at all, for it would seem that they would get tangled with one another. Fortunately, fleeing is not the animal's means of escape from an enemy; instead it curls up in a circle when alarmed or disturbed, leaving the hard upper surface exposed as a protection for the legs and the soft underside. Each of the segments contains a pair of scent glands that secrete an objectionable fluid used in defense. They are located on the sides of the body and can be seen with the lens. You can also see the spiracles—the external openings of the air tubes—near each leg. The antennae are provided with numerous hairs that are olfactory in function.

Centipedes (Fig. 19B), or "hundred legs," have flattened bodies with fewer legs than millipedes. They have only one pair of legs to a segment, and unlike the legs of the millipedes, these carry the centipede rapidly over the ground. Centipedes live in much the same kinds of places as millipedes and are often found in association with them.

Look closely at a centipede with your lens; the first pair of legs has been modified to form a pair of hooklike jaws, called the maxillipeds, which project forward beneath the head. They are prehensile organs, and their terminal joints, which are very sharp, contain ducts of poison glands. The poison they secrete is used to kill insects, worms, and other small animals for food, because centipedes differ from millipedes in being carnivorous. Their poison has little effect on us—except perhaps to leave a red spot on the skin—but there are some centipedes in tropical countries that reach a foot in length and whose bite is not only painful but dangerous.

Although the prints of the raccoon may be seen occasionally, March weather is usually not to his liking, and except for short trips he remains in his winter retreat. His expeditions abroad during the early spring days are actually trial trips and usually of short duration, and it is not until spring has really taken possession of the woods that he finally emerges to search among the sodden leaves and debris left by melting snows for newly awakened snakes and beetles.

Unlike the raccoon, the skunk is now abroad rather regularly. At first he confines his activities to the woods, where he feeds on whatever insects he can find and such other small animals and birds that he can succeed in catching. But when the spring thaw sets in and the snow gradually disappears, he abandons the woods and thickets for more open land, where he hunts meadow mice on the newly exposed areas of turf or snakes that may have been lured from their winter quarters by the sunshine.

Some time during the month the woodchuck will emerge from his underground burrow, although he may already have come out for short spells in February. Female black bears, followed by their cubs, will leave their dens and search for food, feeding on buds and twigs for the most part. About the middle of the month, we should see the chipmunk on a log or tree root and hear his loud chirpy "chuck-chuck-chuck." Other chipmunks, deep in their burrows, respond to the clarion call and with a rush scamper out to add their loud and vigorous notes in a spring salute.

In chipmunk land, spring is the time for lovemaking; before the month is over the males go in search of mates. Other animals feel the same urge, and from isolated woods may be heard the barking of foxes or the noisy caterwaulings of wildcats as they seek their mates. Meadow mice are more silent.

Few birds mate this early, though sometimes bluebirds, robins, and crows will try to get a head start. The results are often tragic, however, for the weather is not yet warm enough for nest building and family raising.

If the winter has been a mild one, the downy woodpecker will begin to advertise for a mate by drumming loudly on a resonant tree or pole. English sparrows and barred owls, however, are indifferent to the temper of the elements; and as the owls engage in their grotesque lovemaking, weird sounds fall upon the night air.

I know of no month that offers so many delightful and unexpected surprises as March. Early in the morning I may awaken to the gay carol of a song sparrow and then later in the day find the spring beauty in a hollow and the fairy shrimp in a woodland pool. Like its cousin the gaudy portulaca, the spring beauty (See Fig. 15c), as if mindful of economy, opens its starry blossoms only in the sunshine, when the bees and flies are about, and closes them at other times to protect the store of pollen and nectar from unwanted pilferers. It is a charming little flower, pink with veinings of a deeper pink, though sometimes white with pink veinings. The blossoms, on long slender fragile stems, are turned mostly in one direction. Pick them, and the petals close and the whole plant droops as if in rebellion against such an indignity.

Examine the flower with your lens and note that there are five petals but only two sepals. Observe, too, that the golden stamens develop before the stigma matures, so that self-fertilization cannot take place. When the anthers have shed their pollen, then and only then do the three stigmatic arms open outward to receive pollen dust from a younger flower. This is carried to them by female bumblebees and the little brown bombylius as well as other friendly visitors which, in fair exchange, seek the nectar to which the pink pathways guide them.

I never know when or where I will find the fairy shrimp (Fig.

20A). The distribution of this unusual animal is freakish and uncertain. It may be abundant in one pool and entirely absent from another nearby, which to all appearances is exactly like the first. For several years it may appear in the same pool as regularly as the seasons, and then it may not be seen there again for four or five years even though the conditions appear to be the same. Or it may be abundant one season and then not reappear for several years.

FIGURE 20

Adult Male and Larva
of Fairy Shrimp

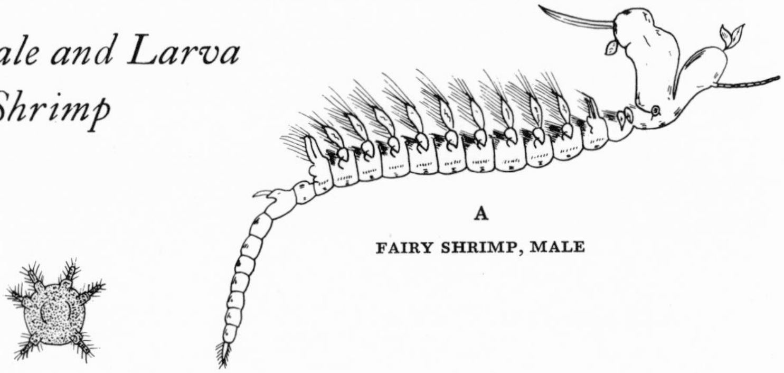

A

FAIRY SHRIMP, MALE

B

NAUPLIUS

The fairy shrimp is not a shrimp, though it is distantly related to shrimps and resembles them in form. It measures about an inch long, but may sometimes reach a length of 4 inches, and is colored with iridescent tints of red, blue, green, and bronze. The odd little animal always swims on its back, which is so transparent that it is possible to see its beating heart. It propels itself through the water with eleven pairs of leaflike appendages. For this reason it is known as a phyllopod, which means leaf-footed, and belongs to a group of crustaceans known as the Phyllopoda. These appendages, or leaf-feet, are actually gill feet, since they serve as respiratory organs as well; moreover, they are provided with "chewing bases" that help manipulate the food. The gill feet are borne on the body segments posterior to the head. In swimming, each pair in succession, from the hindmost forward, is pressed back against the water. At the same time, currents of water pass over the gills and convey microscopic plants and animals to the mouth. When the fairy shrimp swims, the waving plumes of the gill feet are the most conspicuous part of the animal, although hardly less so than the rich body colors that shimmer in the sunlight.

It is rather difficult to observe the fairy shrimp at close range in a pool; it is better to catch several specimens with a water net and transfer them to an aquarium at home. Be sure, however, that the temperature of the aquarium water is kept the same as the temperature of the water in which they are found.

Note that the fairy shrimp has a large head and two black eyes elevated on the ends of short stalks, and that the hind part of the body is slender without appendages and is brightly colored by the red hemoglobin of the blood. Examine the gill feet closely with the lens and you will see a flattened plate at a point near the spot where they are attached to the body. This plate is used in breathing.

When the ice breaks up in the spring, adults, spiderlike young (nauplii; see Fig. 20B) recently hatched from the eggs, and maturing young or larvae may all be found swimming in ponds or temporary pools formed by melting snow or by rain. As this is the mating season, many pairs may be seen swimming about together, always on their backs, the male above the female in close embrace. The sexes are easily distinguished—the male by his modified antennae, called claspers, which are used to hold the female, and the female by the prominent brood pouch. Just behind the gills, on the eleventh segment of the body, are tubelike appendages by which the male transfers the sperm cells to the female. You can see them with the lens if you collect a male.

Since fairy shrimps can live only in cold water, mating is accomplished in a matter of weeks. Then, as the water begins to warm up, the adults fall to the bottom and die. The eggs which the females have meanwhile produced also fall to the bottom, where they lie dormant either in the wet mud of permanent ponds or in the bottom mud of pools that dry up.

The eggs can resist long periods of desiccation. As a matter of fact, it appears that they must lie dormant for a long time—at least the entire summer; indeed, some eggs even appear to need drying before they can hatch. As a rule they do not hatch until sometime in late winter, but as they are sensitive to temperature changes, they may do so in the fall. Young fairy shrimps are sometimes found in ponds during mild weather following a very early autumn freeze. However, such young are killed by warm spells that follow. This may help to explain why the distribution of the fairy shrimp is freakish and irregular, for in the ponds where young are observed in the autumn, few if any will be seen the following spring.

For the first few days of March the birds we see are those that have wintered with us, but as the days pass others from the South appear on the scene. Some of them stay for a few days or a few weeks and then go on to more northern breeding grounds, while others remain through the summer until the days begin to grow noticeably shorter and the chill of approaching autumn warns them that it is time to return to warmer regions. In March, a day or two after the bluebird appears I am sure to see a robin on the lawn or to hear his sharp clucking call from a nearby tree; or perhaps I may see the robin first. For some reason the first robin of spring always seems to be alone, and yet a day or two later I may see several of them together, as if the others had been close at hand but timid about showing themselves. This explanation seems unlikely, however, for the robin seems to know we mean him no harm and walks unconcernedly about our lawn in search of worms. Sometimes his trusting nature will lead him to select a nesting site as close to the house as the cedar tree outside my window, where one year his mate built her nest while I watched her.

It may be days before the robin begins to sing; then early some morning his simple song falls upon our ears. But even before this, the song sparrow sounds his gay carol from every quarter. What does he care what the weather is? Once he starts to sing, he will sing on the brightest morning or bleakest day; even though sharp winds may blow and snow and ice still decorate the landscape, he will pour out his liveliest carol in a spirit of optimism and in defiance of the elements, as if to speed departing winter on its way.

Sometime during the month we hear the familiar note of the phoebe about the barn, in the orchard, or along a rushing stream. The strange reedlike sound of the redwing issues from the distant marsh as he sits perched on a swaying rush, his scarlet epaulets flashing in the bright sunshine. And suddenly one day grackles, whirling into a leafless tree, iridescent plumage twinkling greens and purples in the dancing sunbeams, advertise their return by discordant chatterings.

Before the month is out we should see a marsh hawk sail out of the blue sky in his buoyant and unhurried manner and a cowbird fly out from the shadows and strut over the ground like some silent shade. And where the leaves have gathered in the thicket, the fox sparrow announces his presence by noisy scratchings or by a few notes that are like the soft tinkling of tiny silver bells, while the meadowlark calls from a still brown field, and the vesper sparrow flirts his tail where the whitlow grass blooms.

Whitlow grass is a small plant, almost miniature in proportion, and rather insignificant. We would ordinarily pass it by, but we should take a good look at it since it will serve as an introduction to the large mustard family which includes such familiar plants as the pepperwort, pennycress, shepherd's purse, rock cress, watercress, sweet alyssum, candytuft, stock, mustard, cabbage, radish, turnip, cauliflower, and Brussels sprouts.

A common plant and widely distributed, whitlow grass grows in waste lands, sandy fields and roadsides. The flowers, so small that they almost escape notice, develop on stalks 1 to 5 inches high that rise from a rosette of hairy, lance-shaped, toothed leaves. Look closely at a flower and you will observe that it consists of four petals arranged in the form of a cross (see Fig. 15D). But since the petals are two-cleft, the crosslike effect is not so much in evidence as it is in other members of the family. The lens show that there are six stamens, two of which are shorter than the others. Although each flower secretes four drops of nectar, few insects visit the flowers. Hence the anthers of the four long stamens shed their pollen directly upon the stigma below them, leaving to such insects as may visit the flower the task of transferring the pollen from the two short stamens, which they must touch to reach the nectar.

So March draws to an end. Whether it will go out as a lion or as a lamb is anyone's guess; perhaps the first of April will be a day of celebration for those of us who thought spring was well on its way. But spring cannot be too long in coming, though snowflakes may fly on the morrow, for all signs point to spring, and nature cannot be entirely wrong.

APRIL

"When wake the violets, Winter dies;
When sprout the elm-buds, Spring is Near."

OLIVER WENDELL HOLMES

"Spring Has Come"

APRIL

IN THE NORTHERN LATITUDES WE CAN EX-
pect April to be wet and chilly, and yet there will be
days when it is sunny and mild and fragrant with the
smell of spring—days that makes us forget there are
still patches of snow in the woods and shady corners. But the snow
soon disappears under the combined influence of April showers and
the sun's warming rays. Countless shoots, with early leaves snuggled
close, begin to push their way through rain-moistened soil to tinge
the earth with vernal green, and the buds of trees begin to open and
clothe the woodlands with a singular and evanescent beauty. In the
South, where one may "lean his cheek against the air as if it were
velvet," the weather is most delightful, and growing things are well
advanced.

I am afield as often as I can escape from the invisible fetters
that chain me to a pattern of living, partly of my own doing and
partly the result of the times in which we live. I may only step
outside my house or I may wander half a mile, or even further if
time permits. But it is not distance in feet or miles that matters; it is
just getting outdoors to keep pace with an awakening nature world,
lest I fail to glimpse an event that on the morrow will have passed
and I will have to wait for another year.

Early in the month I like to get into the woods and poke among
the dead leaves for the wild ginger, a curious woodland plant that
hides its small bell-like flower under woolly leaves, as if in apology
for having blossomed. We may wonder why this flower of sober
purplish brown should bloom so close to the ground, virtually concealed among the leaves, its color not infrequently resembling the
leaf mold just beneath it. Perhaps it is a nice refinement to woo the

fungus gnats and early flesh flies, brought into active life from the maggots that have spent the winter beneath the dead leaves and the decaying bark of fallen logs; for in its cosy cup they can find both food and shelter from the chilling winds of early spring.

Wherever the turf is sodden and uninviting, in moist woodlands and along the brooksides, I usually find the dainty yellow adder's-tongue (Fig. 21A) or dogtooth violet. We can accept its name of adder's-tongue, for when the sharp purplish point of a young plant pushes its way above the ground, one does not need much imagination to note the fancied resemblance to a serpent's tongue. As for the name of dogtooth violet, we can only conjecture; it neither looks like a dog's tooth nor is even a violet. Indeed, merely a glance at the solitary russet-yellow, bell-shaped flower reveals its kinship to the tulip, lily, and trillium. Examine the flower more closely and you will find that the perianth consists of six similar petal-like segments and that each of the six stamens stands before each of the segments. These are structural characters of a family of plants whose members are remarkable for the simplicity and beauty of their flowers.

In spite of their striking external multiformity, flowers are comparatively simple and uniform in structure. Figure 21B is a diagrammatic sketch of a typical flower with which you can compare almost any blossom. The receptacle is the tip of the floral stem, from which the floral organs—sepals, petals, stamens, and pistils—grow. The outer or lowest of these floral organs are the sepals, usually small green leaflike structures that enclose and protect the other floral organs in the bud before they are fully developed. Collectively they form the calyx. Above and inside the sepals are the usually conspicuous, often brightly colored petals, collectively known as the corolla. The number of petals is usually the same as that of the sepals, or sometimes it is a multiple of the sepal number. Their purpose is to attract the insects whose visits are important in the reproductive functions of flowers. Flowers draw insects to them in several ways: by bright colors; by a sweet liquid, secreted in glands called nectaries, which is much desired by bees and other kinds of flower-visiting insects; and by odors from essential oils and other substances produced by special modified cells.

Inside and above the petals are located the male (stamens) and

FIGURE **21**

Representative
Spring Wild Flowers

C

PURPLE TRILLIUM

D

MARSH MARIGOLD,
FLOWER AND LEAF

A

YELLOW ADDER'S TONGUE

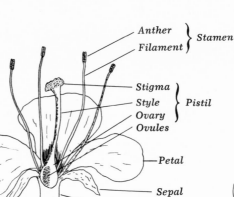

Anther
Filament } *Stamen*

Stigma
Style } *Pistil*
Ovary
Ovules

Petal

Sepal
Receptacle
Peduncle

B

DIAGRAMMATIC SKETCH
OF A TYPICAL FLOWER

G

BLUET, FORM A

E

CINQUEFOIL

F

GROUND IVY

I

FLOWER OF VIOLET,
INTERNAL STRUCTURE

H

BLUET, FORM B

female (pistil) reproductive organs. A stamen consists of a slender stalk or filament, surmounted at its apex by a single enlarged, often more or less cylindrical or ovoid anther in which develop the pollen grains that later lead to the formation of the male reproductive cells or sperms. The pistil is situated in the center of the flower. It consists of three fairly distinct parts: an enlarged globose ovary containing the immature seeds or ovules in which the female reproductive cells or eggs are formed; an elongated style which rises from the top of the ovary; and at the top of the style, a slightly enlarged stigma upon which the pollen grains fall, or to which they are brought by wind or insects. Stigmas are frequently very rough or bristly and sometimes are covered with a sticky fluid so that the pollen grains will adhere to them. A pistil is also composed of one or several carpels or seed-bearing organs. If a pistil is made up of one carpel, it is called a simple pistil; if composed of two, three, or more carpels fused together, a compound pistil.

The sepals and petals are considered accessory parts of a flower because they are not directly concerned with the reproductive processes, whereas the stamens and pistils are considered essential parts, being directly involved in the formation of seeds. A flower in which all the floral structures are present is said to be complete; a flower that lacks one or more is incomplete. Flowers that have both stamens and pistils are perfect flowers; those that have only stamens or pistils are imperfect. Imperfect flowers with stamens are called staminate flowers; imperfect flowers with pistils, pistillate.

Sometimes I find the purple trillium or ill-scented wake-robin (Fig. 21c) growing with the gilded lilies of the adder's-tongue. Bees and butterflies do not visit carrion-scented flowers—not because they are fastidious in their taste, but because they cannot obtain nectar from them. Since these insects are not available for cross-pollination, the plant must depend on the carrion or flesh flies, which are doubtless attracted to the flowers by their resemblance in color and odor to decaying flesh. The flesh flies lay their eggs in the flowers and, in return for a monopoly of the pollen food, which probably tastes as it smells, transport the pollen grains.

There are some twenty species of trilliums, and all are handsome woodland plants with stout stems, ruddy purple at the base, and perfect flowers of three sepals that remain until the plant withers and three larger petals that are white, purplish, or purplish red, depending on the species. Not all of them blossom in April; some wait until May, including the painted trillium, the loveliest of them all.

The skunk cabbage remains a relic of March, but its mottled spathes have given way to leafy crowns that will soon blend with the leaves of the hellebore[1] now beginning to unfold. In woodland brooks, where rushing waters from melting snows course their way, marsh marigolds (Fig. 21D) huddle on little islands and open their golden flowers to provide a festive board for bees and brilliant flowerflies. The marsh marigold is not a true marigold, nor even a cowslip, which it is sometimes called, but more of a buttercup. Its scientific name, *Caltha palustris* (*Caltha*—cup, *palus*—a marsh), means marsh cup, which is perhaps more fitting. A marshy plant it is, at home in swamps, low meadows, river banks, and even ditches, its bright yellow flowers a welcome relief among the dull hues of the early spring landscape.

Examine the flower closely and you will find that there are no petals but instead petal-like sepals. Note, too, that although the anthers and stigmas mature simultaneously, the anthers open outwardly and the outer ones, or those farthest from the stigmas, open first—a nice adjustment to ensure cross-pollination.

The famed arbutus prefers drier ground—hillsides and borders of rocky woods. It also grows in the vicinity of evergreens, and it is in such places that I usually find its open chalices scenting the air of the spring woods with a delicious spicy fragrance that blends with the smell of pine and of damp soil being warmed into life. It seems a sacrilege[2] to pick the dainty blossoms which look as if they have been placed in our early spring woods as messengers of hope and gladness. So they must have appeared to the pilgrims after that first winter on the bleak and rugged New England coast.

In hidden copse and shaded thicket, in woodland border and

---

[1] A familiar plant of Eastern swamplands found growing with the skunk cabbage. The coarse roots, often mistaken for those of harmless plants, are very poisonous, and horses and cattle have been killed by cropping the young leaves in spring. Even the seeds are fatal to poultry.

[2] —And in some states it is illegal.

low hillside, the solitary buds of the bloodroot slowly emerge in the warm sunshine from the embrace of silvery-green leaf cloaks. They expand into white-petaled, golden-centered flowers of evanescent beauty, for on the morrow the delicate blossoms, unable to withstand the spring winds, fall and are gone.

Observe that the golden-orange anthers mature after the two-lobed stigma, which shrivels before the pollen is ripe, and that the outer stamens are somewhat shorter than the inner ones in an advanced flower. Also that the stigma is prominent in a newly opened flower so that cross-pollination may be effected by such insect visitors as hive bees, bumblebees, mining bees (Fig. 22A), and the bee-like flies. Puncture not only the root but any part of the plant and there flows a red-orange juice which the Indians found serviceable as a war paint, for it stains whatever it touches.

Of greater permanence are the flowers of the early saxifrage. Explore the woodland or hillside in early spring and you are sure to find rosettes or fresh green leaves rooted in the clefts of rocks. If you examine the rosettes carefully, you should find small fine-haired balls in the center of the leafy tuffets. These little balls soon expand into branching downy stems, bearing many little white starlike flowers that are visited by the early bees and such butterflies as the mourning cloak and tortoise shell. Why the downy stems? The hairs are sticky and thus guard the flowers against unwanted pilferers, such as the crawling ants whose feet become ensnared in them.

Sometimes we find the early saxifrage crowding the fernlike leaves of the Dutchman's-breeches, for both plants delight in rocky situations. Despite its rather inelegant name, Dutchman's-breeches has a certain air of refinement; indeed, the finely dissected leaves and dainty heart-shaped blossoms, pendant from a trembling stem as a gem hangs from a lady's ear, suggest the feminine rather than the masculine. The tongue of the honeybee, as can be seen when it is examined with the lens (Fig. 22B), is too short to reach the nectar secreted in the two deep spurs. Butterflies have a tongue that is long enough, but it is difficult for them to cling to the pendulous blossoms. Hence cross-fertilization of Dutchman's-breeches is effected mainly by the early bumblebees.

By its pale foliage and attenuated saclike blossoms the corydalis at once shows its kinship to the Dutchman's-breeches and the bleeding heart of our gardens. The dainty little pink sacs, golden yellow at the mouth, hang upside down along the graceful stems and sway with every passing breeze. The corydalis, too, favors rocky woodlands where great boulders are covered with lovely little forest gardens.

Perhaps no plant is more typical of early spring than the wood anemone or wind flower, whose tremulous starlike blossoms quiver in the slightest breeze along the woodland border and on shaded hillside. It has a slender (though tough and pliable) stem and a horizontal rootstock that firmly anchors it in the ground. Its solitary white flower with many stamens and carpels, and with petal-like sepals (the petals, with a view toward economy, being dispensed

FIGURE 22

Bees, Caterpillars, and Insect Structures

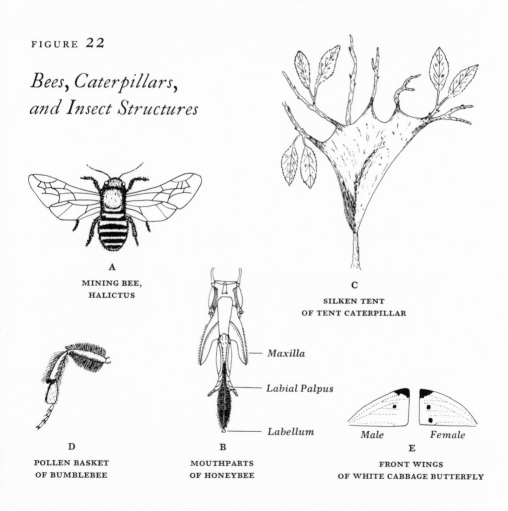

A

MINING BEE,
HALICTUS

C

SILKEN TENT
OF TENT CATERPILLAR

— *Maxilla*

— *Labial Palpus*

— *Labellum*   *Male*   *Female*

D

POLLEN BASKET
OF BUMBLEBEE

B

MOUTHPARTS
OF HONEYBEE

E

FRONT WINGS
OF WHITE CABBAGE BUTTERFLY

with), is set off by a background of whorled leaves, the better to advertise its wares to the insects that visit it. The flower also nods in cloudy weather, a refinement to ensure fertilization by wind-carried pollen in case insects fail to do so. Thus the dainty anemone is as well equipped to survive in the never-ceasing struggle for existence as the more familiar and highly successful dandelion.

It will be noticed that plants having thick rootstocks, corms, and bulbs that store up food for the winter are prepared to rush into blossom far earlier in spring than the fibrous-rooted species that must accumulate nourishment after the season has begun. The snowdrop, crocus, and grape hyacinth of our gardens are cases in point.

As a rule, too, our early spring flowers are small, shy, and delicate-appearing, and we find them by fortunate accident or by painstaking search. Of course, there are exceptions such as the yellow rocket, which sends upward from a single root crown a dozen or more sturdy stems a foot high, each bearing one or more showy spikes of small, brilliant yellow flowers that brighten meadows and the banks of neglected runlets. And we need only visit a field or rocky pasture to find the wild strawberry, cinquefoil, and ground ivy trailing their vines over the ground and opening their blossoms to the sky. Here, too, the pussytoes, or early everlastings, unfold their little clustered heads, tufts of silver-white silk on stems rising from charming rosettes that could have been found throughout the winter. And everywhere the pestiferous chickweed is in flower.

Look at a flower of the yellow rocket and observe the four petals arranged in the form of a cross, as in whitlow grass. Hence this showy plant is also a member of the mustard family, or Cruciferae. The word crucifer means crossbearer, but the crucifers are no martyrs. They are, on the contrary, a group of vigorous plants adapted to succeed in the struggle for existence.

Consider, for instance, the shepherd's purse. One stem serves for many flowers, none of which develops at the tip, since to do so would stop its upward growth. The flowers, though small, are not entirely inconspicuous, for they are clustered together and, thus massed, attract the insects that aid in pollination and that might pass them by if they were placed separately on the stem. Here is one reason why the plant has been able to march around the world and to compete successfully with other plants. The shepherd's purse also continues to blossom until frost covers the ground, and by extending its flowering season far beyond that of any native flower, it avoids the fierce competition for insect trade which it would encounter were it to flower during a shorter period. In your daily journeyings note the places where you find it growing; it is not a proud plant but will take root wherever it may, being satisfied with unoccupied waste land and locations where other plants refuse to grow.

We need only to glance at one of the little yellow flowers of the cinquefoil (Fig. 21E) to observe its resemblance to the rose; indeed, with the magnification of the hand lens we could very well take it for a yellow rose. Compare it with any rose, wild or cultivated, and you will see they have much in common: five sepals, five petals, and many stamens in multiples of five. There are some three hundred species of cinquefoil found everywhere in the north temperate and subarctic regions. Our most common species is the five-finger, a name alluding to the five leaflets that spread like the fingers of a hand.

The strawberry always brings to mind the luscious berrylike fruit that is such a delight to our taste, but the flowers appeal no less to our sense of the aesthetic: the many orange-colored stamens provide a delightful contrast to the white of the petals, especially when seen with a lens. Note the many pistils; they will form what is called an aggregate fruit, as we shall see when we come to examine it a little later.

The corolla of the bluish-purple flower of the ground ivy (Fig. 21F) differs from a conventional flower in that it is tubular in shape, with a two-lobed upper lip and a three-lobed lower lip, the middle lobe being the largest. Look at several of these flowers and you will find that some are larger than others and that the larger contain both stamens and pistils, whereas the smaller contain only a pistil. Why both kinds of flowers? Perhaps the small flowers are visited by insects which have already become dusted with pollen from the larger ones, thus ensuring cross-pollination. As for the larger flowers, the stamens mature first, and later, when they are past maturity, the pistil develops and receives pollen from the younger flowers. In this way the larger flowers protect themselves against self-fertilization.

The ground ivy is a member of the mint family, a large group of aromatic herbs with square stems, tubular flowers, and foliage covered with tiny glands that secrete an essential oil. Hence many of them are cultivated for use in medicine, confections, cosmetics, and cooking. Spearmint, peppermint, sage, thyme, and rosemary all belong to this family.

To my way of thinking, one of the most delightful of the early spring flowers is the bluet. By itself it is not a showy flower, but when massed together in countless numbers, bluets trace a milky path beautiful to see in fields and meadows. A bluet may be light blue, pale lilac, or nearly white. It has a yellowish center and is funnel-shaped, with four pointed, petal-like oval lobes. However, it is not the structure of the flower which warrants our looking closely at it, but rather the fact that there are two kinds of flowers which, strangely enough, do not occur in the same patch. Should you select a flower from one patch and examine it with your lens, you may find that the stamens are situated in the lower part of the corolla tube, with the stigmas exserted. But a flower from another patch may have the stamens elevated to the mouth of the corolla, with stigmas below the opening—in other words, just the reverse of the first flower.

The reason for the two kinds of flowers is to secure cross-pollination. An insect visiting form A (Fig. 21G) becomes dusted with pollen from the anthers in the middle of the tube, and this pollen is brushed off by the stigma of form B (Fig. 21H). Conversely an insect visiting form B gets dusted with pollen which is removed by the protruding stigma of form A.

Before the month is out, the violets will be in flower everywhere—

> *Half hidden from the eye;*
> *Fair as a star, when only one*
> *Is shining in the sky.*

Always perennial favorites, the violets are probably the most beloved of all our wild flowers—and perhaps of our garden flowers as well, for the pansy is actually a violet. These delightful flowers are found in a wide range of habitats, dry, sandy plains, rich woods, wet meadows, and swamps; but as a rule, any one species prefers only one of these situations. There are some three hundred members of the violet family. We know them as small, low-growing herbaceous plants that grow in groups, with blossoms that peep out from the foliage. In the tropics, however, violets become shrubs and small trees.

Violets are somewhat unusual in their floral structure and in the fact that there are two kinds of flowers. The conspicuous flowers have five sepals, five petals, and five stamens, and when viewed with the lens, the lower petal is seen to extend backward to the stem as a spur or sac which forms the nectary and which gives the violet its characteristic shape (Fig. 211). The lateral pair of petals is narrower than the upper pair, and two of the stamens have appendages that project back into the sac of the odd petal. The other less conspicuous (cleistogamous) flowers are produced lower down among the leaves; they bear rudimentary petals and are fertilized in the bud without the flowers opening, and it is from these that the seeds are produced.

If March were mild and springlike, the catkins of the alder opened then, but usually it is not until April that the scales open and the long plumed, pendant tassels emerge to wave like pennants in every transient breeze.

Such tassels are called catkins from their fancied resemblance to a cat's tail. Botanically they are also known as aments. Whichever we call them, they are actually flower clusters. Flowers such as the hepatica, yellow adder's-tongue, and bloodroot are borne singly at the end of a long stalk and are called single or solitary flowers. In most flowering plants the flowers are borne in groups or clusters. A cluster results from the branching of the main stem and is known as an inflorescence. There are different kinds of inflorescences, such as the raceme, spike, cyme, corymb, head, umbel, and catkin, depending on the manner in which the flowers are arranged.

The familiar alder of New England is the speckled alder, a small tree or shrub, so named because its stems are sprinkled with numerous conspicuous light-gray spots that are actually breathing pores called lenticels. It is essentially a plant of the Northern states, but a related species, the smooth alder, is widely distributed throughout the South. The two, which may often be found growing together where their ranges overlap, are very much alike in habit and have similar catkins.

FIGURE 23

Catkins

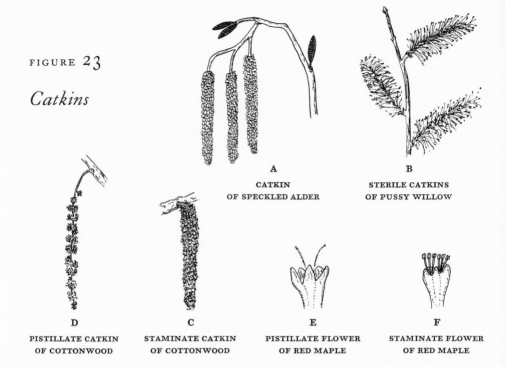

A
**CATKIN
OF SPECKLED ALDER**

B
**STERILE CATKINS
OF PUSSY WILLOW**

D
**PISTILLATE CATKIN
OF COTTONWOOD**

C
**STAMINATE CATKIN
OF COTTONWOOD**

E
**PISTILLATE FLOWER
OF RED MAPLE**

F
**STAMINATE FLOWER
OF RED MAPLE**

When examined with a lens, the catkin (Fig. 23A) is found to consist of brown and purple scales surmounting a central axis. The scales are set on short stalks, and beneath each scale are three flowers, each having a three- to five-lobed calyx cup and three to five stamens whose anthers are covered with yellow pollen. Pistils seem to be lacking. But if one of the shorter erect, deep-purple catkins is examined, every fleshy scale will be seen to enclose two flowers, each having a pistil with a scarlet style. Hence there are two kinds of catkins on the alder, staminate and pistillate; and since both are found on the same plant, the alder is said to be monoecious.

Because the pollen grains are transferred from one flower to another by the wind, the alder long ago dispensed with petals. These would only prevent the wind from picking up the grains and acting as the agent of pollination, an adaptation of distinct advantage. Once pollination has been carried out and fertilization completed, the pistils develop into small woody structures that resemble miniature pine cones. The scales protect the seeds formed beneath them, and when the seeds mature the scales open and release them to be scattered by the wind.

The hazelnut responds to the warmth of spring in much the same manner as the alder. During the winter the staminate catkins hang stiff and rigid, but under the influence of the spring sun they relax, develop their pollen, and fade away. The flowers consist of four stamens with two-cleft filaments, each fork bearing an anther (Fig. 24A). The pistillate flowers are little starlike tufts of crimson stigmas of surprising beauty when viewed through the lens (Fig. 24B). The nuts, enclosed in a leafy husk, are sweet and rival the filbert in quality. The chipmunk is extremely fond of them, as well he might be: they are not only good eating but fairly accessible,

FIGURE 24

Flowers
and Fruiting Capsules

A
STAMINATE FLOWER
OF HAZELNUT

B
PISTILLATE FLOWER
OF HAZELNUT

C
STAMINATE FLOWER
OF PUSSY WILLOW

D
PISTILLATE FLOWER
OF PUSSY WILLOW

E
CAPSULES
OF PUSSY WILLOW

F
STAMINATE FLOWER
OF ASPEN

G
PISTILLATE FLOWER
OF ASPEN

since the hazel is a shrub of rather low growth. This little animal is not a good climber, so that the nuts of the tall trees are not available until they fall.

Most of us may pass by the alder in the swamp and the hazelnut in the thicket; but not so the pussy willow by the roadside, for its yellow catkins (Fig. 23B) are sure to catch the eye. The yellow color is the color of countless pollen grains that await the wind to carry them to sticky stigmas which may be near at hand or more distant, for the pussy willow is dioecious and the staminate (Fig. 24C) and pistillate (Fig. 24D) flowers occur on separate plants. Keep some of the pistillate flowers (which are softer and silkier in appearance than the staminate ones, and less yellow in tone) under observation for a few days and they will develop into conic-shaped capsules (Fig. 24E) that contain undeveloped seeds. When the seeds are fully matured the capsules open, the seeds are released, and, since they are furnished with long, silky down that functions as an effective parachute, are carried by the wind, perhaps to lodge eventually in some faraway place.

Aspens and poplars closely follow the alder, hazelnut, and willow in bringing out their catkins. The aspen is said to be the most widely distributed tree in North America, and though quite common, it has little value,[3] is rarely planted, and is often an undergrowth in an oak wood where it is likely to remain unnoticed. Yet it frequently forms a little thicket of its own on a gravelly bank by the roadside or on the border of a swamp, though it is not a water-loving plant.

Its catkins, appearing before the leaves, are furry and show a touch of pink. With the lens note how the scales of the staminate flowers are deeply cut into three or four linear divisions and are fringed with long, soft, gray hairs (Fig. 24F). There are six to twelve stamens. The pistillate flowers, found on separate trees, have a two-lobed stigma with an ovary surrounded with a broad oblique disk (Fig. 24G).

Except where it has been planted as a shade tree, the eastern cottonwood is not common in the Northeast, but elsewhere it covers a wide range, being especially abundant in the alluvial soils of the plains and prairie states. The staminate trees are usually densely

---

[3] Except perhaps to the beaver, the tree being one of this animal's favorite food trees.

flowered, with catkins (Fig. 23C) 3 to 4 inches long and ½ inch thick. Viewed with a lens the flowers are seen to contain as many as sixty or even more stamens with large dark-red anthers. The pistillate flowers (Fig. 23D) are less numerous and have a somewhat globular ovary surrounded at its base with a cup-shaped disk and three or four greatly dilated or lobed stigmas.

It is the law of the wildwood that forest trees shall bring forth their flowers before their leaves. Poplars and aspens, the American elm, and silver and red maples do so in early spring; birches, oaks, and hickories somewhat later. The elm, which can trace its lineage to the Miocene, is one of the first trees to accept the challenge of March with its meteorological vagaries, and even when the snow is still on the ground the flower buds begin to swell and shine. Before April is well on its way, these buds shake off their brown scales and the tree appears as in a coppery mist. Look closely at a twig and you will find eight to twenty tiny flowers in umbel-like clusters. Viewed through the lens, the flowers are most attractive with their bell-shaped, reddish-green calyx cups, bright-red anthers, and pale-green pistils. Just before the lopsided leaves appear a few weeks later, the blossoms will have given way to little flat, oval, green samaras winged all around for flight on the air currents. We can find them on the ground, but for the most part they lie there unnoticed.

Even before the elm has blossomed, the silver maple puts forth its greenish-yellow flowers, and soon after the red maple follows. Have you ever noticed how some red maples appear very red when in blossom while others appear yellowish? The difference is due to the color of the flowers. The red maple is essentially dioecious, but this rule is not a hard and fast one, for a branch with staminate flowers can be found on a tree that is pistillate, and pistillate clusters may occur on a tree that is staminate. Even perfect flowers may be found. Examine the flowers closely and you will see that the reddish flowers are the fertile ones (Fig. 23E); the sterile ones (Fig. 23F), fringed with yellow stamens, are orange-colored.

From my window I can look out across the field where the sparrows found good cheer not very long ago, and see the shadbush on the woodland border, its silvery-white chandeliers brilliant against the background of leafless trees. And when I take the path that cuts through the field into the woods beyond, and follow the woodland trail with all its twists and turns for about half a mile or so, I come upon the spicebush, its knots of golden-yellow blossoms gilding the swampy April woods. If you want to know why it is called spicebush, break a twig and you will detect an aromatic odor.

Some bushes are slightly more yellow than others. Ordinarily, staminate and pistillate flowers are found on separate plants. When both are examined with the lens, the staminate flowers will be found to have not only a yellow calyx but yellow anthers as well, which makes them a brighter yellow than the pistillate flowers, since these have only the yellow sepals. The six-parted calyx takes the place of the corolla, which is absent. There are also minor details that you may discover for yourself.

April is not simply the month of early spring flowers and blossoming poplars, elms, and maples; it is also the month when many insects emerge from their winter retreats and the eggs of others hatch into nymphs, grubs, and caterpillars. Within the dark recesses of their burrows, tree borers resume their nefarious activities, and in the ground the larvae of click and May beetles begin to feed on various roots. Ladybird beetles come out of retirement and wander about in search of food; carabid beetles roam at night to look for prey and hide by day beneath stones and logs; ants undertake to repair their colonies; woolly bears uncurl from beneath old boards; and Polistes wasps start building their nests.

In April the wingless females of the spring cankerworm climb up the trunks of trees, where the winged males search them out. A few other moths also appear, and some butterflies too, and together with bees and wasps and various flies, seek the nectar hidden away in countless blossoms, while industrious ichneumons chase them all.

Sometime during the month the eggs of the tent caterpillar hatch, and soon the small silken nests (Fig. 22c) appear on the branches of the wild cherry. The caterpillars are most interesting to keep under observation from the moment they appear until they leave the "nest" to search for a place in which to pupate. If you have a few minutes each day to spend with them, I suggest you find an egg cluster early in the month and keep an eye on it. A reading glass will serve your purpose better than a hand lens.

It often crosses my mind how eagerly we await the arrival of the first bluebird or robin or search excitedly for an early spring flower, but how casually we accept the appearance of the first bumblebee. Yet the bumblebee, in her rich velvety costume of black and gold, her wings as yet untorn from long foraging flights, strikes just as much of a welcome note as she emerges from her winter resting place and hums her way over field and meadow.

The bumblebees we see in April are queens, the only survivors of last year's colonies. Their mission in life is to establish new colonies, but instead of getting to work immediately, they fly about for a week or more sipping the nectar of early spring flowers and filling the "baskets" on their hind legs with pollen grains. Such activity is not without purpose, for after a fast of eight or nine months they must store up energy in preparation for their domestic duties, and they must obtain the food necessary to feed the growing grubs of the new colony.

Capture a queen with the insect net, and after having killed it in the killing jar, examine with your lens the outer surface of a hind leg to find the pollen basket (Fig. 22D). It is merely a smooth, shining hollow with long overcurling hairs on its sides. The outer surface of the male's hind leg does not have the long side hairs because the males do not collect pollen; anyway, you will not find the males at this time of year.

Most of the moths and butterflies that appear in early spring are small, dull-colored forms—the advance guard of the brilliant hosts of midsummer—but there are exceptions, such as the common white or cabbage butterfly we so often see flying over fields, meadows, and gardens, stopping now and then to sip the nectar from an early spring flower. If you were to examine several specimens of the butterfly, you would most likely find that some of them have only one conspicuous black spot on the upper side of each front wing, while others have two. As with many insects, the sexes may be distinguished by differences in color or color patterns; in the cabbage butterfly the male has one black spot, the female two (Fig. 22E). Such differences, known as colorational antigeny, occur among insects in general but are most conspicuous among the butterflies and moths.

Do such sexual differences serve a useful purpose? Probably

not. It is doubtful if they have any protective value, but they may, in some species, serve as recognition marks by which the sexes can locate one another. However, in most instances they are usually so trivial and variable as to be negligible. Furthermore, as far as we know, insects are unable to perceive colors except in the broadest sense. If such sex-linked color differences do not enable the males and females to recognize each other, at least they help us to do so.

As the April days succeed each other almost too quickly, or so it seems, various forms of water insects begin to appear in our ponds and streams to join the company of diving beetles, whirligigs, and water striders. Look in the shallows along the edge of any pond, stream, or even spring pool, and you will probably find what appear to be bits of rubbish or bundles of sand grains or pieces of leaves moving as if carried along with the water currents. But remove them from the water and examine them and you will see that they are portable cases inhabited by wormlike creatures.

These wormlike creatures are known as caddisworms (Fig. 25A). They are not worms but the larval form of the caddis fly, an insect which resembles a moth except that its body is more slender and more delicately built. Caddisworms look like caterpillars, with chewing mouthparts and well-developed thoracic legs, but they have only one pair of abdominal legs or prolegs. They live for the most part on vegetable matter and move about with only their heads and legs exposed, ever on the alert to withdraw into their cases when disturbed. Hence it is that when you remove a case from the water you will most likely not see the insect at first. Let it remain undisturbed for a few moments, however, and its dark-colored head will soon emerge, followed by the six legs. Note that the case is open at both ends.

Caddisworms may be seen with the naked eye, but the lens is needed to find out how their cases are put together and to observe a few details concerning the insects themselves. They may readily be picked up with the fingers or scooped out of the water with a pan or some similar container. To examine the caddisworm, pull it gently out of its case and with the lens observe the little tassels of short, threadlike white gills along the sides of the abdomen. The water enters through the opening of the case at one end and passes out at the other, propelled by undulating movements of the caddis-worm's body. That the water may circulate freely over the gills and through the case, the insect is provided with three tubercles, called "spacing humps," located on the first abdominal segment, one on the back and one on each side. Look at the prolegs: they are furnished with hooks, called drag hooks, that anchor the caddis-worm to its case.

Unlike the snail, the caddisworm is not grown fast to its porta-ble dwelling. This can be demonstrated by a simple experiment. Using your finger, hold the case down on a flat surface with its occupant wrong side up. After a few struggles the insect will suc-ceed in turning itself over.

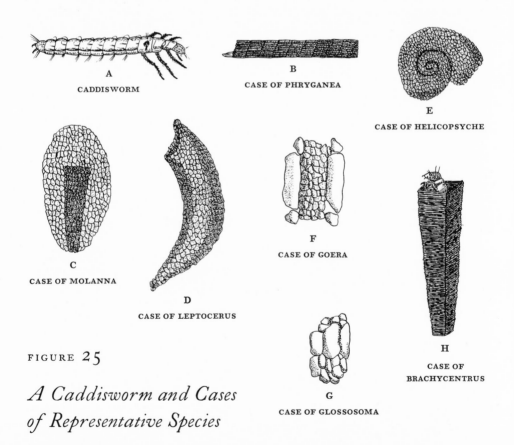

A

CADDISWORM

B

CASE OF PHRYGANEA

E

CASE OF HELICOPSYCHE

C

CASE OF MOLANNA

D

CASE OF LEPTOCERUS

F

CASE OF GOERA

H

CASE OF BRACHYCENTRUS

G

CASE OF GLOSSOSOMA

FIGURE 25

A Caddisworm and Cases of Representative Species

We can watch a caddisworm build its house by removing it from its case, placing it in a pan or a photographic developing tray containing some water, and providing it with the same materials its case was made from. It will immediately begin to build a new one. To follow its building activities use a reading glass instead of the hand lens. Should you wonder what a caddisworm would do if dispossessed of its case and deprived of its normal building materials, substitute other materials and see what happens.

There are some four hundred species of caddisworms in North America. They live in practically every kind of aquatic situation and can be found throughout most of the year, a few even in winter. To keep the record straight, not all build portable houses, not all have tubercles, and not all have abdominal gills; but those most of us are likely to find do. The cases of different species (Fig. 25B,C,D,E,F,G) differ greatly in form and in materials used, but each species builds its own type of case and uses a specific kind of material. Particles of dead leaves, sticks, pebbles, sand grains, and other substances are utilized and cemented together with a silk secreted by modified salivary glands. The silk is not spun into a thread but is poured forth in a gluelike sheet upon the materials to be cemented.

A most remarkable case is one made of minute twigs, roots, fibers, and fragments of wood cut to proper lengths to form even, straight sides. The case diverges toward the anterior end. During the first weeks of its existence, its maker, Brachycentrus, lives in the side waters of brooks, where it forages along the banks. Then it moves out into the center of the stream and attaches one edge of the case's larger end to a submerged rock or stone. Securely cemented to its support, the insect henceforth lives a sedentary life. It always selects an exposed situation and faces up current in a position shown in Figure 25H, with its head projecting slightly and its legs extended upstream. It is thus able to grasp and devour small larvae and bits of vegetation that float within its reach.

In addition to the insects, other forms of water life become active in April. Sponge gemmules, the winter buds of the freshwater sponge, germinate and colonies begin to form; freshwater mussels spawn; and painted and spotted turtles, having emerged from the bottom mud of the ponds in which they hibernated, swim about or bask on a protruding stone or partly submerged log, ever alert to plunge into the water upon the approach of danger, real or fancied.

As worms, insects, snails, and other small aquatic animals become abundant in the ponds and streams, the fish, which were on short rations through the winter, feast and soon grow fat and brisk.

FIGURE 26

Female Newt and Eggs

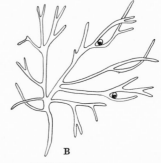

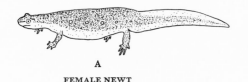

A
FEMALE NEWT

B
EGGS OF NEWT
AMONG LEAVES OF HORNWORT

The yellow perch has already begun to spawn, and the agile, blood-thirsty pickerel and pugnacious brook stickleback are getting ready to discharge the duty of egg laying so that the young may hatch and get a fair start for growth during the summer. And as our native fish recover from the stupor of winter and once again swim briskly through the waters, the migratory species—herrings, alewives, and shad—crowd to our watercourses from the sea. Meanwhile, the spotted salamander and the common newt have begun to mate and to lay their eggs in ponds and pools.

The female newt (Fig. 26A) lays her eggs one at a time, each encased in a capsule of jelly and fastened to the stem or leaf of a water plant (Fig. 26B). Under the lens the eggs are found to be brown at one end, creamy or light green at the other. Since as many as several hundred may be laid by one female, and since only a few are deposited each day, egg laying is a drawn-out process.

Sometime during the month you should find the spring peeper, whose mysterious piping voice has been the only evidence of his presence. Walk slowly toward the edge of a pond and you will most likely see him swim vigorously through the water and then climb up on a floating twig. He probably won't remain there long but will plunge into the water and swim to the protective cover of floating leaves. Here he will begin to sing, and you can see his swollen throat gleaming like a great white bubble. Another peeper may join him, and a third, a fourth, and still others, until many have joined the chorus. Make a noise, ever so slight, and at once there is complete silence. Later in the day when the sun begins to sink in the western sky, they sing in earnest, and throughout the night their high-pitched chorus, which reminds me of sleigh bells, can be heard for half a mile or more.

The singing of the frogs and toads in the ponds and marshes is as much a part of our spring as the appearance of the first wild flowers, the return of the migratory birds, and the blossoming of the trees, and is the prelude to mating and egg laying. To the uninitiated the singing of one species might sound much like that of another, but actually each species has its own distinctive call as well as several other calls. I can still remember when I first heard the chorus of wood frogs in a woodland pool and mistook their cluckings for the quacking of ducks.

Actually, the call of the wood frog is a short, sharp, snappy clack; the call of the leopard frog—the familiar frog of the marshes, ponds, and cattail swamps—is a low, guttural note.

The call of the pickerel frog is similar to that of the leopard frog but shorter, somewhat higher-pitched, with a distinct snoring quality. It resembles the sound produced when you tear a piece of resistant paper. Green frogs, although they appear in ponds fairly early, do not begin to call until May. The call is an explosive sound and once heard, is not quickly forgotten. Low-pitched and prolonged, it is likely to be repeated five or six times in succession. Twang a rubber band stretched tightly over an open box and you will have an idea of what it is like. Sometimes when it is given with less than its usual force, the call resembles the drumming of a woodpecker.

About the last of April or the first of May, we begin to hear the isolated calls of the common tree frog. The call is a loud, resonant trill, ending abruptly. The frog, who calls ten or eleven times in half a minute, begins in the early afternoon, especially on a warm, moist day, and continues into the night. I can still remember the time I visited a pond at night and with a flashlight spotted the gnomelike male by the water's edge, his vocal sac distended into a small, pearly, translucent balloon. He didn't seem to mind the glare of the flashlight and kept right on singing, unmindful of my presence.

One of the most musical sounds in nature is, perhaps rather surprisingly, the call of the toad. Toads emerge from their winter burrows in early April, and although an occasional sweet, tremulous call may be heard from a pool by the middle of the month, it is not until the end of the month that the chorus is in full swing. The male does all the vocalizing, producing a long, sustained, musical, high-pitched trill.

It might be interesting to collect a few eggs of some of these amphibians and examine them with the lens, for they all differ in size and color. For instance, the eggs of the spring peeper are white or creamy and black or brownish, the eggs of the common tree frog are brown and cream and yellow, the eggs of the leopard frog are black and white, and those of the pickerel frog are brown and bright yellow. Surprisingly, the eggs of some of the smaller species, such as the robber frogs and little chorus frogs, are the largest, while the smallest are those of the large bullfrog. Some species, like the peeper, lay each egg separately; others lay several eggs at a time. Frogs, such as the leopard and pickerel frogs, lay their eggs in gelatinous masses; toads lay their eggs in long, curling tubes of jelly containing a single row of black eggs.

April is the natal month for many wild animals. The naked young of the white-footed mouse are born in a warm nest of leaves and shredded bark; the kittens of wildcats and young mink are brought forth in hollow logs and burrows; the flying squirrel introduces her young ones to the world in a tree cavity; the red fox gives birth to pups in a secluded den; and baby raccoons are born in chambers in hollow trees.

Now that domestic cares have been thrust upon him, the raccoon can no longer idle his time away, for food is not yet plentiful and he is often compelled to go hungry. In the mad scramble for food, however, he has a decided advantage over the woodchuck and other vegetable eaters, since he will eat almost anything. But despite his omnivorous tastes, I have never known him to eat mushrooms. In general I have often wondered why more animals do not eat these lowly plant forms. If we can eat them without injury—that is, the nonpoisonous kinds—certainly many of the herbivorous mammals should be able to do so. Of course, there are a few animals that do include them in their diet. The common slug, whenever it chances upon a mushroom, usually stops to gourmandise until there is practically nothing left of the plant. Insects are especially fond of a mushroom diet, and I dare say that the larvae of certain kinds are among the happiest creatures on earth, for the mother is particularly careful when depositing her eggs to seek out only the tastiest mushrooms to serve as a food bed for her offspring. The tortoise, too, will often stop in its wanderings to nibble at a mushroom, and deer and cattle have been known to feed on them. But the prime mushroom eater is the red squirrel. This little animal is especially fond of them and even stores them in the forked branches of trees for future use. Though April is not a mushroom month, several species may be found, such as the glistening

FIGURE 27

The Coprinus,
Sponge Mushroom,
and Horsetail

A

GLISTENING COPRINUS

B

SPONGE MUSHROOM, MOREL

C

HORSETAIL,
FERTILE SHOOT

coprinus and the sponge mushroom or edible morel, the latter being one of the most sought after.

The glistening coprinus (Fig. 27A) may be found at the base of trees or on covered stumps, particularly after a shower, and the sponge mushroom (Fig. 27B) or morel in orchards and woodland areas. With your lens you will observe the cap of the former often glistening with minute shining particles and that of the latter covered with a network of blunt ridges enclosing irregular depressed areas.

Before the month is out, countless horsetails will shoot up in sandy and gravelly places, such as along roadsides, in waste places, and along railway embankments. Although these plants, as shown in Figure 27C, do not seem to bear any resemblance to a horse's tail, the green vegetative shoots that follow later do so if we use a little imagination. Horsetails are of no economic value today, but if we can judge from the fossil remains, which show a considerable number and variety of these plants, they probably were at one time an important element in the flora of the earth. Like the ferns and club mosses, they seem to have contributed their vegetative parts and spores to the formation of coal during the Carboniferous period.

At a casual glance the horsetail stem, which is the same diameter from top to bottom, is pale and rather weird in appearance. With the lens it is seen to be ornamented at intervals with slender black pointed scales. These scales, which are much reduced leaves that long ago lost the ability to carry on photosynthesis, point upward, are united at the base, and encircle the stem in a slightly bulging ring. At the tip of the stem is a cone-shaped whitish structure called a strobilus. As shown by the lens, it is made up of tiny disks (sporangiophores) that remind us of miniature mushrooms, each of which bears five to ten spore-producing structures or sporangia. After the spores have been released, the sporangia remain on the sporangiophores in torn scallops.

They do not remain there for very long, however, because once the spores have been released, the stem dies and is followed by a slender green stem with numerous branches set in whorls at intervals or nodes. At one time the stem probably had a whorl of leaves at each node, but since there are now so many green branches, the leaves have long since been reduced to mere points and appear to be only a sort of trimming. Both the stem and the branches are composed entirely of segments. The bottom of each segment is set into a socket of the segment below it, and each is easily pulled out. Handle a horsetail and you find it is rough and harsh in texture. The reason is that it is impregnated with silica. Because of its abrasive quality the horsetail was once used for cleaning and polishing metal utensils and was called a scouring rush.

In the North at this time of year, few grasses have advanced to the point of flowering, though the sweet vernal grass (Fig. 28A) is likely to have pushed up its compact, spikelike panicles, and perhaps has expanded them with open blossoms of violet anthers. In the South there is rarely a month when the low spear grass (Fig. 28B) is not in flower, but in Northern latitudes it is not apt to blossom until spring is well on its way. In April, however, it will send up tufts and so change the hillsides from brown to green.

Since the grasses have not yet grown too high, we can easily see the wolf spiders running over the ground. Sometimes one of them can be seen dragging a tiny globular parchmentlike affair covered with silken threads. This is the egg sac. When the eggs hatch, the spiderlings climb upon the body of the mother and are carried about, papoose fashion, for some time.

These spiders capture their prey by stalking it or chasing it in the manner of wolves; hence their name. There are a number of species, some of which have acquired the habit of digging burrows in the ground to which they retire when not out hunting. A few merely excavate a shallow depression beneath a log or stone and line it with silk, but others dig a vertical tube, going down as much as a foot or more. Some of the latter species build around the opening a circular wall of earth and pebbles which they bring up from below, or a turret of grass and dirt which they fasten together with silk or bits of twigs.

The turret is not an ornament but is used as a watchtower. The spiders spend much of the day perched on the turret with their heads projecting just far enough so that they can view the immediate neighborhood. From such a vantage point they can detect their prey more readily, and can equally well observe the approach of an enemy, from which they can escape by dropping into the burrow.

FIGURE 28

*Two Early
Spring Grasses*

A
SWEET VERNAL GRASS

B
LOW SPEAR GRASS

During the month garter snakes may be seen basking in the sunshine, for by now most of them have emerged from their winter quarters. Lizards, too, may be appearing, depending on the region. In New England these animals are scarce, and as far as I know only one species, the five-lined skink, has extended its range into this section of the country. But more about these animals later.

April, of course, is the month when there is a great influx of birds from the South. Indeed, April has hardly arrived when coots, rails, and members of the heron family put in an appearance on the coast and in the marshes, taking the place of the ducks that have journeyed northward. About the second week I usually see the Savannah sparrow searching among the emerging grasses for his insect food, and a few days later I hear the trill of the chipping sparrow early in the morning. And one day—it may be early or late in the month—I catch a glimpse of the yellow palm warbler in a thicket, flitting from bush to bush, wagging his tail with the same unhurried motion as the phoebe, or hear the scratching of the brown thrasher as he rustles the dead leaves, but I do not see him often, for he is shy and furtive; it is only when he calls to his mate in loud clear notes from a sapling that I see him silhouetted against the sky.

The swallows appear when you least expect them, and in a newly plowed field you can see the killdeer searching for grubs.

From almost any thicket you can hear the peculiar whistle of the white-throated sparrow, and if you go into the woods on a warm, sunny April day you are likely to come upon the towhee in a sun-kissed clearing, scratching among the leaves like a fox sparrow. Perhaps you will also see the myrtle warblers by the brookside, chasing flying insects.

Certainly you will hear the scream of the red-shouldered hawk as it circles high in the sky. The flight of the hawk always fascinates me, and I invariably stop to watch it sail through the air on out-stretched wings, ascending and descending, balancing on the cool currents high above the ground, sometimes stationary on motion-less wings, sometimes climbing an invisible staircase until it is a mere speck in the sky, then suddenly lifting its wings and plunging toward the earth with meteoric speed.

By now the robins and bluebirds have begun to turn their thoughts to domesticity, and the chickadees, that have been around the house all winter, have scattered to the seclusion of the nearby woods to prepare their nests. The male house wrens, too, are busy collecting sticks and building nests as they await the arrival of their mates; but they could save themselves the trouble, for the females will use neither the materials they have collected nor the nests they have built, and sometimes not even the site they have selected. But you could never convince them that their labor is wasted, since it is all a part of their courtship ceremony.

Actually it is a little too early for nest building. But several birds, among them the meadowlark, have begun to court their mates, and almost any day I expect to see the catbirds perform their courtship antics in the seclusion of the lilac bushes outside my window. Meanwhile other birds are arriving in great numbers, display-ing themselves joyously to each other, singing loudly in rivalry or for pure delight, and staking out claims to nesting areas; the trees are getting ready to unfold their young, tender leaves; the vibur-nums and dogwoods are beginning to show signs of blossoming; and everywhere ferns are breaking through the soil in watchspring fashion, the interrupted and cinnamon ferns wrapped in brownish wool, others less warmly clad. May is about on us and spring is well on its way.

MAY

"Then bursts the song from every leafy glade."

OLIVER WENDELL HOLMES

"Spring"

MAY

AS APRIL PASSES INTO MAY, THE WEATHER becomes more settled and we can expect the days to be warm, when the winds blow gently, the sun shines brightly, and cottony clouds build mountains in the sky. But there can be some cold days and a night or two of frost, when a cold front pushes its way down from Canada; such days, however, are not a part of May, or shouldn't be. The weather in the North is freakish anyway, but elsewhere—such as in the south—May is summery and the landscape is a pageant of colors from countless wildings and garden blossoms.

In the South the trees are in summer foliage: not only the evergreens that are in leaf throughout the year, but also the deciduous trees, for they replace their leaves early. Or should we say that spring comes early though the four seasons in the South are not so distinct and different in character as they are farther north. In the North it is not until the first week in May, or perhaps a little earlier or a little later, depending on the weather, that the deciduous trees begin to open their buds. One day they may be quite naked, and on the morrow, or so it seems, are dressed in leafy splendor. Of course, all do not bring forth their leaves at the same time but gradually, as the month progresses.

When the leaves first emerge they are tender and fresh, a delightful green, as yet unmarred by the disfigurements of summer. But not all are green; the leaves of the white oak come out a bright red, fade to a soft pink, then become a silvery white and finally a yellow green. And as the young leaves are covered with a soft down, the entire tree acquires a frosty, misty appearance quite pleasing to the eye, though unfortunately this effect lasts only a few days.

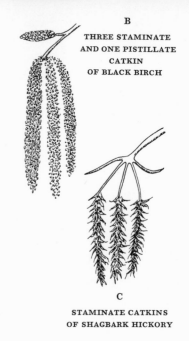

FIGURE 29

Catkins of Scarlet Oak,
Black Birch,
and Shagbark Hickory

B

THREE STAMINATE
AND ONE PISTILLATE
CATKIN
OF BLACK BIRCH

A

STAMINATE CATKIN
OF SCARLET OAK

C

STAMINATE CATKINS
OF SHAGBARK HICKORY

Willows, poplars, and some maples have already blossomed, but oaks, birches, hickories, and a host of other trees wait until May to decorate the landscape with countless blossoms that provide a festive board for bees, butterflies, and a horde of other insects. The staminate flowers of the oak (Fig. 29A) are borne in fringelike catkins, the pistillate in few-flowered clusters. We might look briefly at the flowers of the scarlet oak, but any oak will do. The staminate flowers have a bright yellow calyx and yellow anthers (Fig. 30A); the pistillate involucral scales are reddish, the stigmas a bright red (Fig. 30B); perhaps not the most beautiful of flowers, but a pleasant surprise when looked at with the naked eye and then with the lens.

The hop hornbeam, a lonely forest tree, and its cousin the hornbeam or blue beech, common on the borders of streams and swamps, also bear flowers in catkins. So, too, do the birches. There are nine different kinds of birches in North America. Six are trees, and five of them are found east of the Rocky Mountains. Here again, any birch will do if you want to examine the flowers. I am rather partial to the black birch, perhaps because its inner bark is fragrant and has a pleasant spicy taste, but every birch has something to commend it. The scales of the staminate flowers (Fig. 29B) are bright red-brown above the middle, pale brown below, a color scheme lost to everyone unless he is of an inquiring nature and has a lens in his hand. The anthers are a bright yellow. The colors of

FIGURE 30

Flowers of Various Trees

F
STAMINATE FLOWER
OF BEECH

A
STAMINATE FLOWER
OF SCARLET OAK

C
STAMINATE FLOWER
OF SYCAMORE

G
PISTILLATE FLOWER
OF BEECH

B
PISTILLATE FLOWER
OF SCARLET OAK

H
FLOWER
OF TULIP TREE

I
STAMINATE FLOWER
OF WHITE ASH

D
STAMINATE FLOWER
OF SUGAR MAPLE

E
PISTILLATE FLOWER
OF SUGAR MAPLE

K
FLOWER OF
FLOWERING DOGWOOD

J
PISTILLATE FLOWER
OF WHITE ASH

the pistillate flowers (Fig. 29B) are somewhat more subdued; the scales are pale green, the styles a pale pink.

As far as I know, the hickories are strictly American trees; not one species is found elsewhere. They are closely allied to the walnuts, the chief botanic distinction being in the husk; and like the oaks, they too have taken a firm hold of the earth. And also like the oaks, their flowers are borne in catkins, the staminate in groups of three (Fig. 29c), the pistillate in clusters of two to ten. Everyone who roams the woods knows the shagbark, a ruggedly picturesque tree, its bark broken into long, unsightly pieces sometimes 3 feet or more in length, loose at the edges but clinging by the center as if reluctant to let go. All trees, of course, shed their bark; they have to so that the trunk may increase in width. But most trees do so unobtrusively, the bark falling off in plates or scales of varying size. The shagbark has no compunction in advertising the process, the silver maple hardly less so, and the sycamore not at all, for the bark of this tree flakes off in great irregular masses, leaving the surface a crazy patchwork of greenish white and gray and brown; the smaller branches sometimes appear as if they had been given a coat of whitewash.

The sycamore also blossoms in May, both the staminate and pistillate flowers in dense heads (Fig. 30c), the staminate dark red on axillary peduncles, the pistillate light green tinged with red on longer terminal peduncles. The sugar maple, its sugary sap a highly prized article of commerce, blossoms somewhat later than its cousins the silver and red maples, its greenish-yellow flowers (Figs. 30D, E) in hairy, thick clusters appearing sometime during the month. The flowers of the beech, too, appear in clusters, the staminate (Fig. 30F) a yellow-green in pendant balls, the pistillate (Fig. 30G) solitary or paired, and surrounded by numerous awl-shaped bractlets.

It is difficult to refrain from comparing the small flowers of these woodland and forest trees with the large showy blossoms of the magnolia and its cousin the tulip tree. The flowers of magnolias are often 6 inches, sometimes a foot, in diameter, and in most species deliciously fragrant. Where they are planted as ornamentals, as in the South, they perfume the air with a scent quite overpowering. When you examine a blossom note the texture of the petals: they are thick, waxy, and lustrous, and, of course, the coloring is exquisite. The leaves also are most attractive—in many species shining, leathery, and evergreen, sometimes exceeding a yard in length. Their foliage is most luxuriant and tropical; yet magnificent as these trees are, they now are but a shadow of their former greatness, for eons ago in preglacial times, forests of these trees covered the midcontinental plains of Europe and America and even extended northward to the Arctic Circle.

Magnolias are essentially Southern trees, but the cucumber tree, the hardiest of them all, is found as far north as the Canadian border. Somehow it seems out of place in our northern forests, for its large green leaves betray the tree's tropical character and are a sharp contrast to the smaller leaves of the oaks, elms, and maples. Although the yellowish-green flowers become fairly large, measuring about 2 inches long, they are virtually lost among the leaves. Unlike the flowers of most magnolias, they are neither beautiful nor fragrant. The fruit at first is green and shaped like a cucumber, but it soon flushes pink and later turns red as autumn approaches. Then, as the fruit matures, the red berries within break through the skin and hang for a time on long white threads, eventually to be eaten by the birds or to be torn off by the wind.

Place a garden tulip on a branch of the tulip tree and it could almost pass for the blossom of the tree itself. Large and showy, with dashes of red and orange on its greenish-yellow tulip-shaped corolla (Fig. 30H), the tulip blossom in no wise fails in comparison with the cultivated *Tulipa*. But we have to look closely at the flower to appreciate its distinctive charm—something we are not apt to do, for usually it not very accessible and we have to stretch a bit to reach it. In the forest the tree reaches a height of 190 feet, a magnificent size; but when it stands alone it attains its finest growth, for then its trunk rises like a Corinthian column into a long, narrow, pyramidal crown of pleasing symmetry. We don't realize how tall it can get until we see one outlined against the sky and then compare it with neighboring oaks, elms, or maples.

I first became acquainted with ashes in the Arnold Arboretum[1]

---

[1] Which belongs to Harvard University and is located in Jamaica Plain, Massachusetts.

years ago. I was a novitiate in botany and often visited the place for a practical exercise. I can still recall the day in May when the ashes were in bloom and I examined the flowers for the first time. Perhaps it was because they intrigued me so that the occasion is still fresh in my mind.

There are a number of species, but the white ash is generally considered the most beautiful of our American ashes. The flowers appear before the leaves, like most of our forest trees, the staminate on one plant (Fig. 30I), the pistillate on another (Fig. 30J). At first the panicles are compact; later they become long and loose. The staminate flowers come out purple, then they turn to yellow; the pistillate ones are vase-shaped and purple, with an elongated style and a spreading divided stigma. The naked eye cannot see them adequately; we need the lens to do them justice.

Probably no plant satisfies our sense of the aesthetic more than the wild crabapple in blossom, for then the tree is invested in a rose-colored bloom and fills the air with a spicy, stimulating fragrance. Linnaeus probably never saw the tree, but he must have possessed a degree of omniscience when he named it *coronaria*.

The cultivated apple may not have the same appeal as its wild counterpart, but an apple orchard that extends as far as the eye can see is a breathtaking spectacle when in blossom. And no less so is a peach orchard in the South when countless blossoms veil the land-scape in a rose mist. The pear, the plum, the cherries, and the quince, too, blossom in May—the pear and plum in thickets and woodland borders, the wild black cherry along the roadside, the red cherry in burned-over lands, and the quince in old gardens. Ex-amine any or all of the flowers and you will find that they are all built on the same plan: five sepals, five petals, and many stamens in multiples of five that make them kin to the rose. To be sure, there may be minor differences in flower structure, but they do not in-fringe on the general floral pattern.

I doubt if there is a more decorative ornament in the wildwood than the flowering dogwood when its naked branches become ornamented with a multitude of white or pink-flushed blossoms. Strictly speaking, I should not call them blossoms; the actual flow-ers are small and greenish and somewhat tubular. What then are the four large white petal-like structures that catch our eye as we wander along a woodland trail or stroll down a country lane? They are merely involucral bracts or scales, reduced leaves designed to attract the mining bees and other insects with similar appetites to the nectar in the floral tubes. Look at the flowers with your lens; they are in clusters of ten to thirty and are made on a plan of four: four lobes to the calyx, four divisions of the corolla, and four stamens (Fig. 30K). There is only a single pistil. Later it will become an egg-shaped scarlet drupe.

Though a few shrubs have blossomed in April, most of them wait until May and throughout the month they parade through the land with all the colors of the artist's palette: forsythia, lilac, bridal wreath, mock orange, weigela and barberry in the garden; prickly ash along a river bank; beach plum and downy hudsonia by the seashore; black alder and bladdernut in moist woodlands; chokeberries on rocky uplands; smooth winterberry and swamp laurel in swampy places; Labrador tea in cool bogs; bayberry wherever it can find room to grow; and dogwoods and viburnums and a host of others—the list is almost interminable.

In the South the azaleas may still be in full bloom, and where they are planted in gardens and parks furnish a riot of color. We have azaleas in the North too, the cultivated kinds as well as the wild ones such as the pinxter flower or pink azalea (sometimes erroneously called the wild honeysuckle, though it is not a honeysuckle at all). Sometimes I find it in blossom in April, but it is usually in May that I find a sunny glen aglow with its fragrant rosy masses. Merely a glance at its long protruding style (Fig. 31A) tells us that it forms a convenient landing place for a visiting bee.

Few wild flowers are pollinated by the wind except incidentally; most of them depend on insects and by design or accident have become modified the better to attract and make use of them. Somehow the early meadow rue never did get around to doing so. In some parts of its range the plant blossoms in April, but I usually don't find it flowering until May in rocky woods where its feathery flowers hang in clusters like fleecy clouds. The flowers are small and rather insignificant, yet are not unattractive when seen with the lens. They lack a corolla and hence have no way of attracting insects, but since they do not have any nectar there is no reason why they should do so. Undoubtedly insects do visit the flowers and occasionally must transfer some of the pollen, but this primitive plant nevertheless depends almost wholly on the wind.

With its feathery plume of small white flowers the wild spikenard, or false Solomon's seal, seems more a plant of summer than one of spring. It is true that it extends its flowering season into July, but it is essentially a flower of the spring woods and thickets. The flowers are slightly fragrant, though probably not enough so to attract the bees and other insect visitors were they set sparingly on the stem. Clustered together as they are, however, they collectively advertise their wares to every passing insect, though their chief guests are the mining bees of the Halictus tribe.

Where the spikenard grows we also find the bellwort, its pale yellow bell hanging from the tip of the branch and almost hidden from view by the leaves, so that we often fail to see it unless we actually look for it. Examine the flower closely and that of the wild spikenard and you will see their relationship to the lily. The lily family is well represented among May flowers, with the Canada mayflower, a tiny woodland plant; the little star of Bethlemen of grassy meadows; the Indian cucumber of swampy woods; and the lily of the valley in the garden. Another is the greenbrier that zigzags its prickly way through tangled underbrush and over stone walls and fences which line the roadside, its yellow-green flowers fewer in number than those of its cousin, the carrion flower of foul odor.

Take almost any woodland trail and you will find the frosty-white little starflower with golden anthers poised airily above a whorl of leaves. And midst the brown carpet of the woodland floor, if you are lucky, you will see the striped pinkish pouches of the moccasin flower, or stemless lady's slipper, swinging balloonlike in the murmuring wind. There are other lady slippers, some more showy and perhaps more beautiful, but the moccasin flower is supposed to be the most common of them all. I am not too sure that it is, since many thoughtless people have picked it indiscriminately and it is rapidly vanishing from the scene. But if you can find it, observe how ingeniously it compels insects to carry its pollen. For once an insect has feasted in the large banquet chamber, it must first rub along the sticky overhanging stigma with its rigid, sharply pointed papillae that comb out the pollen it has brought from another flower, and then it must

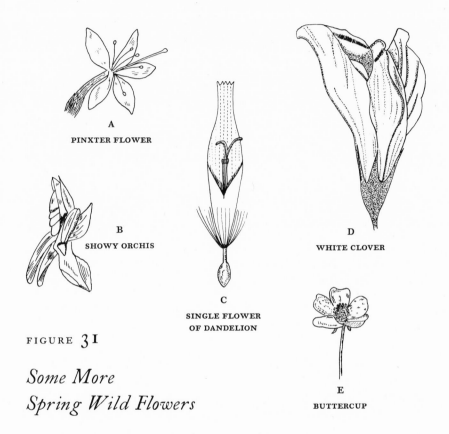

A
PINXTER FLOWER

B
SHOWY ORCHIS

C
SINGLE FLOWER
OF DANDELION

D
WHITE CLOVER

E
BUTTERCUP

FIGURE 31

Some More
Spring Wild Flowers

pass an anther that, drawn downward on a hinge, covers it with more pollen before it can emerge into the open.

The orchids, often called the elite of the plant kingdom, probably show the most advanced development of floral structure for insect pollination, the flowers having been modified to function as spring traps, adhesive structures, and hair triggers attached to explosive shells of pollen. Indeed, in some instances, the flowers have become so modified that they look like insects themselves.

An orchid flower consists of three similar sepals, two lateral petals and a third differentiated into a conspicuous lip that is often spurred and always conspicuously colored and that secretes nectar. In most orchids there is only one stamen, united with the style into a single fleshy body known as the column, which is placed at the axil of the flower and facing the lip. The style, instead of ending in the stigma as in most flowers, terminates in a beak, the rostellum, at the base of the anther or between the two anther cells. The stigma, which is gummy or viscid, is located directly below the rostellum. Two stemmed, pear-shaped pollen clusters called pollinia are produced in the anther cells. Each of the pollinia is composed of several packets of pollen that are tied together by elastic threads which run together and form a stem ending in a glandular or sticky disc contained within the rostellum. It is the disc that becomes attached to an insect and ensures the transfer of the pollen to the stigma of another flower.

If you can find the showy orchis (Fig. 31B) at the moment a bumblebee is visiting it, you can see just how an orchid manages to secure cross-pollination by an insect. Or perhaps it would be better to find the plant first and then wait for a bumblebee. It requires a little patience, of course; unfortunately we cannot command the insects to perform when we want them to. Incidentally, a reading glass would be most useful in this case.

The showy orchis grows in moist, rich woods; I most frequently find it in hemlock groves. The flowers, few in number on the stem, have the sepals and petals lightly united in a magenta or dark rich crimson hood with a conspicuous, almost white, lip prolonged into a spur in which the nectar is secreted. When the bumblebee, generally a queen, alights on the projecting lip, she thrusts her head into the spur, and as she sips the nectar, brushes past the rostellum at the top of the column, thus rupturing the thin membrane covering it and exposing the sticky round discs attached to the pollinia. The discs immediately adhere to the bee's face or forehead and are carried away by her as she flies to another orchis blossom, where the pollen grains are scraped off on the viscid stigma.

The cultivated columbines I have in my garden, though considerably larger and more showy than the wild varieties, somehow lack the appeal of those I find in the woods, where their scarlet and yellow cornucopias dance in the breeze with an elfin charm. I think this is true with most flowers, or at least with the wild ones, for whenever I have transported them from the fields or woods to the garden, many of them do not seem to belong there. There are some who will disagree with me, but I much prefer to see them in their natural surroundings. Of course, many of them are garden escapes, but presumably all of them were originally wildings.

As I write, my thoughts turn to the fringed polygala, a dainty and charming little flower that I often find hidden in the woods or along a stone wall by the roadside. Purplish rose, magenta, or crimson magenta, with two winglike sepals highly colored like the petals and the three petals united in a tube and fringed at the end, the flower looks like a tiny butterfly that has settled on the ground. The fringe has a practical purpose; it serves as a landing platform for a visiting bee that, upon alighting, depresses the platform. The stamens and rigid pistil, which are contained within the tubular petals to protect them from rain and useless visitors, are thereby forced out and come in contact with the insect, when an exchange of pollen takes place.

The beautiful little naked broom-rape comes to mind too. It is a rather curious plant and a parasite, fastening itself on the roots of other plants in woods and thickets and drawing its nourishment from these roots. There is but one flower to a stem; it is purplish or violet in color and delicately fragrant. I recall, too, the wood sorrel, a dainty woodland plant whose leaves close at evening, and the May-apple and the Indian turnip, perhaps better known as jack-in-the-pulpit. The May-apple is a handsome woodland plant with white waxy flowers that are all but hidden by the umbrella-like leaves. These are remarkable for their size, sometimes measuring as much as a foot in width.

Merely a glance at the Indian turnip shows its affinity to the skunk cabbage. The minute greenish-yellow flowers are clustered on the lower part of a smooth, club-shaped, slender spadix within a spathe that curves in a broad-pointed flap above it. As revealed by the lens, the flowers are usually staminate on one plant, pistillate on another; hence the plant must depend on insects for cross-pollination. These are small insects, the fungus gnats of the genus Mycetophila, which may frequently be found imprisoned in the narrow confines between the bases of the spadix and the spathe. Sometimes both kinds of flowers may be found on one plant; this is the exception rather than the rule, but it suggests the possibility that the plant's dependence on insects for fertilization may be a recent development. The curiously shaped green and maroon or whitish striped spathe is somewhat variable in depth of color. In the open and exposed to the sunlight, it is paler than when found in the deep woods, where it is dark purple. The Indians utilized the farinaceous root (corm), which is shaped somewhat like a turnip, as an article of food. I would advise against trying it, for unless properly prepared it can produce blisters in the mouth. Both the skunk cabbage and the Indian turnip belong to the arum family, which also includes such water-loving plants as the dragon arum, arrow arum, water arum, golden club, and sweet flag.

Although the blue-eyed grass may be found blooming during the greater part of the summer, it first appears in May, and you will find it blossoming among buttercups, clovers, and dandelions in almost any field and meadow. It is not a grass in spite of its name, though it is rather grasslike in appearance with its linear pale blue-green leaves; as a matter of fact, it is related to the blue flag of pond margins and to the iris of the garden. It is a dainty flower with a perianth of six spreading divisions, violet-blue or purple in color and each tipped with a thornlike point. Its center is beautifully marked by a six-pointed star accented with bright golden yellow, each of the star points penetrating the petal-like divisions. Only for a day, and the day must be a sunny one, does the flower open; and as the day draws to a close it folds together never to open again. But on the morrow a new bud will open, and a new flower will take its place. Pick a flower and it will close immediately.

The yellow star grass also has a grasslike name, but it, too, is not a grass. It is a plant of the fields and meadows, and one would think that its yellow flowers would be lost among the buttercups and dandelions. But somehow it manages to get its starlike blossoms high enough to call itself to our attention; more important, the smaller bees and smaller butterflies find it readily and visit it often, for the flower produces an abundant supply of pollen as if to make sure of their continued visits.

Perhaps no flower has devised such an ingenious way to prevent self-fertilization as the wild geranium, since the pollen becomes mature and the anthers fall away before the stigma becomes receptive. Look as you may, and though the flower is perfect, you will not find the two reproductive structures present at the same time. The wild geranium is a delicate flower, pale or deeper magenta, purplish-pink, or light purple or lavender, and is common in open woods, thickets, and shady roadsides. During cold, rainy, or cloudy weather it may retain its anthers for several days before the stigma becomes mature, while on a warm, sunny day, when insects are flying, the change may take place within a few hours. The blossoms are cross-fertilized mostly by honeybees, the smaller bees of the Halictus tribe, and the flowerflies (Fig. 32A).

It is human nature to view the common and familiar with a certain amount of indifference, and sometimes this indifference may border on the contemptuous. I don't think I have ever heard anyone speak well of the dandelion, simply because it is a weed; were it a rare exotic everyone would extol its virtues. Yet it is those very virtues that makes it a weed, if we can define a weed as a plant that grows where we do not wish it to grow, for the dandelion seems able to grow almost anywhere. That is one of the reasons why it is one of the most successful of plants; it can meet the fierce competition for existence and survive.

Look at a dandelion dispassionately and you will see why. Note how deeply the stocky root penetrates into the ground, far below where heat and drought can affect it or where nibbling rabbits, moles, and grubs can break through and feast. Watch the winds buffet and bend the stem, and though it is a hollow tube, how invincibly strong it must be, since no harm befalls it. As any engineer will tell you, a hollow tube is stronger than a solid one. Why are grazing cattle not tempted by it even though they devour other succulent plants indiscriminately? Is it because the rosettes of leaves secrete bitter juices?

Examine the golden-yellow flower head with the lens and what will be revealed? Not one flower but often three hundred minute, perfect florets, all cooperating to ensure cross-pollination from small bees, wasps, flies, and other insects that come seeking the nectar, secreted in each little tube, and the abundant pollen, both of which are greatly appreciated in early spring when food is scarce. And after flowering, the golden head is transformed into a globular white, airy mass of tiny parachutes, each one a seed, and each one ready to sail away on the slightest breeze, to be carried, perhaps, untold distances before finding a resting place.

The dandelion represents a type of inflorescence known as a head. Such an inflorescence consists of a number of (usually) small flowers that are sessile (without a pedicel or stem) and crowded together on a flattened or convex-shaped receptacle. The bunching together of many small flowers intensifies the color and scent attractive to insects and thus lures them more efficiently than a number of small individual flowers would be able to do. Many plants accomplish this objective by grouping their flowers closely together in the form of a raceme, spike, umbel, corymb, cyme, and so on; but the most effective is the head, as found in the dandelion and other members of the composite family: sunflower, daisy, aster, chicory, thistle, and goldenrod. In many of these species there are two kinds of flowers: tubular and strap-shaped, as we shall see later. In the dandelion they are all strap-shaped (Fig. 31c). Once upon a time, the flowers may have been five-petaled blossoms, for the five teeth at the top of the corolla and the five lines descending from them would seem to indicate that once distinct parts had been fused together to form a more showy and suitable corolla. Observe the five anthers that create a tube from which the pistil extends with its two-lobed stigma. Also note that the florets may be in various stages of development; those in the outer row of the dandelion head blossom first. After a corolla has opened, the anther tube first appears and then later the pistil, which gradually rises out of the anther tube and extends above it, when the stigma lobes quirl back.

The familiar red and white clovers (Fig. 31D), appearing during the month in fields and meadows and along the roadside, are also heads of small individual blossoms. The lens shows that each one is perched at the tip of a little stalk and that each has a tubular calyx with five delicate points and a corolla that reminds us of the sweet pea. The upper petal, known as the standard or bearer, more or less encloses the two lateral ones, called wings, and the two

G

COCOON OF BLACKFLY

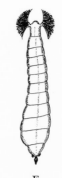

F

LARVA OF BLACKFLY

FIGURE 32

Stages in Life History of the Blackfly; Other Insects

A

FLOWERFLY

H

BLACKFLY

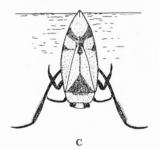

C

BACKSWIMMER

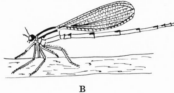

B

DAMSELFLY

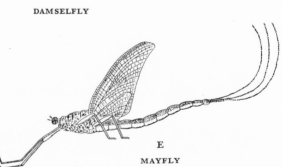

E

MAYFLY

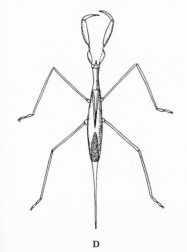

D

WATER SCORPION

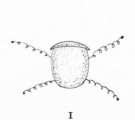

I

EGG OF MAYFLY

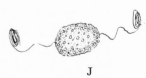

J

EGG OF MAYFLY

lower ones are united to form the keel. There are ten stamens, nine of which are united, and one pistil.

In the red clover the individual florets are closely packed together in a somewhat pyramidal globular cluster that ranges through crimson or magenta to paler tints of the same colors. Each floret contains an abundance of nectar, but it is of little use to the honeybees, for their tongues are too short to reach down into the nectar wells.[2] The butterflies with their long tongues can do so easily and often drain the nectaries without giving any service in return, for they are not sufficiently heavy to depress the keel and thus expose the anthers. Only the burly bumblebees can do so and hence are the plant's chief benefactors. Years ago, farmers in Australia planted red clover and had a bounteous crop, but failed to get any seed to plant for the following year simply because they had neglected to import the bumblebees. Once they had taken care of the oversight, all went well.

If there is any one flower that we truly associate with May, I would say it is the buttercup. I think we can trace this association back to our childhood days, when we found a special delight in the bright-yellow blossoms. There is an early buttercup that blossoms in April, but the one we know so well appears in May. The scientific name of the flower is Ranunculas, which means "little frog" in allusion to the wet places where buttercups like to grow. But though there are some species that prefer moist habitats, others are common in the woods, fields, and roadside banks.

There is nothing complex about a buttercup's blossom (Fig. 31E). Five pale-yellow sepals with brownish tips, though they are green in the bud; five petals, pale beneath but a bright yellow above, and shining as if they had been varnished; and numerous stamens and several pistils; this about tells the story. Each petal is wedged-shaped, with its broad outer edge curved to form a cuplike flower. Remove a petal and examine it with your lens, and you will find a small scale at its base. It covers the nectariferous pit.

Find a buttercup that has just opened and you will see the anthers huddled in the center; later they form a fringy ring about the pale green pistils, each pistil having a short yellowish stigma.

[2] It has been shown, however, that honeybees can be induced to pollinate red clover.

Note how the anthers open away from the pistils to prevent self-fertilization; note also that they shed much of their pollen before the stigmas are ready to receive it.

In such a galaxy of flowers as the hepatica, wood anemone, early meadow rue, marsh marigold, columbine, and buttercup, only the marsh marigold and buttercup would seem to have anything in common, and yet all are of the same family—a family considered by botanists a primitive representative of the angiosperms. But who cares whether they are primitive or no; suffice that they are among the most delightful flowers of the wildwood and bring cheer and joy to those of us that travel the byways of the nature world.

Many years ago Charles Darwin, in writing about earthworms,[3] said that "it may be doubted if there are any other animals which have played such an important part in the history of the world as these lowly organized creatures." It is rather a far-reaching statement and one might be tempted to question it, but on reflection there is a great deal of truth in it. To many of us the earthworm is merely of value as bait for catching fish. But because of its underground activities it is of inestimable value to the farmer and to those of us who have a garden, for it works unseen day and night, harrowing and fertilizing the soil for our benefit. It burrows 12 to 18 inches into the ground and brings the subsoil to the surface; it grinds the soil in its gizzard and changes it into a soil of finer texture; and it even fertilizes the soil by secreting lime that neutralizes the acids in it.

The earthworm is not only a tiller of the soil; it is an agriculturist as well, for it plants fallen seeds by covering them with soil and cares for the growing plants by cultivating the soil around the roots. Furthermore, it enriches the soil by burying the bones of dead animals, shells, leaves, twigs, and other organic matter that, upon decaying, furnish the essential minerals to the plants. It even provides drainage by boring holes to carry off the surplus water, and by doing so promotes aeration.

The changing character of the landscape and much of the beauty of our fields and forests can be attributed to the labors of this diminutive workman; the earthworm can also be credited with hav-

[3] "The Formation of Vegetable Mold through the Action of Worms."

ing preserved many ancient ruins and works of art by covering them with earth. The familiar mounds of black earth or castings that we so often see on the ground or on our lawns are particles of soil swallowed in their burrows and brought to the surface. Since there may be as many as fifty thousand worms in an acre of ground,

FIGURE 33

Earthworm, Hairworm, and Multiple Tails of Hairworm

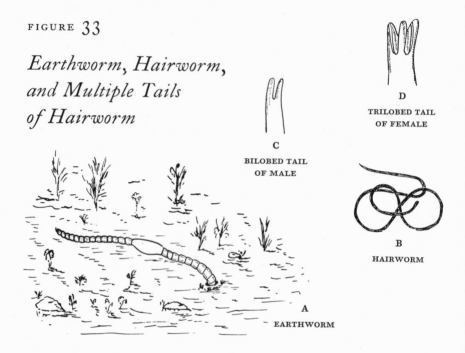

D

TRILOBED TAIL
OF FEMALE

C

BILOBED TAIL
OF MALE

B

HAIRWORM

A

EARTHWORM

Darwin estimated that more than 18 tons of earthy castings may be carried to the surface in a single year, and in twenty years a layer 3 inches thick would be transferred from the subsoil to the surface.

Earthworms are strictly nocturnal animals and are not found outside their burrows during the day unless "drowned out" by a heavy rain. During the hours of sunshine they remain stretched out in their burrows with their heads—or rather anterior ends, for they do not have heads in the strictest meaning of the word—near the surface. For some reason they cannot seem to find their way back to their burrows if they leave them, so they anchor themselves to the walls as they stretch themselves over the surface of the ground in search of food (Fig. 33A). You have doubtless seen a robin tugging away at a protesting worm, and if you have ever tried to pull one out of the ground you found it was not easy.

Look at the earthworm through your lens and you will see stiff bristles projecting from the body. There are four pairs on every segment of the body except the first three and the last. The bristles, or setae, protrude from small sacs in the body wall and can be extended or retracted by special muscles. When the earthworm wants to remain fixed in its burrow, it simply extends the setae out beyond the surface of the body and into the sides of the burrow, and it will be securely anchored, or relatively so. Then when the animal wants to change its position, it retracts the bristles and is free to move. Of course, the earthworm is not so securely fixed in place that it cannot be dislodged by the use of superior force, but then it will most likely be injured or even killed.

The presence of the setae explains in part how the earthworm moves within its burrow or over the ground when a heavy rain forces it out. Place the animal on a stretch of bare ground and observe that it moves by first extending the anterior part of its body, anchoring this part by means of the setae, then drawing up the posterior end. The setae on this end are then extended into the ground, those of the anterior region are retracted, and this part is extended lengthwise. All this is accomplished by two sets of powerful muscles, one set running circularly around the body, the other set lengthwise. You can see the action of these muscles if you look closely.

Partway between the lateral setae and the ventral setae, there is a small pore or opening visible with the lens. This is the external opening (nephridiopore) of a coiled tube called a nephridium that functions as a kidney. There is a pair of nephridia in every segment except the first three and the last. While handling the earthworm you will probably notice two swollen areas on the lower surface of segment 15 (the segments are numbered consecutively from the anterior end and are easily discerned by the grooves extending around the body). Small openings in the ridges are the openings of the sperm ducts. The openings of the oviducts are small round pores, one on each side of segment 14; eggs pass out of the body through them. (An earthworm has both male and female sex organs.) Two pairs of minute pores concealed within the grooves which separate segments 9 and 10 and 10 and 11 are the openings of the seminal receptacles into which sperms are transferred from one to another during the process of copulation.

On the upper surface a reddish-purple line is visible just beneath the integument or skin. This is the main blood vessel, which functions as a heart. Blood is forced forward by visible wavelike contractions of this blood vessel that begin at the posterior end and move quite rapidly toward the anterior. At the anterior end of the earthworm you will note a fleshy flap or lip called the prostomium. It pushes food into the mouth. If you want to see how it works, place an earthworm in a glass dish containing some soil, give it a small piece of lettuce or cabbage leaf, and then with a reading glass watch what happens.

An earthworm has decided preferences when it comes to food and must, therefore, have a fairly well developed sense of taste. In many ways it is an exceptional sort of animal; for instance, it doesn't have any eyes, and yet it is sensitive to light. Actually it has sense cells (photoreceptors) in the epidermis that respond to various wavelengths of the spectrum. Some of them are most sensitive to the blue wavelength and are the ones that send the earthworm back into its burrow at the first approach of dawn; others are sensitive to yellow; but strangely, the animal doesn't have any photoreceptors that respond to red light. So if you want to collect some worms at night for a fishing expedition on the morrow, use a flashlight shaded with a red filter.

Earthworms respond to sound vibrations and other stimuli by means of sense organs that are essentially groups of sensory cells. These cells connect with the central nervous system and communicate with the outside world through sense hairs that penetrate the cuticle, a thin transparent membrane that covers the body and protects the animal from physical and chemical injury. The sense of touch is also well developed, for the animal apparently likes to have its body in contact with solid objects. But I cannot subscribe to the view that it is perfectly at home on a hook, as one distinguished scientific authority once remarked.

And don't regard the earthworm as altogether stupid, for experiments have shown that it is capable of storing impressions until such a time as they may be useful—what the psychologists call "latent memory." May is the mating season for earthworms, and if you go out on a moist night with a flashlight you will probably find many mating pairs on your lawn or in your garden. The eggs are laid in yellowish-brown capsules shaped like a football and about the size of an apple seed.

Writing about the earthworm brings to mind an entirely different kind of worm, one which is something of an oddity; and that is about all we can say for it, though it has a rather interesting life history. Perhaps you remember having seen a roadside ditch partially filled with water what appeared to be "horsehairs" twisting and wiggling about. These "horsehairs" are hairworms (Fig. 33B) and occur not only in roadside ditches but in ponds, brooks, and springs, where they lie on the bottom, sometimes singly, but more often in groups twisted together and resembling a snarl of twine, a loose-coiled wire, or a clump of matted roots.

These worms measure a foot or more long, sometimes as much as 2 feet, and are completely covered with a thin layer of brown chitin (the material which forms the outer covering of insects). Though not completely rigid, they have a certain amount of stiffness, so that when they slowly coil and uncoil they appear as if pieces of wire had suddenly come to life. The body tapers to a point at the anterior end, but at the opposite end, as may be seen with the lens, it splits into two (Fig. 33C) or three (Fig. 33D) parts or may even be blunt, according to the sex and species.

During May the winter bands of the white-tailed deer break up, and in secluded thickets the does nurse spotted fawns. Early in the month muskrats are born, and the short-tailed shrew has a litter of young. And on a warm, sunny day, as the cottontails play along the woodland border and the chipmunks romp in rocky uplands, young woodchucks, now about a month old, are taken by their mother to the entrance of the burrow and, with wondering brown eyes, gaze for the first time on the outside world.

In May the northward migration of birds reaches its height, and on any sunny morning you can see countless gaily dressed warblers flit among the treetops. You can watch a redstart whirl and dash like a will-o'-the-wisp in a shaded thicket, or a yellow warbler cavorting like an animated sunbeam among the leafy branches of a brushy swamp, or you can listen to the wood thrush tune his lyre amid the misty greenery of a shady nook. As you approach some tangled shrubbery along the roadside, the yellowthroat nervously voices his alarm with scolding chirps and chattering notes as, high above on the branch of a nearby tree, the peewee gives voice to his plaintive song. And where the water of the brook flows to its distant destination, you can see the water thrush teetering on a partly sub-merged log, wagging his tail in the usual thrush-like manner, or you can catch a glimpse of the scarlet tanager in the distance as he flashes red against the blue sky. Too, you can hear the oriole singing his song of joy as he weaves his way among the apple blossoms, the vireo in the woodland grove, and the mewing of the catbirds as they perform their courtship ritual in the privacy of the lilac bushes.

Although a few birds mated in April and began building their nests before the month was out, nest building doesn't really begin in earnest until May. How birds can build a nest, following the same design and for the most part using the same materials as their ancestors, without being shown is still something of a mystery; we call it instinctive behavior, yet I wonder if a certain amount of intelligence doesn't modify their actions.

We are so accustomed to associate nest building with birds that we are inclined to think of it as an avian monopoly and to lose sight of the fact that other animals engage in a similar activity—as certain fish, for instance. True, the nests made by the large- and small-mouthed bass and sunfish are merely circular depressions excavated in the bottom sand—not much better than the hollow nests scooped out by terns. But the dainty nest of the stickleback rivals some of the finest examples of avian architecture.

Like other animal forms, a fish has become modified in various ways to live in a certain kind of habitat, an aquatic one. The arrow-shaped or spindle-shaped configuration of its body that permits it to pass easily through the water, the gills for extracting dissolved oxy-gen, and the scales which serve as a protective covering are only three features common to most fish. Fish are not unique in having gills and scales, since these structures are also found in other ani-mals; nor do they have an exclusive patent on the form of the body. Yet when we think of a fish, these are the features that come in-stantly to mind.

FIGURE 34

Fish Scales

C

CYCLOID SCALE
OF BROOK TROUT

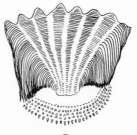

D

CTENOID SCALE
OF YELLOW PERCH

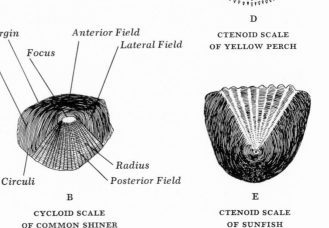

Anterior Margin

Anterior Field

Lateral Field

Lateral Field

Focus

Radius

Circuli

Posterior Field

A

GANOID SCALE
OF PIKE

B

CYCLOID SCALE
OF COMMON SHINER

E

CTENOID SCALE
OF SUNFISH

A fish scale is not an uncommon or rare object, yet few of us know just what one looks like or what it actually is. It is formed by certain scale-forming cells (scleroblasts) in the dermis, the inner layer of the skin. These cells lay down two layers of different substances, an outer bony layer and an inner threadlike layer containing calcareous deposits. As a fish grows, the scales increase in thickness and size by successive additions of bony material, these additions being indicated by lines of growth. Because periods of growth alternate with periods of inactivity, owing to seasonal variation of food and other conditions, it is possible to estimate the relative age of a fish by examining these diarylike lines, much as we can determine the age of a tree by counting the rings in a cross section of the trunk.

There are three principal types of scales: ganoid, cycloid, and ctenoid. Ganoid scales (Fig. 34A) are usually rhomboid or diamond-shaped and occur in such fish as gars, pikes, and sturgeons. Cycloid scales (Figs. 34B,C) are usually circular, with concentric rings about a central point, and are found in trout, minnows, and most other soft-rayed fish. Ctenoid scales (Figs. 34D,E), which are characteristic of perch, bass, sunfish, and most other spiny-rayed fish, are similar to cycloid scales except that the posterior margin bears small spines or teeth. Both cycloid and ctenoid scales are arranged on the fish in oblique rows and overlap like the shingles on the roof of a house.

There is more to a scale than one might suspect. As shown in Figure 34B, we can distinguish in a cycloid scale such parts as lateral field, anterior margin, circuli, and so on; but such features are of interest mainly to the specialist and the ichthyologist. Most of us will be satisfied to examine a scale or two. Needless to say, the lens is imperative, and even a microscope will be very useful. I might add that a scale can be taken from a fish with a knife or a pair of tweezers without killing the fish. After a scale has been removed, the fish can be returned to the water. A further thought also occurs to me: people collect all kinds of things as a hobby: postage stamps, coins, buttons—the list is endless. A collection of fish scales may have no monetary value and may serve no useful purpose other than providing a hobby for someone seeking something new. They are fairly easy to obtain and, when cleaned and mounted on small cards with correct labels, can be kept indefinitely.

Mushrooms are still not too common in May; a few species may be found, however. The shapely little brownie cap (Fig. 35A) frequently springs up on our lawns and pastures, and I have often found the hedgehog mushroom on living oaks, locusts, and beech trees. The uncertain hypholoma sometimes appears as early as May, its white, fragile caps quite conspicuous among the grasses of a lawn, pasture, or roadside. The reddish caps of the waxy laccaria may be seen in a woodland grove, although it also occurs in swamps and wet places. And if we have a few days of wet weather or a heavy rain, the fairy-ring mushroom may also appear during the month in odd ringlike formations. When viewed through the lens, all these mushrooms appear quite different than when seen with the naked eye.

Mushrooms belong to a group of plants called fungi. The fungi lack leaves, flowers, and chlorophyll, among other things, and include the familiar molds and mildews that so often appear on our foodstuffs and clothing, as well as the less familiar smuts and rusts. We shall consider these unusual plants on another occasion, but at the present time we might turn to one of the rusts that makes its presence known at this time of year in the form of brown swellings which may be seen on cedar trees and from which extend numerous long, thin, bright-orange tendrils, or horns, that twist about like petals of a flower (Fig. 35B). Known as cedar apples, these curious formations excite interest whenever they are seen.

Cedar apples, which are actually galls, represent one stage in the rather complex life history of the plant known as the apple rust, which is, incidentally, a disease of the apple. The tendrils, or horns, which are small at first but grow longer with every spring shower, produce spores called teliospores. These teliospores in turn develop within the tendrils into basidiospores. When fully matured, the basidiospores are released and carried about by air currents. If they fall on apple or crab-apple leaves and the conditions are favorable, they germinate, penetrate the leaves, and produce fibers (mycelium) that ramify throughout the leaf tissues. The presence of these fibers is revealed by the appearance of yellow spots on the leaf surface. The yellow spots resemble rust, hence the name of the

FIGURE 35

Plant Structures

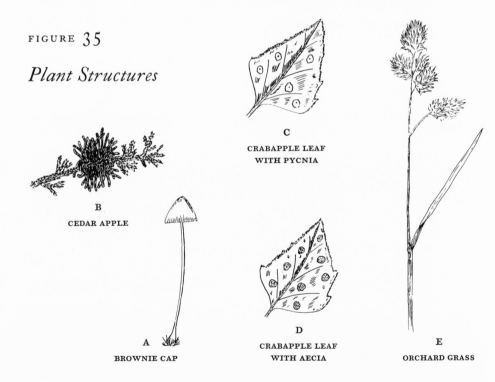

C
CRABAPPLE LEAF
WITH PYCNIA

B
CEDAR APPLE

A
BROWNIE CAP

D
CRABAPPLE LEAF
WITH AECIA

E
ORCHARD GRASS

plant; if we examine them with the lens, we find them to be amber blisters. These blisters contain flask-shaped structures, pycnia, that increase in size and exude a sticky substance which in turn produces pycniospores.

The pycniospores are of both sexes, and a pycniospore of one sex must reach one of the other sex for fertilization to occur. This is effected by insects that are attracted to the exudation. After fertilization, the pycnia change into black dots (Fig. 35c), surrounded by a reddish circle. The fungus now grows through the leaf, forming structures that appear, under the lens, as small cups with recurved and fringed or toothed margins (Fig. 35D). These cups, called aecia, produce aeciospores.

The aeciospores mature in July and August and, like the basidiospores, are scattered by the wind. If they land on cedar leaves, they germinate and by the following June produce small, greenish-brown swellings or galls. The galls gradually increase in size until by autumn they have become chocolate brown and kidney-shaped, with small circular depressions. They remain in this form until the following April when, under the influence of spring showers, they begin to put forth the tendrils or horns. After successive showers, the horns grow longer and longer, until the galls or cedar apples covered with these horns sometimes get to be the size of a small orange. Another crop of teliospores is now produced, thus completing the life cycle in something like two years. It is worth noting that the rust cannot spread from cedar to cedar or from apple to apple, but must alternate between the two hosts.

Although most of the grasses do not begin blooming until June, a few blossom in April and a few more in May. Although the blue-green leaves of the orchard grass (Fig. 35E) may be seen in April along the wayside and in fields and similar places, it is not until the clovers, buttercups, and dandelions bloom that the panicles of this grass become painted with large anthers of purple and yellow, terracotta, and pink. About the same time, the silvery hair-grass lifts its spreading panicles from the ground and opens its purplish spikelets for a day. It is a grass of waste places, and in a similar situation we also find the early bunchgrass. In bright sunlight its spikelike panicles often become tinged with greenish purple, but this soon fades, and beneath the hot sun of summer its slender stems become a shining yellow slightly tinged with pink. The slender sphenopholis prefers open woods, and in woodlands generally the white-grained mountain rice sends up slender stems with short, narrow panicles of a few pale-colored flowers. The meadow foxtail, too, blooms in fields and meadows, and along railway embankments and dry road-sides the downy brome grass with its panicles of drooping spikelets reminds us of oats, both cultivated and wild.

As the earth gradually warms up, ferns become increasingly abundant. Countless little croziers of the cinnamon fern, protected from the cold by wrappings of rusty wool, appear like wraiths in the dim-lit swampy woodlands, and from its growing rootstalks scattered fronds of the sensitive fern rise into the air along the roadside and in the wet meadow. Here, too, the delicate little croziers of the royal fern uncurl with dainty grace, and the interrupted fern sends up the fronds from which it gets its name. On sunny hillsides and in open woods the bracken strikes a springlike note with its light-green color, and by the brookside the common bladder fern clings to moist rocks or, in the deeper woods, nestles among the spreading roots of some great forest tree. The rusty woodsia, now a soft, silvery green, but brown-haired later in old age, thus reversing the usual order of things, appears in masses luxuriant to the eye and velvety to the touch. And in the shaded hollow the charming but aloof maidenhair fern echoes the spirit of the woods, even as the marsh fern with its bright green balls tempers the black muck of the swampland.

One advantage of living in the country is that all sorts of ecological habitats are near at hand or at least not too distant, such as the brook that meanders across the countryside less than a mile from where I live. I often go there, sometimes merely to watch the water flow and ripple against the brookside or a partly submerged rock. For it is a slowly moving stream, not the bubbling brook of the poets, but rather a lazy one, as if in no hurry to reach its destination. A favorite spot of mine is where the hobblebush throws its straggling branches across the water, and from this vantage point I can watch the bees and flowerflies—gorgeous little creatures that seem to favor blossoms as gaily colored as their own lustrous bodies

—while they feast on the marsh marigolds that grow in the brook bottom. Other insects, too, engage my attention, such as the dragonflies and damselflies. The dragonflies, as is their custom, are almost constantly on the wing, darting swiftly to and fro, but alighting occasionally on a leafy blade; the damselflies (Fig. 32B) are less on the wing, and, when they take to the air, flutter about with the uncertainty of butterflies; many of them descend among the forget-me-nots that blossom by the brookside and lose themselves among the flowers.

Water striders are always present where there is a surface to skate on, trying their best to avoid the ripples sent out by the whirligig beetles; and as these whirling dervishes dance their merry dance, a backswimmer (Fig. 32C) dives to the bottom. Few brooks are without a water boatman or two, and I can always see water beetles crawling along the bottom. Occasionally what appears to be a brown waterlogged stick turns out to be a water scorpion (Fig. 32D). But there are other members of the brook society beside insects: snails, pill clams, and planarians, for instance, and briefly a scud appears and disappears with an agility that would put a magician to shame. And sometimes a black-nose dace glides into the sunshine.

As I watch the dace swim about, I cannot help thinking how well the little minnow illustrates the principle of countershading, with the olive-green of its back and the silvery-white of the underside. The backswimmer also exemplifies the same principle, for the back of the insect is a pearly color and the lower surface is dark. This would seem to be just the opposite of what it should be, but actually the insect swims upside down, so that the lower surface is uppermost and blends with the bottom mud and sand when the insect is seen from above; while the lighter back, when seen from below, blends with the sky. Possibly this coloration helps the insect to capture its food and to escape being a victim itself.

The backswimmer is a common inhabitant of our ponds and streams. We usually first see it resting at the surface, floating head downward, with the tip of its abdomen extended above the surface into the air and its long hind legs outstretched like oars or sweeps, ready to propel the insect through the water for food, or to safety, should danger threaten. It is an expert diver, and when it descends to the bottom it carries a silvery film of air down with it. Since it is lighter than water, it has to hold on to a plant stem or some other object, but its front legs are admirably modified for grasping either a foothold or prey. Scoop up one of these insects, and with your lens you can see how well the front legs are adapted for either of these two objectives. You might also like to look at the strong suctorial beak with which it sucks the body juices of its victims. The beak can inflict a burning sting, so use care in handling the insect. The effect of the sting may last for some time, particularly if you are susceptible to insect poisons.

I also saw a few mayflies flying about the streamside. During the month mayfly naiads transform into adults, and quite frequently clouds of these flying insects are seen in the vicinity of a pond or stream. If you don't live too far away, they will appear about your porch lights at night and on your window screens during the day. They are beautiful, fragile insects (Fig. 32E), translucent, soft gray and brown, with large front wings, small hind ones, and two or three long tails that extend backward from the tip of the body.

Most adult mayflies live only a few hours, or at most a day or two, although a few species live several weeks. Even if they wanted to live longer, they would be unable to: they would starve to death, since they are unable to eat. Examine their mouthparts through your lens and you will find that they are shrunken, if not actually vestigial, and useless. Moreover, their digestive tract has been transformed into a sort of balloon inflated with air that helps them to fly. Their legs are delicate structures and poorly adapted for walking.

All this is for a purpose; their main objective in life is to mate, and once mating has taken place and the females have deposited their eggs in the water, both males and females have no further reason for living; hence they all die. Sometimes the surfaces of ponds and streams are strewn with their bodies, which become food for eager fish.

There are few phenomena of the insect world more striking than the mating flight of the mayflies, which usually occurs in the cooler hours of the day or in the evening just after sunset. Thousands may participate in the mating flight, swinging up and down through the air in a joyous, rhythmic dance. They move up and down together, swinging downward in a swift descent toward the surface of the pond or stream and then bounding upward as if they were thistledown wafted by a gentle breeze.

Most of them die an heirless death, for there are many males but few females. The females lay their eggs almost immediately, the short-lived species depositing them in clusters that rapidly disintegrate upon reaching the water, whereupon the eggs sink to the bottom; the less perishable species dropping the eggs one at a time. The latter females either alight on the surface of the water at intervals to wash off the eggs, or creep down into the water enclosed within a film of air to lay them on the undersides of stones and then float up to the surface and fly away. Should you feel sorry for these insects because they live for only a few hours, do not waste your sympathy, for actually the naiads live for several years. Indeed, the mayflies live longer than most insects, although their life span is exceeded by some wood-boring beetles that take as many as ten years to mature and by the periodical cicada that takes even longer.

Mayflies are not the only insects which mate during the month, for on some predetermined day the winged males and females of all termite colonies in a given locality rise in dark funnel-like swarms on their nuptial flight. Birds and predacious insects take a large toll, but those that remain drop to the ground and pair. The paired males and females, known as kings and queens, at once seek nesting sites and begin the work of excavating the galleries of a nest in the ground or in wood. They do not mate immediately, but they remain together indefinitely and later mate repeatedly. We shall have occasion to look more closely at the termites further on.

I often think how remarkably well some animals are adapted to live in an environment or situation in which at first we might wonder how they can live at all. Consider the small, seemingly delicate blackflies that live among the rocks of a waterfall or in the rapids of a rushing stream. Or more accurately, it is the larvae that live in such places. How they can manage to do so without being carried downstream or hurled against the rocks would seem to border on the miraculous, until we come to examine them. If they lived separately, it would be like looking for the proverbial needle in a haystack to find one. However, they live together by the thousands and collectively appear as a swaying mass of greenish "black moss" on a rock, stick, or leaf. Scoop up a handful of this "black moss" and you will collect hundreds of them.

They are too small to see any of their adaptive details with the naked eye, and the lens is needed to magnify, for instance, the disklike sucker at the hind end of the insect (Fig. 32F). This sucker is fringed with little hooks which enable it to cling to a rock, stick, or other support. At the opposite end, just behind the head, is a fleshy proleg, also provided with a sucker. By means of these two suckers the insect can walk over any kind of surface with a sort of looping gait. But what happens if the larva is dislodged by the rushing water, which happens frequently? It simply spins a silken thread from its salivary glands, attaches one end to an anchorage, and clings to the other end by the hooks on the hind sucker. Actually the insect does not necessarily wait for an emergency to spin the silken thread, but spins a number of them intermittently throughout its larval existence. It fastens them in all directions from one rock to another or from one leaf to another, and they serve as a sort of labyrinth along which it can travel from place to place in search, perhaps, of a better feeding spot. Quite frequently when suddenly dislodged or frightened, though it is not easily alarmed, it simply slides along one of these previously spun threads until it can manage to secure a fresh foothold. It can quickly spin a new thread, however, if the need arises.

At the anterior or head end are two fan-shaped brushes. The larva uses them to collect its food, which consists essentially of algae and diatoms. On the dorsal side of the last abdominal segment are three retractile blood gills by which respiration is carried on. Blood gills are merely expansions of the body wall filled with blood. Oxygen passes from the water through the thin covering in much the same manner as it does in the gills of a fish.

When full grown, the larva spins a boot-shaped cocoon, which it fastens to a rock or other object but more often to other cocoons, since it usually lives in company with other larvae. Sometimes the cocoons are so thickly clustered on the rocks that they form golden-brown blankets which are even more mosslike than those of the larvae. The pupa (Fig. 32G) breathes by means of tracheal gills that project from the top of the cocoon and wave about in the water. When the transformation is completed, the adult fly emerges from its pupal skin in a bubble of air that carries it to the surface. On reaching the atmosphere, the bubble bursts, liberating the fly which immediately takes flight.

Blackflies, which are stout, humpbacked, short-legged insects (Fig. 32H), mate and lay their eggs soon after emerging from their pupal skins. Should you visit a waterfall or rapids when they are laying their eggs, you will see them dart in and out of the water and attach their eggs to the stones. If the water is shallow and not too swift, they crawl about on the surface of the rocks and then desposit their eggs. I might add a word of caution: the flies can inflict a painful bite, and it is advisable to smear blackfly cream or a similar concoction on your hands, face, and other exposed parts of the body as a precautionary measure.

Riffle beetles, which also lay their eggs on stones in swiftly flowing streams, have contrived a rather ingenious way of doing so. The insects are small, about a quarter of an inch long, and their bodies are covered with silken hairs which are capable of retaining a film of air. On hot days the females settle on stones projecting above the water and then, enveloped in air, climb down over the water-washed stones and lay their eggs in the swiftest part of the stream. Certain mayflies have copied the idea, or perhaps the beetles got it from the mayflies; the mayflies, however, use their wings instead of hairs to entrap a film of air. One species of mayfly lays her eggs with threads attached to them (Fig. 32I); another with floats (Fig. 32J), so that the eggs may be suspended or float in the water.

The larva of the riffle beetle is a curious insect and not readily recognized as an insect—it was originally described as a crustacean. Very small, flat, circular in outline, copper-colored, and called a water penny (Fig. 36A), it is usually found clinging to the lower surface of a stone. I say this with a certain amount of reservation, however, for even when a stone is overturned and the water penny is right beneath your eyes, you probably won't see it. But if

A
WATER PENNY

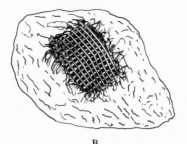

B
NET OF HYDROPSYCHE

C
EGGS
OF CODLING MOTH

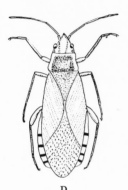

D
SQUASH BUG

E
EGGS OF SQUASH BUG
ON LOWER SURFACE OF LEAF

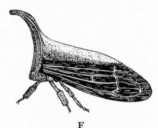

F
TREEHOPPER

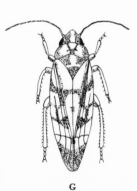

G
LEAFHOPPER

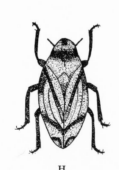

H
FROGHOPPER,
LEPYRONIA QUADRANGULARIS

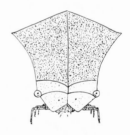

I
TREEHOPPER

FIGURE 36

*Bugs, Hoppers,
and Moths*

J
SPITTLE MASS
OF FROGHOPPER

K
LOWER SURFACES
OF HIND WINGS
OF PEARL CRESCENT

you do, you have to turn it over on its back to see its legs and five pairs of glistening white gills.

I have already mentioned that not all caddisworms build portable houses; some species that live in waterfalls, riffles, and rapid streams dwell in tubes composed of silk and debris which are permanently attached to a rock or other support. But in addition to these tubes, caddisworms also spin nets adjacent to the tubes, with which they capture their food. The net of Hydropsyche, for instance, is funnel-shaped (Fig. 36B) and provided with a strainer at the downstream end. Sometimes the nets of Hydropsyche are spun in crevices between stones, but usually they are built up from a flat surface and then take the form of semi-elliptical cups kept distended by the flowing water. Food materials such as algae, insect larvae, and various small animals that pass through the net are trapped by the strainer.

Several mayfly naiads occupy rapid currents, swift eddies, and waterfalls and show some nice refinements for living in such places. For instance, the naiads of Epeorus have flattened bodies, knifeblade legs, and grappling claws, all adjustments for living under the constant pressure of swiftly flowing water. Probably the most interesting of all mayfly naiads is Chirotenetes. A rich chocolate brown with a light median stripe over the head and thorax, it lives in the rushing water of stony creeks, where it leaps and dashes with considerable agility. But it spends much of its time merely sitting on its middle and hind pairs of legs with its head up and the first pair of legs, that are fringed with long hairs, held forward. As the current flows by, the long hairs spread out and function as a basket to capture whatever food the current carries. Chirotenetes feeds on a mixed diet of algae and midge larvae, a rare habit among mayflies.

Probably no insect is better equipped to be a member of a waterfall society than the larva of the net-veined midge, which is found only in the swift-flowing streams of mountainous or hilly regions. Few of us ever see it, and I mention it only because it is a curious-looking insect and because the adult, which is mosquitolike in form, differs from all other insects in having wings marked by a network of fine lines. The larva is less than half an inch long and looks very much like a row of six black beads strung on a thread. The first bead represents the head and thorax that have been fused together and the remaining five the abdomen. Instead of legs the larva has a row of six sucking disks along the ventral surface that are highly efficient for walking and for clinging to a rock.

The net-veined midge is also of interest for another reason. Normally when a winged insect emerges from its pupal skin, it has to wait for its wings to expand before taking flight. During this interval the insect is in a helpless condition and subject to danger from various sources. In certain caddis flies that emerge from swiftly flowing water, the time required for the wings to expand is much less than that of most insects; but in the net-veined midge the wings have reached their full expansion before the insect escapes from its pupal skin, so that when it reaches the surface of the water it has only to unfold its wings and fly away, a matter of a second or two.

As we have seen, there is more or less a give and take between plants and insects, and in many instances nice adjustments have been made by both so that they may profit from each other. There are also many instances, however, where one profits at the expense of the other. But even then adjustments must be made, as with the codling moth that emerges from its cocoon when the apples blossom so that it may lay its eggs in time for the larvae to enter the young developing apples. The larvae enter at the blossom end and tunnel to the core, where they live for about a month, feeding on the pulp and destroying the beauty and quality of the fruit. Then when full grown, they burrow out through the side of the apple and crawl to the ground, where they pupate. But we are not much concerned with the insect itself or its life history, except for one stage in its life cycle—the egg. The egg is a most curious object (Fig. 36c) and if you have never seen it (and few of us have), you should look for one. You may not find it, for it is flat and scalelike, about half the size of a pinhead, and so extremely thin and transparent that it is barely visible to the naked eye, and then only by reflected light. But look on the leaves and branches of an apple tree anyway—and be sure to use the lens or a reading glass, for the eggs are laid singly, which complicates matters.

The codling moth is a major pest of the apple, and so is the spring cankerworm, though to a lesser degree. We have already met its relative the fall cankerworm, or more specifically, we have seen its eggs, which as you recall, resemble miniature flowerpots. The eggs of the spring cankerworm are no less interesting; they are ovoid, slightly ridged, and iridescent purple, a rather unusual color for an egg, but then insect eggs come in all colors. The eggs are laid in groups in crevices of the bark and are somewhat easier to find than those of the codling moth. The caterpillars belong to a family called the Geometridae, and like other members of the family, have the habit of spinning a long silken thread which they attach to a twig and often use to let themselves partway down to the ground. Then they climb up again in the manner of a sailor going up a rope. Watch them with a reading glass and see if you can discover what happens to the thread as they make their way upward.

Sometime during the month the squash bug (Fig. 36D) puts in an appearance in my garden, and in spite of my efforts to get rid of it will probably remain there until the first frosts. Then it will disappear to spend the winter in some nook or cranny, only to reappear next spring. I don't mind having a few squash bugs around, but when they become numerous my squashes and cucumbers suffer, the vines first wilting and then turning black. Apparently this condition is not due so much to the plant juices that the insects suck with their long beaks as to a toxic substance that they inject while feeding.

Some authors describe the adult squash bug as a dark, sordid, ugly, ill-smelling creature, but I can't entirely agree with them. It does give off a disagreeable odor when crushed, but aside from this I don't find it unattractive. To the naked eye it appears blackish-brown above and a dirty yellow beneath, but the ground color is actually ochre-yellow, darkened by numerous minute black punctures, with crimson beneath the wings. Seen through the lens these colors take on a different hue, and if the insect were not such a pest I think we would regard it in a different light.

In color the nymphs are unlike the adults; when first hatched they are a light green with beautiful rose-colored legs, antennae, and beak; later they become a somber grayish-white with dark legs.

The eggs are laid in clusters (Fig. 36E), though each egg is separated from the others. They are somewhat flattened on three sides and have a yellowish-brown or dark-bronze color. Seen with the naked eye they are pretty little objects, but they are even prettier under the lens. They can usually be found attached to the undersides of the leaves of various members of the Cucurbitaceae: squash, cucumber, pumpkin, melon.

From now on until the cold weather sets in, we shall see any number of small insects hopping around on shrubs, trees, and various forms of herbage. Some of them are treehoppers (Fig. 36F,I), some are leafhoppers (Fig. 36G), some are froghoppers (Fig. 36H). There are some twenty-odd species of froghoppers; one of the more common and very widely distributed is the species *Lepyronia quadrangularis* (Fig. 36H). It is brownish in color, densely covered with microscopic hairs, and has wings marked with two oblique brown bands. The name of the insect is not inappropriate, because it looks rather like a miniature frog. But you have to catch one and look at it with your lens to see the resemblance.

Many of us may not be acquainted with the froghoppers, but few have not seen the white frothy masses that, as the spring advances, are gradually becoming quite conspicuous on grass stems and other vegetation (Fig. 36J). These frothy masses are produced by young froghoppers (nymphs), and they live within them until they become adult hoppers and start jumping around. Usually one young froghopper is found within a frothy mass, but in some cases there may be as many as four or five. The froth or spittle is a viscid liquid which the insects expel. Actually, the liquid is a mixture of two substances derived from different sources which the nymphs beat up into a froth by whisking it around with their bodies. At one time it was an open question what purpose the frothy mass serves, but it is now generally believed that it offers protection against parasites and other enemies. Apparently the name froghopper was originally bestowed upon the insects not because they look like frogs, but because the froth was thought to be "frog spittle" voided from the mouths of tree frogs. The froghopper is also called the spittle insect for the same reason.

Although butterflies are not so much a part of spring as they are of summer, as the month progresses and the days get warmer many of them are lured from their winter retreats or their mummylike cases. We may catch a glimpse of the painted lady, the red admiral, the gray comma, the hop merchant, the gray hairstreak, the black and tiger swallowtails, the pearl crescent,[4] and several others, but the time for butterflies is not yet upon us, and we shall defer further treatment of them until a later chapter.

In the nature world, the month of May is an active and busy one, and all too quickly it draws to a close and the tempo slows down. The sexual urge of the mammals, birds, amphibians, and others has passed, mating has been accomplished, and among the mammals and birds, domestic chores command their attention. Developing tadpoles in ponds and streams are well on their way to becoming adult frogs, toads, and salamanders. Most of the trees and shrubs have blossomed and are in the process of developing their fruits. Many wild flowers have appeared, though there are others still to come; and though some insects have laid their eggs, many have yet to do so. And so we come to June, the leafy month of June, as some have called it, and we continue our rambles with an observant eye and our lens in the pocket, for there are many a byway and pathway left to explore.

[4] A word about the pearl crescent may not be amiss since the butterfly has two different forms, one in the spring and one in summer, which vary so much that once they were considered separate species. The butterfly is small, with a wing expanse of only about an inch and a quarter. The wings are a reddish-brown and more or less marked with black wavy lines and dots. In the spring form, the hind wings are heavily and diffusely marked on the lower surface with strongly contrasting colors; but in the summer form, which is somewhat larger, the wings are plain and only faintly marked (see Fig. 36K). A number of years ago, Edwards showed by some interesting experiments that the smaller, darker spring form is the result of low temperatures. He placed a number of chrysalids that normally would have produced the summer form upon ice, and found that these specimens produced the spring form.

JUNE

"Now is the high-tide of the year."

JAMES RUSSELL LOWELL

"The Vision of Sir Launfal"

JUNE

LOWELL ONCE WROTE THAT "JUNE IS THE pearl of our New England year." I am not sure I entirely agree with him, though June is undoubtedly a most delightful month. The thermometer has not yet reached the heights that it does in July and August, nor does it descend to the low points of April or May, for the cold air masses that have their origin in Canada rarely come down as far as New England during this month. Showers are infrequent, too, so the days are invariably pleasant and comfortable. We can make plans for any outdoor activity with confidence, though sometimes our confidence in the weatherman may be misplaced and we will have a day or two of rain. But it will be a warm, soft rain and we don't mind it, for we know it will help our gardens and be a boon to the farmer. Needless to say, the plants of the wildwood benefit as well.

By June most of the birds either have a family to engage their attention or are joyously awaiting the arrival of their young ones. A few have delayed the task of nest building, but not for much longer; any day the marsh wrens will begin construction of their hollow ball-like nests among the grasses and reeds of a swamp, and the red-eyed vireo will weave her dainty, handsome little basket in the fork of a sapling. Cuckoos, warblers, and indigo buntings will also get to work; but the goldfinches, now in their breeding costumes of gold and sable, will wait awhile.

The mammals, too, are concerned with family cares. The spotted fawns of the white-tailed deer receive lessons in woodcraft as they follow the does to their feeding places; the young of weasels, mink, and wildcats move about with their mothers and learn the art of hunting. Young porcupines come down from the hollows in the trees to seek the edges of ponds, where they gorge themselves on succulent water plants; and on some warm night when insects are plentiful, young skunks are led forth by their mothers to be shown the where and how of getting a living.

Insect life becomes increasingly abundant in June, and countless caterpillars, grubs, and flying adults appear everywhere, many of them to serve as food for birds and insectivorous mammals and many others to invade our gardens, orchards, fields, and woodlands. Insects are concerned not so much with domestic chores as with perpetuating the species, and from now throughout the summer, mating and egg laying become a ritual followed as scrupulously as it is by the more advanced animals.

Should we visit a pond, stream, or lake on a sunny day and watch the dragonflies darting here and there, now alighting on a water plant to rest a moment, now streaking off in pursuit of some other flying insect, we might see one of them skim the surface of the water, then suddenly swoop down and touch it. Or we might see another, poised in the air momentarily, descend to the surface in a swift curving movement, hover above the submerged leaf of a water plant, and then fly quickly upward, only to descend again to the submerged leaf. Or we might see a third dragonfly alight on the stem of an emergent water plant rather near the surface of the water and curve its body below it.

Somewhere between 625,000 and 1,250,000 different kinds of insects have been named and described. We don't know the exact number, because there are so many different species that it is impossible to keep accurate count. And these appear to be only a fraction of the total number that actually exist. Aside from the huge number of insects, which in itself is enough to stimulate anyone's imagination, the habits of no two species are identical, for each shows some difference in food, structure, or habits. Thus each of the three dragonflies described above was following her own behavioral pattern in laying her eggs—for that is what they were doing.

Insect eggs are not laid at random wherever the female happens to be when it comes time for her to lay them. She deposits them where she is reasonably sure that her young will have an ample supply of food when they emerge. (This may seem to involve a certain amount of intelligence, but actually it is merely instinct.) Thus dragonflies, whose young live in the water, lay their eggs where the newly hatched naiads will be in a habitat to which they are adapted and where they can readily find food. Even the dragonflies that fly over distant fields return to the water to lay their eggs.

We have seen that the eggs of the tent caterpillar (Fig. 37A) are clustered on the twigs and branches of the wild cherry so that the caterpillars when they emerge have access to fresh cherry leaves, the hatching of the eggs being timed to correspond with the appearance of the leaves. The milkweed butterfly lays her eggs on the leaves of the milkweed plant because the caterpillars will eat only milkweed leaves,[1] and the black swallowtail butterfly seeks out members of the carrot family. How these insects and others find their specific host plants is rather a mystery. Doubtless the senses of touch, taste, smell, and sight are all involved, but there is still much we do not understand. Of course not all insects are so restricted in food plants; in captivity the gypsy moth, for instance, will feed on over 450 species of plants.

Terrestrial insects have relatively little trouble laying their eggs where the young will have a nearby food supply, since both adults and young live in the same kind of environment. But insects that spend their adult lives on land and their nymphal or naiadal lives in the water are faced with the problem of getting their eggs into a habitat which, as winged insects, they are not well adapted to enter. The problem has been solved in various ways. Some species of mosquitoes, for instance, lay their eggs in dry places, where they remain until rains or melting snows provide the necessary moisture. Horseflies, dobson flies, alderflies, and certain caddis flies lay their eggs upon tree branches or stones that overhang the water or on emergent water plants, so that all the young have to do when they emerge from the eggs is to drop into the water. And we have already seen how blackflies, riffle beetles, mayflies, and net-veined midges accomplish their objective.

To the uninitiated it might seem that egg laying is a simple

---

[1] Sometimes the eggs are laid on the spreading dogbane and the green milkweed. Both these plants, however, are closely related to the milkweed.

FIGURE 37

Life Stages of Eggs of Borers and Bugs

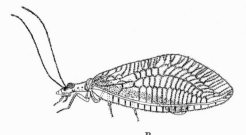

B

LACEWING

A

EGG MASS
OF TENT CATERPILLAR

E

COCOON OF APHIS LION

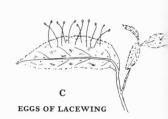

C

EGGS OF LACEWING

F

PLUM CURCULIO

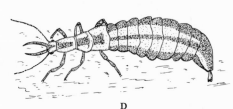

D

APHIS LION

G

EGG SCARS
OF PLUM CURCULI

H

EGG SCARS AND GIRDLING
OF RASPBERRY-CANE BORER

J

ELECTRIC LIGHT BUG

K

EGGS OF
ELECTRIC
LIGHT BUG

I

MALE WATER BUG
CARRYING EGGS

matter, but it isn't as simple as it may appear. The urge to safeguard their eggs and to make ample provision for their young is so strong that many insects have developed unusual habits to ensure their survival. Consider, for instance, the lacewing or golden-eyes. Look for the insect (Fig. 37B) when you are outdoors, and carry a reading glass with you so you can watch the female lay her eggs if you happen to come upon her when she is doing so. From the tip of her body she ejects a drop of sticky fluid on the surface of a leaf. Then she lifts her abdomen and spins the drop into a thread half an inch or so long. This hardens almost immediately on exposure to the air. She next lays an oblong egg about the size of a pinhead on the tip of the stalk. The egg firmly in place, she spins another stalk on which she lays an egg, and continues to repeat the performance until she has laid her full complement (Fig. 37c).

The lacewing has a good reason for taking all this trouble. She usually lays her eggs in the midst of a colony of aphids or plant lice, and since insects lay their eggs within easy reach of a food supply, it follows that her young feed on these soft-bodied insects. They are, as a matter of fact, highly carnivorous and eat a variety of small insects, but especially aphids; hence they are known as aphis lions. To a carnivorous animal, anything in the form of another animal is usually included in its diet if it is able to subdue it; and since eggs are of animal origin, they are equally tempting. Accordingly, if the lacewing were to lay her eggs in a cluster, the first aphis lion to emerge would probably feed on the eggs. So instead of having half a dozen or more offspring, the mother lacewing would probably end up by having only one. Hence she places the eggs beyond reach. Of course, there is nothing to prevent the first aphis lion from climbing the stalks; but why go to all this work when soft, juicy aphids are right at hand?

If you look carefully at the eggs with your lens, you should be able to detect the little doubled-up and still unhatched "lions." You may even find one about ready to emerge. If you do, watch closely and see the jaws thrust through the shell, opening it, as it were, for a peephole. Then the head gradually appears, followed by the legs and finally by the spindle-shaped body.

Once the "lion" has emerged, it is not in a hurry to descend the stalk but clings to the eggshell while it views its surroundings. Then, apparently satisfied with what it sees, it begins to move about, but as the eggshell limits its sphere of action and there is no place else to go but down the stalk, the "lion" finally grasps the stalk with its first pair of legs and, with the help of the other two pairs, begins a careful descent. This is a feat for a creature only a few minutes old with no previous experience at gymnastics. At last it arrives safely on the leaf and pauses for a moment to look around. Spying the nearby aphids, it proceeds to satisfy its hunger without further ado.

The aphis lion (Fig. 37D) is not a particularly prepossessing creature when seen with the naked eye, and it is less so when seen through the lens. It has an extremely large head, and the sides of its body are armed with immense curved hairs that give it a ferocious appearance. But perhaps its most outstanding feature are the long, pointed, sickle-shaped jaws. They are effectual instruments for grasping the soft-bodied aphids and, as they also form a hollow tube with an opening at each end, the opening at the base leading into the "lion's" throat, they also serve as an efficient sucking apparatus.

Although the aphis lion bears little resemblance to its namesake of the jungle, it is well named if we think of it in predatory terms. The aphids, which are usually distended or swollen with sap, provide an easily accessible food supply, and the "lion" preys almost exclusively on them, sucking their blood until there is nothing left of the victim but a shriveled mass of skin. If you use a reading glass to watch a "lion" feed, you will observe how it grasps an aphid between its long jaws and then rolls it first one way and then another, at the same time extracting the body juices. An aphis lion always seems hungry, and the number of aphids it will consume seems limited only by the number available. Try counting the number of aphids that a lion will eat at one sitting, and you will probably tire of the exercise before the lion will. When full grown, the lion spins a white spherical cocoon (Fig. 37E), usually on the lower surface of a leaf. Later the winged adult escapes through a circular hole previously cut by the insect before transforming. The piece cut out remains attached to the rim of the hole like a lid.

Insects lay their eggs in all sorts of places. Some are exposed to view on tree trunks, branches, fences, and the like. Held in place by a viscid substance that hardens into a cement upon drying, they are not easily dislodged even though they may be buffeted by rain, wind, sleet, and snow. Other eggs are hidden away in the ground or inside stems, tree trunks, fruits, seeds, and roots. A few insects which make incisions, punctures, scars, and holes in plants and in the ground have an egg-laying device extending from the posterior part of the body with which to do so. This instrument, called an ovipositor, is essentially a sharp piercing organ. It may be simple or elaborate in structure. It is also quite strong: drilling a hole in a tree is not easy, for example; yet some ichneumon flies seem to do so rather effortlessly.

When an incision or hole has been made and an egg or eggs laid in it, most insects fly away and leave the eggs to hatch; they do not even take the trouble to attach them with cement. There are, however, a few insects that take certain measures to safeguard their eggs. Actually these are cases where such measures are needed, for the insects lay their eggs in places where they may be damaged by the growing plant. Consider the plum curculio, an insect highly injurious to plums, cherries, and other stone fruits. It is a rather odd-looking insect (Fig. 37F), but then so are all snout beetles. Snout beetles puncture or make incisions in plant tissues with a long curved beak that, when examined with the lens, is seen to be a curious modification of the head. There are a number of species—the imbricated snout beetle, the strawberry crown girdler, the black vine weevil, the apple curculio, and the notorious cotton boll weevil, to name a few—so they are not difficult to find.

But to get back to the plum curculio. When the female is ready to lay her eggs, she makes an incision in the fruit and deposits an egg in it. With her beak she pushes the egg down into the cavity which she has excavated and then, in front of the incision, she cuts out a crescent-shaped slit (Fig. 37G) so that the egg rests in a flap of tissue. This protects the egg from being crushed by the developing fruit.

The raspberry-cane borer, although not a snout beetle, has a somewhat similar habit. Before depositing her eggs in the new growth of a raspberry cane, the female first cuts two ringlike girdles about a half inch apart around the stem a few inches below the tip. Then midway between the two rings she punctures the stem and deposits an egg (Fig. 37H). The two ringlike girdles cause the top of the stem to wilt and fall over, thus arresting the stem growth, a necessary precaution because an enlarging stem would have crushed the egg. How the plum curculio and raspberry-cane borer developed such nice refinements in the art of egg laying leads us to ponder on the subtle designs of nature.

We might also speculate on how or why the female of a certain species of water bug decided, once upon a time, to attach her eggs to the back of the male instead of to water plants, as all normal water bugs do. Or why the female of a certain water boatman took to attaching her eggs to the crayfish. In the case of the water bug, the male doesn't think much of the idea, but what can a poor male do when the female has other ideas? Willy-nilly, the eggs are attached to his back (Fig. 37I) and he has to swim around with them until they hatch, which takes about ten days or so. You might watch for him as he swims, and if you can catch him, look at the eggs with your lens. There are about a hundred or so, securely attached with a waterproof glue so he couldn't dislodge them if he wanted to.

In almost any pond or stream we can find a relative of this water bug. It is considerably larger and is called the giant water bug. There are a number of species of giant water bugs; all are wide, flat-bodied, oval, brownish or grayish insects, and very predaceous. Their front legs are raptorial and admirably suited for catching prey; the middle and hind legs, flattened and oarlike, are adapted for swimming; and at the tip of the abdomen is a pair of narrow, straplike, retractable respiratory appendages.

Because of their size we do not expect these insects to fly easily, but they are actually capable flyers and often fly considerable distances. Some species are attracted to electric lights and hence are called electric light bugs (Fig. 37J). The eggs of one of our more common species are curiously streaked or striped, are attached in clusters 2 or 3 inches long to various water plants, and are so big that the emergence of the young can be seen with the naked eye, though a reading glass will help you to watch the hatching process a little better. Each egg has a cap at the free end which the young water bug loosens when ready to emerge. The line along which the cap opens is indicated on the eggshell by a white crescent-shaped streak (Fig. 37K).

Most insect eggs are visible to the naked eye. But even with the best eyesight it is difficult to see the sculpturing and other ornamentation that many of them have. The conical, pale eggs of the white cabbage butterfly (Fig. 38A) are quite visible on cabbage leaves, but their ribbed surface is not apparent unless magnified. Similarly, the eggs of the harlequin bug, found on such plants as the turnip, potato, eggplant, mustard, and radish, appear white with two black bands but are seen to better advantage with the lens, when they look like small barrels (Fig. 38B). The eggs of the spinach leaf miner resemble miniature golf balls.

That insect eggs are anything but egg-shaped seems rather obvious from the few I have mentioned. Actually they seem to come in all shapes: flat (codling moth), conical (white cabbage butterfly), oval (swallowtail butterflies), cylindrical (assassin bugs), elongated (asparagus beetle: Fig. 38C), and curved (tree crickets: Figure 38D). Many of them have appendages, such as those of certain mayflies which we have already noted. The eggs of one species of water scorpion has two long filaments (Fig. 38E) and the egg of another three, while the egg of a third species has a crown of eight or more. The eggs of the stink bugs (Fig. 38F) have a circle of spines around the upper edge. Some eggs have a cap that the young can remove in order to escape, such as that of the water bug; the eggs of stoneflies (Fig. 38G) have a similar cap, often ornamented with raylike extensions. But the most unusual of all insect eggs, and perhaps of all eggs, is the egg of the poultry louse. It is white in color, elliptical in shape, and covered with glasslike spines with a long, lashlike whip at the apex (Fig. 38H). Anyone seeing it for the first time and not knowing what it is could very well be excused if he failed to identify it. The eggs of this insect are attached to chicken feathers, if you want to look for them.

With the earth now covered with tender green grass blades as yet unburned and made harsh and crisp by the hot dry weather of summer, grasshoppers have a field day. Every garden, roadside, field, and meadow is a restaurant, and the grasshoppers dine luxuriously. Watching a grasshopper eat at close range, we rather suspect that he gets a great deal of pleasure from doing so, if we can interpret his actions and facial expressions correctly. A reading glass is better for the purpose than the smaller hand lens. Note that his upper jaws or mandibles move sideways instead of up and down. Examined closely with the higher magnification of the lens, they look like a pair of nippers (Fig. 381). Hard and toothed, they are highly efficient biting and chewing tools. Just in front of the mandibles is a hinged flap that moves up and down. This flap is the upper lip or labrum, and the grasshopper uses it to push the pieces of leaf he bites off toward the mandibles. Below the mandibles are the lower jaws or maxillae, each with an appendage or palpus. The function of the maxillae is to hold the pieces of leaf while the mandibles grind it up into smaller bits for swallowing. Directly below and in front of the maxillae is the lower lip or labium; it is provided with a pair of palpi and is used in much the same manner as the labrum. There is also a tongue which is located on the floor of the mouth cavity and is not quite visible unless it is dissected out. As the grasshopper eats, he constantly moves the palpi and taps the leaf with them. Sense organs of taste are located on them and help him to select his food.

Insects vary greatly in their eating habits. The grasshopper, for instance, begins at the leaf margin and eats downward toward the center, making a long even-edged hole in the margin. By repeating the process, he makes the hole deeper or larger and eventually consumes the entire leaf as a well-trained child cleans his plate. Contrast such eating habits with those of the cabbage flea beetle. The insect simply digs right into the leaf wherever the fancy takes it, and after a while a cabbage leaf will look as if it had been peppered with tiny shot. The cabbageworm feeds in much the same manner but makes larger holes.

Some insects feed only on the softer tissues of a leaf and leave the skeleton of the leaf intact. They are known as skeletonizers, and you can find them everywhere. When viewed with the lens, the veins show up like dainty lace. Finally, many external feeders have such characteristic feeding habits, as shown in Figures 38J,K,L that

FIGURE 38

*Insect Eggs
and Larvae*

A
EGG OF WHITE CABBAGE
BUTTERFLY

B
EGG OF HARLEQUIN BUG

C
EGG OF ASPARAGUS BEETLE

D
EGG OF TREE CRICKET

E
EGG OF WATER SCORPION

F
EGG OF STINKBUG

J
FEEDING HABIT
OF SAWFLY LARVA

H
EGG OF POULTRY LOUSE

G
EGG OF STONEFLY

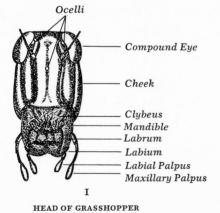

Ocelli

Compound Eye

Cheek

Clybeus
Mandible
Labrum
Labium
Labial Palpus
Maxillary Palpus

I

HEAD OF GRASSHOPPER

K
FEEDING HABIT
OF POPLAR TENTMAKER

L
FEEDING HABIT
OF TORTOISE BEETLE

they may be identified by the type of injury they inflict. The culprits can be recognized after they have long since departed, much as a criminal may be apprehended by his modus operandi.

I doubt if there is anything of plant or animal origin that doesn't serve as food for some species of insect; about the only thing an insect cannot digest is a mineral substance. This doesn't imply that every species can subsist on any plant or animal matter; most of them are restricted to a certain kind of food, though there are some that are fairly catholic in their tastes. Moreover, insects are generally adapted to feed on the kind of food which nature designed for them. Though ants and some beetles are fond of nectar, it is primarily a food for such insects as bees, wasps, flies, butterflies, and moths, that have mouthparts adapted for penetrating to the nectaries and sucking up the liquid.

Watch a butterfly alight on a flower, and then with a reading glass observe how it uncurls its proboscis (normally kept coiled beneath the head, but unwound like a watch spring when brought into use) and extends it into the blossom. When examined with the higher magnification of the lens, the proboscis (Fig. 39A) is seen to consist of the two maxillae, which fit together to form a tube that serves as a passageway for the nectar. The mandibles, which are essentially cutting and chewing tools, are of little value and are either rudimentary or absent. Both the labrum and labium are reduced for the same reason, though the labial palpi are much in evidence. A bulb within the head (you will have to dissect the head to see it), that is alternately dilated and compressed by numerous muscles, creates a partial vacuum within the tube, and the nectar is forced up by atmospheric pressure. The action is like that of a medicine dropper.

The mouthparts of all butterflies and moths are basically the same, with minor variations. In some species, for instance, the maxillary palpi are well developed, and in others the tips of the maxillae have spines which enable the insects to lacerate the tissues of ripe fruits, thus setting the juices free to be sucked up. Some species that do not feed lack the maxillae.

The true flies are a large group of insects with only one pair of wings. The most familiar is the housefly, but there are others: crane

FIGURE 39

Grasshopper Anatomy;
Eggs, Larva, and Pupa
of Familiar Flies

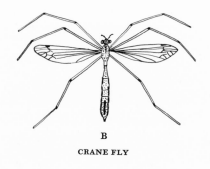

B
CRANE FLY

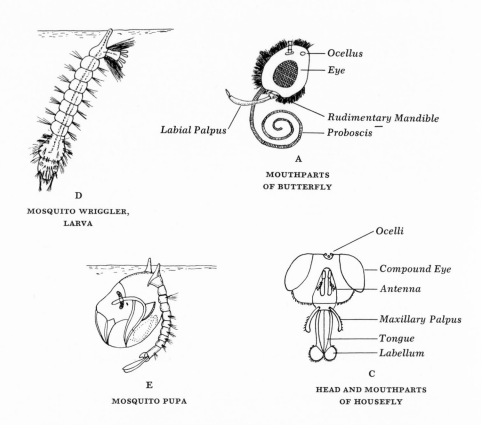

Ocellus
Eye

Labial Palpus
Rudimentary Mandible
Proboscis

A
MOUTHPARTS
OF BUTTERFLY

D
MOSQUITO WRIGGLER,
LARVA

Ocelli
Compound Eye
Antenna
Maxillary Palpus
Tongue
Labellum

C
HEAD AND MOUTHPARTS
OF HOUSEFLY

E
MOSQUITO PUPA

flies (Fig. 39B), flowerflies, blackflies, soldier flies, horseflies, stable flies, and robber flies.[2] The eating habits of these insects vary considerably; hence the mouthparts show various modifications. Essentially, however, they are all adapted either for piercing and sucking or for lapping and sucking, and in the more typical species consist of six bristlelike organs enclosed in a sheath and

[2] Stone-flies, caddis flies, scorpion flies, mayflies, and dragon flies are not true flies.

a pair of jointed palpi. Examine the mouthparts of the housefly (Fig. 39C) with your lens, and you will find that the labium and maxillary palpi are conspicuous, and that at the tip of the labium there is on each side a lobelike appendage, called a labellum, fitted for rasping.

The common and pestiferous mosquito is a fly in the true meaning of the word, and there are not many who have escaped its bite. Yet most of us have never seen what it bites us with. Even the strongest lens may not be powerful enough to show the mouthparts in detail; a microscope may be necessary. They are long and slender, and frequently bunched together so they have to be separated with a needle to be seen. The mandibles and maxillae, the latter barbed distally, are the piercing organs, and though they may appear delicate, can puncture the skin quite effectively. The labrum and epipharynx form the sucking tube; the hypopharynx conducts the saliva; and the labium forms a sheath in which the mouthparts are enclosed when not in use. At the tip of the labium is a pair of labella that are sensory in function. Of course, a bulb or sucking organ is also present, as in all sucking insects.

Both the larva and the pupa of the mosquito are rather odd-looking creatures and bear little resemblance to the adults. Most of us know the larvae (Fig. 39D), the familiar "wrigglers" of roadside puddles, temporary pools, rain barrels, and ponds and marshes. With the hand lens the wriggler is seen to have a large head and thorax and a slender, tapering abdomen. A long straight tube extends from the next-to-last segment, and one or two pairs of tracheal gills occur on the end of the last segment. The tube is a breathing organ with a pair of spiracles armed with a rosette of hairs at the tip. When the wriggler is at rest at the surface of the water, the breathing tube is extended above the surface, and the rosette of hairs is spread out on the surface film to support the animal.

Although the food of the mosquito larva varies with the species, it consists mostly of organic matter suspended in the water or floating on the surface. When feeding, the larva uses a remarkable set of jaws armed with brushes. By moving the brushes rapidly, it creates currents of water that bring food to the mouth. Locomotion through the water is effected partly by undulatory movements of the abdomen and partly by the gills, which take hold of the water and pull the animal backward. The larva, as you will observe, swims "tail first." It does not remain below the surface of the water very long, as it must take in fresh air often.

The pupa (Fig. 39E) is a humped-up little creature that reminds us of a question mark. The head and thorax are greatly enlarged and are not distinctly separated; the abdomen is slender and flexible. During transition from the larval to the pupal stage, a remarkable change takes place in the respiratory system. The single breathing tube of the larva is replaced with two earlike appendages (breathing tubes) on the thorax. They look like a pair of horns. When the pupa is at rest, the head remains upright so the breathing tubes can be extended above the surface of the water. A pair of leaflike appendages at the tip of the abdomen are used in swimming, for unlike the pupae of most other insects, the mosquito pupa is active. However, it usually does not move unless disturbed.

Although many April wild flowers have passed from the scene, most of those that appeared in May are still in bloom, and many others are coming into flower. Indeed, before June is far advanced the showy lady's slipper will bring a touch of the tropics to shady peat bogs where the pale-pink flowers of the cranberry already nod from threadlike stems; the nightshade will unfold its purple pendant blossoms in a hidden nook; and along the wayside the yellow pea-like blossoms of the wild indigo will open to the sky.

It is rather unfortunate that the showy lady's slipper must grow in inaccessible swampy places, for only the most zealous will venture there to glimpse this stately flower which is considered the most beautiful of the Cypripediums. No such problem confronts us in the case of the nightshade, since we find it in almost any thicket and by stone walls and fencerows. The graceful violet-purple flowers are a joy to the eye, but no less so then the drooping cymes of bright-red berries that appear in autumn. It is a most decorative plant, and useful too, for migrating birds find the berries attractive; the hard indigestible seeds pass through them unaltered and are voided many miles distant, a nice refinement to ensure wide distribution. Examine a flower closely and note the deeply five-cleft corolla and the yellow conic center formed by the five anthers that are fused together (Fig. 40A).

The wild indigo is the plant that turns black on withering; we often notice it in late summer along the roadside, when it looks as if it had been scorched by fire. As with all papilionaceous blossoms, the bees are best adapted to fertilize them. The dusky wing finds the indigo a suitable host plant, and we may often see the little butterfly laying her eggs on the leaves.

There are, of course, many other flowers that will engage our attention as the days pass by. Along almost any woodland trail we should expect to find the bunchberry, a smaller edition of the flowering dogwood, the shinleaf and the pyrola with their pretty waxen bells, and their cousin the pipsissewa with its shy, dainty, deliciously fragrant little blossoms and its dark green leaves that retain their color and gloss throughout the winter. Taller than any of these is the

FIGURE 40

Some Summer Flowers, Pollinia, and a Bumblebee Visiting the Iris

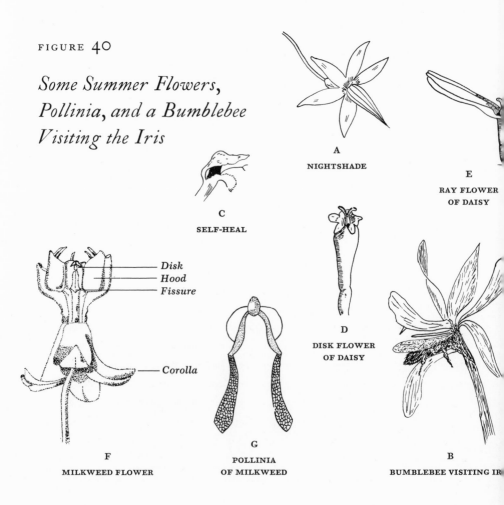

A
NIGHTSHADE

E
RAY FLOWER
OF DAISY

C
SELF-HEAL

Disk
Hood
Fissure

Corolla

D
DISK FLOWER
OF DAISY

F
MILKWEED FLOWER

G
POLLINIA
OF MILKWEED

B
BUMBLEBEE VISITING IRIS

rattlesnake weed, whose yellow flower heads remind us of the dandelion.

In an open glade the delicate four-leaved loosestrife is more than a passing acquaintance, for it blooms throughout the summer, and most of us who ramble through the woodlands know it at least by name. Not many, I am afraid, have ever taken the time to look closely at its little yellow starlike blossoms. Do so sometime, and you will see that they are prettily dotted around the center with terra-cotta red, which may extend in all directions as faint streaks over the corolla lobes. Both the stamens and the pistil project in a cone-shaped cluster, and the stigma extends so far beyond the anthers that self-fertilization occurs but rarely. Each flower is perched at the end of a slender pedicel, and the four leaves (though sometimes there are three or six) are set in circles at intervals on the slender erect stem which, as revealed by the lens, has minute hairs.

Even as May was teetering on the edge of oblivion, the blue flag or iris had begun to tinge the shores of ponds with its royal color, and in peat bogs and swampy ground the pitcher plant is already playing its villainous role of luring insects to their death. Few plants are so curiously fascinating as the pitcher plant. The general form of the flower and its rather peculiar coloring is designed to attract the carrion flies, which are especially fitted to play their part as agents of cross-pollination. It is not the flower but the leaves that draw our attention, for they are hollow and pitcher-shaped, exquisitely designed to trap the insects on which the plant partly depends for its sustenance. Examine one of them and you will find that it holds water and the remains of insects that came to feast but instead found a watery grave. Once inside, the hapless victims find no means of escape, their only avenue to the outside world bristling with downward-pointing stiff hairs that form an effective barrier. Try as they may to fly or crawl out of their prison, every attempt is hopeless, and soon exhausted, they fall into the water to give their lives so the plant may live.[3]

Look in the dictionary for the meaning of the word "iris" and

---

[3] Why does the pitcher plant require an insect diet? It lives in a soil which is usually deficient in nitrogen compounds, and without the insects to make up the deficiency, it could not form proteins.

you will learn why the ancients, always appreciative of the aesthetic, gave the name to the group of plants we know so well. To us the iris is of particular interest because of the ingeniously contrived way it manages to secure cross-pollination by insects—specifically by bees, which seem especially attracted to it. Each of the three drooping sepals forms the floor of an arched passageway that leads to the nectar, while over the entrance to the passageway is a movable and outward-pointing stigma (Fig. 40B). With a reading glass watch a bee when it enters the passageway, and observe how it brushes against the stigma, which scrapes from its back the pollen previously collected from another flower. Guided by the dark veining and golden lines, the bee makes its way to the nectar; and as it moves along the passageway its hairy back rubs against the pollen-laden anther above and becomes covered with a new supply, which it carries to another flower after it has sipped its fill of nectar and gone its way.

In its own manner the frostweed is as unique as any flower you will find. The plant thrives in dry fields and sandy places, and here the solitary flowers with their showy yellow petals open only for a day—and the day must be a bright and sunny one. The next day the petals fall, having served their purpose; the stamens drop too; but the club-shaped pistil, having been dusted with sufficient pollen, remains to develop into a rounded, ovoid pod. Another flower succeeds the first one, and the second is succeeded by a third, and so the succession continues for weeks; then as summer begins to wane, smaller flowers appear which have no petals and which are clustered at the bases of the leaves. The pods of these flowers are no larger than pinheads.

The plant ends its blooming before the first frosts of autumn; why, then, is it called the frostweed? On some cold November morning, examine the plant and you will find ice crystals that might easily be mistaken for bits of glistening quartz about the base of the stem or in the cracked bark of the root where the sap has oozed out and frozen solid. At times the frozen sap assumes a feathery, whimsical form that stimulates the imagination.

Like many of our wild flowers, the self-heal[4] is also an immi-

---

[4] So named because it was at one time considered a cure-all, a specific for all kinds of human diseases.

grant, and like many other immigrants it has become established in spite of fierce competition. One reason for its survival is that it is able to adjust itself to almost any environment; we find it by the roadside, where it is often stunted, dusty, and rather bedraggled; and we find it in fields and meadows, where, conditions being more conducive to vigorous growth, it is a lusty little plant well attended by bumblebees and common sulphur butterflies.

But we are not apt to find the self-heal unless we really look for it, since it is a rather low-growing plant and often lost among the buttercups, daisies, and tall grasses. The tubular purple or violet flowers (Fig. 40c), set in dense spikes that resemble a clover head, are irregularly two-lipped, the upper lip darker and hoodlike, the lower one three-lobed, the middle and largest lobe fringed. There are four twinlike stamens that ascend under the upper lip, the filaments of the lower and longer pair two-toothed at the summit, one of the teeth bearing an anther and the other tooth sterile— a rather odd arrangement. The style is threadlike and shorter than the stamens, and it terminates in a two-cleft stigma.

The yellow hop clover is a quaint little plant we find growing along the roadside and in fields and waste places. In spite of its abundance, however, few of us recognize it as one of the clovers, though its leaves and blossoms resemble those of the more familiar species. The small, pale golden-yellow papilionaceous florets bloom from the base of the flower head upward, and when fertilized, each turns downward and becomes brown. When all have become brown, the withered head looks like a small dried hop. Perhaps it is not a plant to which we would give more than a passing glance, for it bears with many others the onus of being too common. But it deserves a little of your attention, and I would say that were there no other clovers, we would hold it in higher esteem.

June always turns my thoughts back to my boyhood days and how I used to wander through the fields and meadows with all the abandon and carefree exuberance of youth. Today, to recapture that magic freedom, I need only a bright, warm, sunny day, when the blue sky arches over the land, when countless insects hum and buzz about the buttercups and clovers, the daisies and black-eyed Susans, the fiery heads of the Indian paintbrush and the flat-topped clusters of the yarrow, when the music of the bobolinks bubbles up in a cascade of ecstasy from the waving heads of timothy and orchard grass.

You can find much of interest in a June field if you know where and how to look. In January we saw the thrips hiding from the winter cold among the warm woolly leaves of the mullein; now we can watch them threading their way in and out of the tiny florets of the daisy head. For the daisy, like the dandelion, is not a single flower but a cluster of hundreds of tiny tubular yellow florets, each a perfect flower, packed tightly together within a green cup (Fig. 40D). Note how the stamens form a ring around the pistil, how the pistil rises through their midst, and how the two little hair brushes at its tip sweep the pollen from the anthers. As the pistil continues to rise, the pollen is lifted high up where any insect crawling over the florets will be dusted with it. Then, with its pollen gone, the pistil spreads its sticky stigmatic arms, hitherto tightly closed so that self-fertilization could not occur, but now receptive to pollen from another flower.

The ray florets (Fig. 40E) are female and have only the pistil, but this pistil does not have the hair brushes, for it does not need them. Because daisies are so conspicuous and offer so much in the way of liquid refreshment, almost every winged insect eventually finds its way to at least one of them, and with every insect visitor seed production is enhanced. It is no wonder then that there are so many daisies, and that from afar a June field appears as if a blizzard had passed that way.

Unlike the pistillate ray florets of the daisy, the orange-yellow ray flowers of the black-eyed Susan are neutral; that is, they have neither stamens nor pistils. However, the purplish-brown florets that form the conical disk are perfect. The florets at the base of the disk open first, and their pollen forms a yellow circle. Then the next-higher florets on the disk open, forming another yellow circle; and this continues as blossoming circles climb toward the apex. Sometimes we find small caterpillars in the heads of the black-eyed Susan that have the habit of attaching small pieces of the flowers to their backs, keeping them in place with silk. No doubt this bit of camouflage serves them in good stead.

There are eighteen species of milkweeds, but probably few of us are acquainted with more than one or two. With its cloyingly sweet, somewhat pendulous flower clusters, the common milkweed is one of the more familiar of our summer wildflowers, and from now until September we find it blossoming along roadsides and in fields and waste places, its richly nectar-laden flowers a veritable banquet table for all sorts of flying insects.

Milkweeds are excelled only by orchids in the ingeniousness of their flower structure (Fig. 40F). The pollinia (Fig. 40G) are so designed that when an insect steps upon the edge of the flower to sip nectar, its legs slip between the peculiarly shaped nectariferous hoods situated in front of each anther. As it then draws its legs upward, a claw, hair, or spine catches in a V-shaped fissure and is guided along a slit to a notched disk, which becomes attached to the leg. Since the pollinia are each connected to the disk by a stalk, they are carried off when the insect leaves. Upon the insect's arrival at another milkweed flower, the pollinia are easily introduced into the stigmatic chamber. The struggles of the insect at this time break the stalk of the pollinia, and the insect is relieved of its load. Sometimes an insect loses a leg or is permanently entrapped. Anyone can see how it is all done by merely watching an insect on a flower through a reading glass.

Few plants have such a worldwide distribution as the yarrow, and few appear so often in mythology, folklore, and literature. Chiron the Centaur is said to have named the yarrow after Achilles, who is supposed to have used it to heal his wounded soldiers at the siege of Troy. Old books mention its virtues as a love charm, a cure-all for divers ailments, and an ingredient of an intoxicating drink. Aesthetically the plant has mixed qualities, to my way of thinking. The feathery masses of finely dissected leaves have a lacelike character appealing to the eye, but its dusty-looking flower clusters give it a sort of unkempt appearance. When viewed with the lens, the individual flowers belie their collective appearance and are not unattractive. The perfect disk florets are at first yellow but later turn brown; the pistillate ray flowers are white or a grayish white, or in some rare cases crimson-pink.

Two distantly related but dissimilar animals often lie in ambush in the flower clusters. One is a small white crablike spider (Fig.

41A), and the other is a greenish-yellow insect with a broad black band across the expanded part of the abdomen. The spider is of interest because later in the summer it moves to a yellow flower and turns yellow; spiders, as a rule, do not change color. The insect is called the ambush bug and is well named, for both the rather grotesque form of the body and its peculiar coloring simulate the blossoms among which it hides and help to conceal it. The insect is an assassin of the first order and is well equipped for a predatory existence with its powerful beak and grasping front legs (Fig. 41B).

The leg of an insect is composed of several parts: the coxa, trochanter, femur, tibia, and tarsus, as shown in Figure 41C. Most of these parts vary in form in various insects, depending on how the legs are employed; we need only examine the front legs of the ambush bug to see how true this is. The coxa is the part that articulates with the body and is somewhat elongated; the femur is greatly thickened, so that it is half to two-thirds as broad as it is long; and the tibia is sickle-shaped and fits closely upon the broadened curved end of the femur. Both the femur and the tibia, as the lens show, are armed with a series of close-set teeth, so that an unlucky victim is held as firmly as if caught between two saws.

Equally as predatory, if not more so, are the assassin bugs, a group of rather striking insects that are of benefit to us because they prey on cutworms and other harmful species. A common assassin bug is the masked bedbug hunter, so called because it often enters houses where there are bedbugs on which to feed; another is the wheel bug which has a cogwheel crest on the prothorax. Both may be found on various plants and are worthy of close scrutiny.

Predatism occurs in most orders of insects, even among flies which, at first thought, we are inclined to dismiss as nonpredatory, forgetting such biting insects as the blackflies, deerflies, and mosquitoes that can torment us no end. There are others, of course, such as horseflies and robber flies—the latter of proved value because they prey on many injurious species. In June robber flies (Fig. 41D) are quite common in the open fields, where they may be seen swooping down upon their prey in midair or snatching their victims from leaves and carrying them away to a convenient place

FIGURE 41

Insect Miscellany

G
MILKWEED CATERPILLAR

A
CRAB SPIDER

H
SEMI-PUPAL STAGE
OF MILKWEED CATERPILLAR

B
AMBUSH BUG

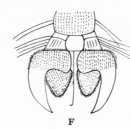

F
STRUCTURE OF AN INSECT'S LEG
(foot of housefly)

I
SEMI-PUPAL STAGE
OF MILKWEED CATERPILLAR

C
EGG OF MILKWEED
BUTTERFLY

D
ROBBER FLY

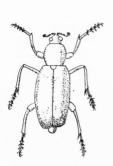

E
ROSE CHAFER

J
SEMI-PUPAL STAGE
OF MILKWEED CATERPILLAR

where they can suck their body juices at leisure. They are fairly large insects with an elongate body and very slender abdomen, though some species are quite stout and resemble bumblebees. It is not known whether this resemblance helps them to get near their prey or serves as a protection against other predators that fear the bee's sting.

Wherever we may go in June, we find insects on the wing or crawling about on the ground; others hidden from view are boring their way in plant tissues. Yellow jackets, bald-faced hornets, mud wasps and digger wasps; mining bees, leaf-cutting bees, mason bees, carpenter bees; stinkbugs, flat bugs, thread-legged bugs, four-lined leaf bugs; treehoppers, leafhoppers; flowerflies, flesh flies, crane flies, peacock flies; buprestid beetles, ladybird beetles, ground beetles; the list is endless. And there are always the rose chafers, which never fail to appear on my rose bushes in such numbers that it seems an impossible task to get rid of them before they damage the blossoms.

The chafers are among the worst insect pests of the garden, because they appear suddenly in great swarms and overrun their food plants almost before we are aware of their presence. They attack not only the rose but the grape as well, at first feeding on the blossoms and then on the newly set fruit and foliage.

They remain for about six weeks and then, having mated, disappear as suddenly as they came. The males fall exhausted to the ground and die; the females burrow into light sandy soil and deposit their eggs. Then they return to the surface, linger on for a few days, and perish, leaving the eggs to hatch and the grubs to feed on the roots of grasses until the following spring, when they change into adult beetles to annoy us once again.

The scientific name of the rose chafer is Macrodactylus, which means "long-fingered," a word that fits it in both a literal and figurative sense. It is a rather beautiful insect, slender in form (Fig. 41E), tapering both forward and behind, and yellowish in color. We would not suspect that the color is due to fine scalelike hairs that cover the insect, for they are hardly visible to the naked eye. They are readily seen with the lens, however. The legs are long, slender, somewhat spiny, and a pale-red color.

Along the woodland border, where the scarlet keys of the red maples hang in drooping clusters and the berries of the shadbush, now ripe, are greedily devoured by hungry birds, brown butterflies, with such delightful names as the little wood satyr, the pearly eye, and the wood nymph, float over the nodding grasses, poise quivering over a nectar-laden blossom, or rest on a leafy plant. Along the roadside the common sulphurs become a familiar sight, and in city park and village yard as well as in the more open field and woodland, the American copper flutters about with careless abandon. The caterpillars of this species feed on the field sorrel, and we often see the butterfly stop to lay its eggs on the leaves and stems. The coppery-red color of the butterfly blends so well with the rusty red of the blossoms that sometimes we don't notice the butterfly on the plant until it moves its wings or suddenly takes to flight.

June is the month that heralds the return of the monarch butterfly from the Southland, though sometimes it may appear earlier.[5] Many of us associate the monarch butterfly with the milkweed, because it is usually the only plant on which the eggs are laid. These eggs are conical, deeply ridged, and pale yellow (Fig. 41F). They hatch within a day or two into small, cylindrical caterpillars with alternate transverse bands of yellow, black, and white (Fig. 41G). The body is divided into a number of segments, and there is a pair of threadlike black "horns" on top of the second segment and a shorter pair on the eleventh. There are also three pairs of thoracic legs and several pairs of abdominal legs and prolegs; as seen with the lens, these have numerous minute hooks, called crochets, that help the caterpillars climb.

As the caterpillars feed and move about on the leaves, the longer pair of "horns" moves back and forth. If the caterpillars are disturbed or frightened, the front horns twitch excitedly; those at the rear are quieter. The horns probably serve to keep away parasitic flies that lay their eggs on the backs of caterpillars.

A caterpillar has but one aim in life—to eat and grow. When a monarch caterpillar reaches its full growth, after having molted several times, it spins a little silken mat on a milkweed leaf. Then,

[5] Every year the butterfly migrates to the South in groups of hundreds, only to straggle back the following spring as individuals.

grasping the mat with the crochets, it lets go of the leaf with its thoracic legs, swings downward, and curves the forward end of its body upward as shown in Figure 41H. It remains in this position for a few hours as the body juices gravitate downward and the lower segments become distended or swollen. Suddenly the skin splits open along the median line of the back and is gradually worked upward by contortions of the body (Fig. 41I).

Meanwhile the caterpillar manages to engage a spiny process (cremaster) at the end of its body in the silken mat; and when firmly engaged, the skin drops off, exposing a strange-looking creature that is broader below than above (Fig. 41J). Gradually the softer outer or new skin hardens into a distinct covering that takes on a beautiful green color, with a number of golden spots distributed over the surface and a few black spots just below the cremaster (Fig. 42A). The insect now passes into a quiescent resting period called the pupal stage or pupa (chrysalis).

During this period, which lasts about two weeks, the wormlike caterpillar with chewing mouthparts is transformed into a beautiful winged butterfly with sucking mouthparts. Some of the external changes may be observed through the covering, which is rather thin and transparent. When the markings on the wings and the outline of the butterfly's body can be distinctly seen, the butterfly is ready to emerge.

The chrysalis of the monarch butterfly has been called "the green house with the golden nails." It is one of the most beautiful natural objects, and when looked at with the lens the colors are even more vivid than when seen with the naked eye. Butterfly chrysalides vary greatly in shape and color (Figs. 42B,C,D). Some are plain, oval, and mummylike. Others are brilliantly colored, often with metallic gold or silver markings, sometimes elaborately sculptured or with long spiny or knobby projections—details visible only with the lens. The chrysalis of the harvester butterfly is most unusual. It resembles a spiral shell (Fig. 42E), although when viewed from above, the anterior half looks like a monkey's face (Fig. 42F). This one is really worth looking for.

Although the pupa of most butterflies is called a chrysalis, some species, such as the hesperids, spin what might loosely be called a cocoon. Many moths spend the pupal stage within a cocoon, but there are many that pupate naked in the ground, such as the tomato hornworm, a strange brown, segmented shell-like object (Fig. 43A) with what appears to be a handle at one end and which we often unearth when we spade our gardens. At this time of the year we may find tent caterpillars, which emerged from the eggs in April and are now full grown, crawling over the ground in search of a place to spin a cocoon. It is a small, tough, silken, oval affair (Fig. 43B), dusted with a yellow powder and generally constructed in a sheltered nook. Under the lens it looks surprisingly different than when seen with the naked eye.

Cocoons differ as much as chrysalides in size, shape, color, and materials. For instance, the apple bucculatrix, a small brown moth only about a tenth of an inch long, spins a small, slender, and strangely ribbed cocoon. Sometimes many of these cocoons are

FIGURE 42

Representative Chrysalids of Butterflies

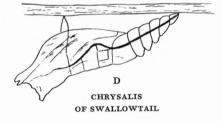

D
CHRYSALIS OF SWALLOWTAIL

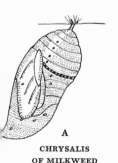

A
CHRYSALIS OF MILKWEED

B
CHRYSALIS OF MOURNING CLOAK

C
CHRYSALIS OF CLOUDLESS SULPHUR

E
CHRYSALIS OF HARVESTER

F
CHRYSALIS OF HARVESTER

clustered on a twig where they may easily be seen (Fig. 43c); indeed, it is these cocoons that may first reveal the presence of the insect in the orchard.

FIGURE 43

Representative Cocoons of Insects

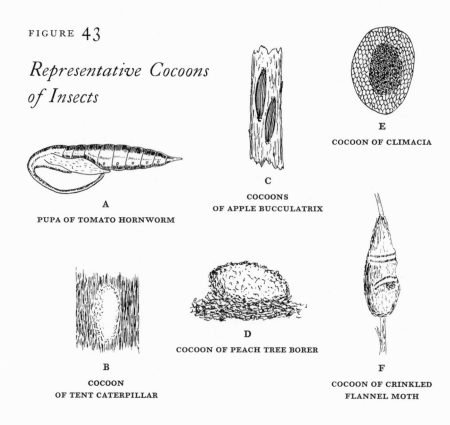

A

PUPA OF TOMATO HORNWORM

C

COCOONS
OF APPLE BUCCULATRIX

E

COCOON OF CLIMACIA

B

COCOON
OF TENT CATERPILLAR

D

COCOON OF PEACH TREE BORER

F

COCOON OF CRINKLED
FLANNEL MOTH

Unlike the trim cocoon of the tent caterpillar, the cocoon of the peach tree borer is an unkempt-looking object, oval and somewhat brown in color, and when examined with the lens, is seen to be made of silk and particles of bark and excrement (Fig. 43D). You will find it on the bark of the peach tree.

Few cocoons are attractive in appearance; most are drab-looking objects, though they are all cleverly designed and constructed to serve the needs of their inhabitants. And not all cocoons are spun by caterpillars; a few beetles, certain flies, and several species of sawflies, such as the yellow-spotted willow slug, the elm sawfly, and the larch sawfly, do so too. The three sawflies construct tough silken cocoons in the duff or debris on the ground. A cocoon is also spun by the spongilla fly, which is not a true fly but belongs to the family Sisyridae, a small group of tiny, smoky-brown insects. They are all called spongilla flies because the larvae live as parasites in the freshwater sponge Spongilla. The species *Climacia dictyona* differs from the rest of the group in being yellowish and in certain minor characters; otherwise it is much like the others. The habits of all are similar. When the larva of Climacia is about to pupate, it leaves the sponge and crawls out of the water to a leaf, twig, or other object. Here it spins a hexagonal meshed net, quite lacelike in appearance when seen with the lens. Beneath this silken covering the larva spins a hard, tough cocoon of closely woven silk. The entire structure is a beautiful object (Fig. 43E).

It is not simple matter for winged insects to escape from a cocoon, but nature in her omniscience usually provides a means of doing so. Insects with chewing mouthparts need only gnaw their way out, but sucking species must have some other way of escaping. A few species secrete a liquid that softens the silk at one end of the cocoon and then, by forcing the strands apart or by breaking them, make an opening. The pupae of some silkworm moths are provided with a pair of large, stout, black spines to slit the cocoons, and the pupa of the white-blotch oak leaf miner has a toothed crest for the same purpose. The caterpillars of the cecropia and promethea moths construct a valvelike structure at one end of the cocoon which separates easily when the adult moths are ready to emerge. And just before the caterpillar of the crinkled flannel moth changes to a pupa, it constucts a hinged partition near one end of the cocoon that functions as a trapdoor through which the adult moth can emerge (Fig. 43F).

On a sunny June day visit almost any pond where the water arum and pickerel weed are in flower, and you will see newts swimming around in the water and countless tadpoles in various stages of development swimming among the plants that grow along the water's edge. You will probably see, too, newly emerged peepers hunting for gnats, mosquitoes, and other small insects. I sometimes wonder if their early escape to land is not a means of survival, for in the water lurk all kinds of enemies. Not the least of these is the

giant bullfrog, whose familiar "jug-o-rum, more rum" is so star-
tlingly weird in the quiet of the night. Toward the end of the month
the bullfrog mates and lays its eggs, some ten to twenty thousand, in
a floating mass that measures nearly 2 feet across. Catfish also
spawn about this time; they fasten their opaque yellow eggs to the
undersides of stones in considerably smaller masses that are only
about 2 inches wide and 1 inch thick.

In the pond you will see backswimmers, water boatmen, and
diving beetles swimming through the water; scavenger beetles and
water scorpions crawling along the bottom; whirligigs gyrating on
the surface, and water striders hurrying out of their way; and in the
air above, thin, long-legged crane flies, damselflies, dragonflies, and
many others, their bodies casting moving shadows on the water.

Watch the dragonflies and you will observe that some of
them—the larger and stronger—keep to the higher regions above
the water, coursing back and forth, passing and repassing the same
point at intervals of a few minutes, while the smaller species are less
constantly on the wing, usually flying in short sallies from one resting place to another or hovering about the water before they alight.
All of them are perfectly harmless, despite popular superstitions,
except to the insects on which they feed and which they capture in
a sort of "basket" formed by their front legs. Actually they are quite
valuable, for they destroy large numbers of mosquitoes.

The aquatic young, or dragonfly naiads, are also predaceous
and perform a like service, feeding on the larvae and pupae of
mosquitoes. They are wholly unlike the adults in appearance, being
dingy little creatures with six queer, spiderlike legs and without
wings, although there are four little wing pads extending down the
back that encase the growing wings. They move along the bottom
rather slowly, but they can lunge and dodge with surprising agility.
To capture their prey they steal up to it and, quickly extending the
hinged lower lip, grasp the victim with the fingerlike pincers at the
end of it (Fig. 44A). Then the lip is folded back and the prey
conveyed to the mouth. In some species the lip, when folded, covers
the face like a mask. The oddly formed lower lip is best seen with
the lens, and the manner in which it is used can be observed at close
range by transferring a naiad to a dishpan of water containing a
little sand and feeding it some water insect.

FIGURE 44

Dragonfly Naiad, Gills of a Damselfly, and a Firefly

B

CAUDAL GILLS
OF DAMSELFLY

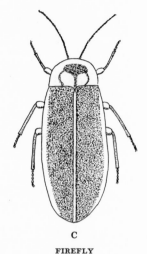

A

NAIAD OF DRAGONFLY
WITH EXTENDED LABIUM

C

FIREFLY

Dragonfly naiads normally crawl about on the bottoms of ponds and streams on their legs, but they are also able to move quickly through the water by making use of Newton's third law of motion: for every action there is an equal and opposite reaction. A toy balloon blown up, pinched by the neck, and then released, is a common application of the principle. In a sense the animal is a living example of what we know as jet propulsion. To understand how it works in the naiad, examine the animal and you will see that it has no external breathing organ. The naiad does possess gills, however; they are located internally, in a special chamber in the rear of the alimentary canal. Known as rectal tracheal gills, they consist of an elaborate lacework of air tubes formed by the infolding of the rectal walls. Water is drawn in and out of this gill chamber through an oral orifice guarded by strainers—specifically, by five pointed spines. At times the water is forcibly ejected by the contraction of the abdominal muscles, and when this occurs the insect is propelled through the water. Thus the breathing apparatus of the naiad serves a double purpose; it is an organ of respiration and of locomotion.

Damselflies are only slightly less interesting than their larger relatives. Smaller, more delicate, and more brilliantly colored, having such names as black-wing and ruby-spot, they fly more leisurely and often rest on the rushes and grasses. Since they are not strong flyers, they often fly in tandem.

Unlike the naiads of the dragonflies, damselfly naiads have three platelike tracheal gills at the end of the abdomen (Fig. 44B). As seen with the lens, the tracheoles are separated from the air in the water only by the delicate wall of the gill, which permits the transfer of gases between the tracheoles and the water. The naiads also use the gills for locomotion but work them in a sort of sculling movement.

Most of our native shrubs have blossomed by the time June appears, and many of them are still in flower: for example, some of the dogwoods—the round-leaved, the panicled, the silky cornel, and the red osier—their small, flat-topped white flower clusters advertising their wares to all who might be interested. The viburnums now come into their own: the dockmackie in dry rocky woods; the nannyberry on the banks of streams; the withe-rod in moist thickets; and the arrowwood in swamps and similar places. To know any one of the viburnums by their flowers is to know them all, for they all spread more or less flattened compound cymes of white flowers; it is their leaves on which we depend to recognize the different species.

Although the high tide of bloom comes in July, even now the elder, "foamed over with blossoms as white as spray," has begun to perfume the roadside with honeylike fragrance and to surpass in beauty and effectiveness the finest of our garden favorites. Pastures, hillsides, and swamps serve equally as a home for the sheep laurel, also known as sheep poison, lambkill, and calfkill—names that attest to the poisonous qualities of the little shrub. The flowers, clustered on the sides of the twigs, are small, five-lobed, and saucer-shaped, and each is a bright pink or crimson. Examine one of the flowers with your lens and you will find that each small saucer has tiny pockets into which the ten red anthers are tucked, the filaments being bent like a spring. When they are touched by foraging insects, the anthers are released with a snap, flinging out the pollen. Touching them with a needle will have the same effect.

In open woodland glades the New Jersey tea of Revolutionary

fame, having appeared in May, is still bringing forth its tiny cream-white flowers in conic clusters, and the sumac is just beginning to open its small greenish flowers in terminal spikes. The sumac's relative, poison ivy, is also beginning to blossom as it climbs over stone walls and up tree trunks. Everywhere the blackberry trails its woody stems with delicate flowers that all but conceal wicked thorns, while its cousin the raspberry, catering only to the small bees, unfolds its flowers on shrubby stems covered with rigid hooked bristles.

Like thrifty housewives of Revolutionary days who were more interested in the leaves of the New Jersey tea than the flowers, we are somewhat inclined to ignore both; and only the small bees seem to have any use for the insignificant flowers of the sumac. Perhaps we might say the same for the flowers of the blackberry and raspberry, though we enjoy their fruits.

Raspberry flowers are smaller than those of the blackberry and are not as effective in attracting insects, though they secrete an ample supply of nectar. The small, erect petals make it difficult for the insects to enter the flowers; only very small ones can do so. Too, these petals prevent the stamens from moving outward away from the stigmas, and as a result self-fertilization often occurs. Just the reverse is true of the blackberry flowers. And who has looked closely at a blueberry blossom? Or is it only the bees and their kind that are concerned with the little bell-shaped flowers?

In some parts of their range, the catalpa, linden, locust, mountain ash, and horse chestnut may open their flower buds in May; in other parts they may wait until June; the catalpa sometimes does not bloom until July. Of all these trees, so diverse in character and appearance, all of them planted more or less as ornamentals in the North, only the linden and mountain ash are native to the northern woodlands.

The white flowers of the catalpa, clustered in panicles often 10 inches high which may cover the tree so thickly that the leaves are all but hidden, suggest the opulence of the tropics. The overall effect of the cluster is one of pure white, but the individual corolla is spotted with purple and gold, the spots forming guideposts for the bees that are attracted by the fragrance of the blossoms.

The creamy-white, papilionaceous flowers of the locust, too, are an invitation to the bees. Like its near relative, the sensitive plant, the leaves fold together; but unlike those of the sensitive plant, they close only at night or on rainy days rather than whenever they are touched. A locust newly in leaf is a beautiful tree; but in winter when the leaves have fallen, it presents an unkempt, ragged appearance, for the trunk is often twisted and the branches, brittle and easily broken, are irregular and twiggy.

Few trees can match the horse chestnut when in flower, for then it is "a pyramid of green supporting a thousand pyramids of white." The flower cluster is what the botanists call a thyrse, a mixed sort of inflorescence in which the main stem is racemose and the secondary ones cymose. It sounds complicated, and it is. We don't have many such clusters; the only other familiar examples that come to mind are the lilac, syringa, and privet. The corolla of each flower in the horse chestnut cluster is white, spotted with red and yellow, and the curving yellow stamens extend beyond the ruffled border. In view of the vast number of blossoms, we would expect a large harvest of nuts; but actually a cluster forms only two or three burs, and some clusters not any. The reason is that only a few of the flowers are fertile; most of them have stamens only, and an aborted pistil.

No matter how isolated a linden may be, the bees are sure to find it, for the flowers are exceptionally fragrant to advertise the rich supply of nectar they provide for their insect visitors. The small yellowish-white flowers are borne in cymose clusters, pendulous at the end of a flower stalk attached, for half its length, to the vein of an oblong leaflike bract as long as the stalk itself.

The mountain ash, with its handsome foliage, clusters of white flowers, and scarlet fruits, is a perennial favorite of mine. The mountain ash of our lawns and parks is the European species; however, our own mountain ash is just as attractive. But we have to travel to the cool mountain woods of New England and Canada or other border states to find it, for it has a distinct preference for cold, unsheltered places where, exposed to the winds, it unfortunately is often stunted. We may find it in the Southern states too, but only in the higher altitudes.

The white pine also blossoms at this late date, the flowers borne in scaly catkinlike clusters, the staminate at the base of the season's shoots, the pistillate at the end of the branches. When seen with the lens, the staminate flowers are oval and yellowish, with six to eight scales at the base; the pistillate are elliptical and pink or purplish.

Whenever our friends in the South go into the woodlands, they have to exercise constant vigilance against poisonous snakes; we in the North are more fortunate. True, we have our own poisonous snakes, but they are few in species and in number; rarely do we encounter one.

All our snakes are now active in our fields and woodlands. Some of them—the milk snake, black snake, and hog-nosed snake —lay their eggs during the month. Those of the black snake are deposited in soft, moist soil on the edge of a meadow. White and leathery, they are covered with coarse chalklike granulations that show up prominently when viewed with the lens. The eggs of all three snakes hatch later in the summer, and as the young eat and grow, they molt at intervals, leaving the cast skins behind as a ghostlike replica of themselves. If you examine one of these cast skins carefully, you will find that it faithfully preserves every minute detail of the outer surface of the snake, including the transparent circular windows that cover the eyes. When looked at with the lens, the skin that covered each scale looks like a tiny lens in itself, surrounded by a raised network of thinner skin that represents the folds which lay between the scales.

Many of our turtles—the painted, spotted, snapping and wood—also lay their eggs in June. Young turtles, snakes, and lizards have the same problem of getting out of the shell that insects do. However, they are usually provided with some sort of device for puncturing or rupturing the eggshell. Young turtles—or we should say embryos, for they are embryos until they have emerged—have a horny caruncle[6] on the tip of the snout, whereas snakes and lizards have an egg tooth protruding from the front of the roof of the mouth. This egg tooth is a true tooth and is usually shed within

[6] However, the caruncle is not always used by embryo turtles in their escape from the shell; in some species it has been shown that the caruncle is quite useless and that the rupturing of the shell is due to the absorption of water, often at a spot away from the hatchling's snout.

a day or two, but the turtle may carry the caruncle for several days; indeed, the snapping turtle may carry it for three or four weeks.

We usually associate turtles with water, but when the spring sun serves notice that tender leaves and berries may be found on nature's menu, the wood turtle leaves the ponds and streams and wanders through pastures, woodlands, and upland fields in search of such tidbits as mushrooms and wild strawberries. In his eagerness to get at the berries, the turtle claws down the plants, and when it has had its fill, lumbers off through the underbrush, his mouth stained with the red juice.

We speak of strawberries as berries, but technically they are not. We shall have occasion to discuss fruits in more detail later on, but for the present we may define a berry as a fleshy fruit containing a number of seeds. If you examine a strawberry (Fig. 45A) with your lens, you will find that it is not actually a berry, though it is fleshy in a way. Compare a strawberry with a tomato, which is a berry, and you will see the difference. In the tomato, the seeds are enclosed in the fleshy fruit, but in the strawberry, the seeds are attached to the surface. Actually, the seeds are a kind of fruit in themselves, called akenes, so that a strawberry in reality is a fruit made up of a number of smaller fruits (Fig. 45B). Botanically it is known as an aggregate fruit.

The wood turtle has no lack of mushrooms on which to feed, for these plants are becoming increasingly abundant in species. In the woodlands we may now find the sheathed amanitopsis of somewhat variable color; the beautiful yellow chanterelle (Fig. 45C) that prefers the deep shade of hemlock groves; the vermilion mushroom, whose cap is often deep red; the oak-loving mushroom, equally at home in a pasture and in a woodland grove, though its habitat seems to be among fallen pine needles and not among oaks, in spite of its name; and the scaly lentinus with its brownish spotlike scales, whose natural place of growth is the decaying wood of evergreens but which also grows on fence posts, bridge timbers, and railroad ties, its destructive effect on the latter having earned for it the name "train wrecker."

Before the month is out, we should also find the common naucoria along the roadside, in pastures, and on our lawns, its homogeneous brownish colors making it fairly easy to identify; the bell-shaped panaeolus of a peculiar leaden tint; and the haymaker's mushroom, essentially a species which likes open, grassy places, such as lawns and city parks. This species is one of the most common of our mushrooms and also one of the easiest to recognize with its conic, smoky-brown cap.

As you wander about the countryside or motor along some highway, observe that a broad ribbon of dull rose often borders winding streams and brooks, though only for a short period—that is, as long as the reed canary grass is in bloom. Even before it flowers, the dense growth of its blue-green leaves is noticed in marked contrast to the lighter color of the marsh. In meadows, fields, and along the wayside, the meadow fescue is also in bloom, its green spikelets tinged with reddish purple, giving delicacy to the one-sided drooping panicles; and overnight the couch grass appears

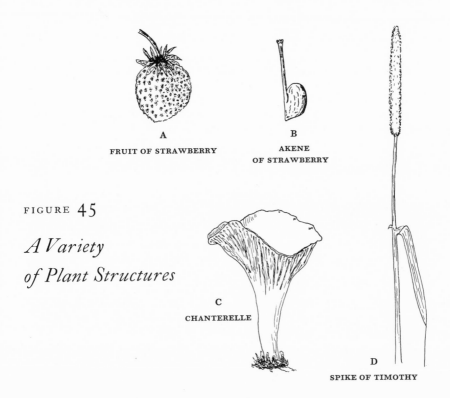

A
FRUIT OF STRAWBERRY

B
AKENE
OF STRAWBERRY

FIGURE 45

*A Variety
of Plant Structures*

C
CHANTERELLE

D
SPIKE OF TIMOTHY

in cultivated lands, fields, pastures, and one's dooryard, too, its stout leafy stems and flattened two-headed spikes shooting up out of the ground with all the energy of the fabled hydra. Cut one stem, and half a dozen will rise in its place.

Look along a narrow brook or ditch or by a damp wayside and you should find the heavy, drooping panicles of the rattlesnake grass. It is one of the more beautiful of our grasses, with its inflated spikelets of pale green and purple, and what is still better, its flowering heads retain their beauty until late fall. Look, too, in the borderland between pasture and marsh and you will see the drooping panicles of the nerved manna grass blooming in tones of purple and green.

And June is not June without the timothy grass (Fig. 45D). In old fields and meadows and along the roadside, its bright-green bayonets rise straight upward, surmounted with cylindrical flowering heads that appear in the sunshine as if gemmed with dewdrops. Sometimes the stalks are 5 feet tall and the flower heads 6 to 10 inches long, but only if the soil is rich; commonly both are much smaller.

With its slender stems and threadlike leaves and its delicate panicles of silvery-pink and -green, the wavy hair grass seems a transient spirit of the wayside that we somehow expect to vanish on the morrow. Grasses, no less than other flowering plants, contribute their share to the perfume of the outdoors, some to a greater degree than others. Few, however, are as fragrant as the vanilla grass. The spikelets are chestnut brown or purple, but it is the leaves that interest us most, for they are very fragrant when dried. Indians use them to weave baskets.

Most of the ferns reach their growth in June. The lady fern is now particularly conspicuous, because its bright pink or reddish stalks are an effective contrast to the green foliage. Later in the summer it loses much of its beauty; the fronds become disfigured and have a rather blotched appearance. The maidenhair fern, still immature, is lovely in its fragility, and its quality of aloofness adds to its charm. Its chosen haunts are dim, moist hollows in the woods and shaded hillsides that slope to the river. Only in limestone regions do we find the slender cliff brake; and if we want to find the rare mountain spleenwort, we must climb the lofty cliff. The Virginia chain fern of swampy places is more accessible. Sometimes we find it standing in deep water.

Fireflies, or lightning bugs, have excited the imagination of many peoples; for instance, they are frequently featured in Japanese art, and they have served as a subject of many musical and poetic compositions, some serious and some in a lighter vein.

The little animated lanterns make a sporadic appearance in early June, when they appear as tiny meteors shooting through the darkness, but a week or two later they are out in full force. We have all seen them, or rather their flashing beacons, for only a few of us are able to recognize them during the daylight hours.

Of what use is the light to the firefly? There are many species of nonluminous fireflies, which live much the same as those that give off light and survive just as well. At one time it was believed that the light serves as a warning to nocturnal birds, bats, and other insectivorous animals—a theory supported by the fact that fireflies are refused by birds in general. But it is now believed that the light is a signal to their mates, since the females of some of the luminous species are wingless. Yet there are wingless females of other insects that do not need light to attract the males; nor are luminous larvae, the so-called glowworms, interested in mating. However, if the light is used to attract mates, the supposition is that they can see each other's light.

We have about half a dozen species of fireflies in the eastern half of the United States, and since each has its own characteristic method of flashing, distinguishable by such features as intensity, duration, number and intervals between flashes, and flight levels, we can learn to identify them with a little effort. One of our most common species is *Photuris pennsylvanicus* (Fig. 44c). The males of the species usually occupy the treetops and flash three, four, or five times with a greenish-blue or pale-blue light; the females remain on the ground and flash one, two, or three times during each period.

Another fairly common species is *Photinus marginellus*. The males usually inhabit low shrubs; the females are closer to the ground. The light of this species is yellow, the flash of the male being somewhat stronger than that of the female. These fireflies start flashing just as twilight ends and tend to stop when darkness has fallen completely, though stragglers may continue to flash well into the night.

The fireflies of *Photinus pyralis* usually begin the evening by flying close to the ground and later fly somewhat higher, though they never attain any great height. They have a rather undulating sort of flight and always flash as they ascend; after flashing, they drop down. They flash with a yellow light, and when the insects are numerous, the flashes appear like brilliant sparks shooting from the ground.[7]

As we watch the fireflies on a warm June evening, the air fragrant with the smell of sweet grass and the silence of the night broken only by the soft whispering of leaves and the faint stirrings of some animal in the underbrush, we are conscious of a change in the tempo of the nature year. The boisterousness and hurly-burly of spring has given way to a quieter and more settled order of things that will prevail for a few weeks, or until a subtle change in temperature and shortening days give warning to all living creatures that it is time to prepare for winter. June passes all too quickly, and we are suddenly aware that summer is about on us when a sphinx moth appears out of the darkness and heads for the lighted window, only to be barred from further progress, like so many other night-flying insects, by the screen that protects us from a full-fledged invasion.

[7] Fireflies, of course, are not the only animals that are luminescent; there are the one-celled animals Noctiluca and Gonyaulax, the crustacean known as the "sea firefly," the parchment worm, and the rock-boring clam. There are also several species of luminescent bacteria and mushrooms. At night remove the bark from a rotting log on which the honey mushroom is growing, and you will probably see what is called fox fire. Or place a few young specimens of the Jack-o-lantern mushroom in a dark room, and they will emit a soft greenish light. I must admit it isn't very strong, but if you are patient and wait until your eyes become adjusted to the darkness, you will see it.

JULY

"The meadow sides are sweet with hay."

JOHN T. TROWBRIDGE

"Midsummer"

JULY

T HE FIRST DAY OF THE MONTH MAY BE A HOT, cool, or rainy one; but whatever it turns out to be, the succeeding days are sure to be hot because the sun is now high in the sky, and its rays come to us more directly. The weather pattern has changed, too; usually the only rain we get is when thunderheads build up during the day and explode in late afternoon. If all the raindrops have fallen before the sun sets, a rainbow may appear in the eastern sky.

In July the time to ramble is late in the day when the afternoon begins to wane, or else in the cool of the morning, before the sun gets high, the deerflies start biting, and walking along the country road or open upland becomes unpleasant and tiring. Sometimes I am out very early, just as a glow signals the dawn of a new day; sometimes I wait and set out on a longer walk after breakfast. The birds are now less vociferous than they were a month ago; occasionally I hear the staccato notes of the scarlet tanager, but for the most part the birds have tapered off in their singing. As the month advances, we hear in the early morning or in the evening only feeble reminders of the brilliant songs of May and June.

Birds are less in evidence, too; yesterday they seemed to be everywhere, but now they appear to have vanished—which is not true, of course; it only seems that way. Yet the orioles have left their swinging nest in the elm tree and with their brood have taken to the woods and thickets to feed on ripening berries. Other birds have too, and the woods are alive with an increasing number of small birds. Within a few short weeks many will be heading south again, and once more we shall see the passing migrants. Even now the male bobolinks, who only a few days ago made the meadow ring with their music, are getting ready to don their autumn dress, and by the end of the month both the young and the old will have assembled in flocks ready to begin their southward journey. They get an early start; other birds wait awhile.

Follow a woodland trail, and you should see the ovenbird; walk along a country road, and you will see swallows perched in a row on a telephone wire. Perhaps a kingbird will dart out from his perch on a fence post or other vantage point and chase a flying insect across the pasture. Then as the day begins to lengthen, the chimney swifts emerge from their quarters in a church steeple and, with quivering wings and shrill twitterings, describe a lacework of disappearing lines in the sky. When night has fallen, the singularly mournful and plaintive wail of the screech owl and the loud sweet notes of the whippoorwill are heard in the distance.

The mammals, always shy, are also seen less often, and except for the sun-loving woodchucks, squirrels, and chipmunks, we must wander afield on moonlit nights see the fox, skunk, raccoon, and cottontail rabbit. After the sun has set and shadows begin to deepen, the bats emerge from their daytime resting places and fly about in search of their evening meal, for they have fasted all day. Like the birds and butterflies, they have their own flight patterns, and one can easily learn to recognize the different species by watching them fly. And as everyone knows, bats find their way by echolocation; that is, they send out supersonic notes that are reflected back, and by relocalizing these reflected waves are able to avoid obstacles.

Though the birds and mammals have temporarily retired behind the scenes, the summer stage is not without its actors. For now the insects have taken their place. Wherever you may wander throughout the countryside, an innumerable company of butterflies dance, rise, or dip in the bright sunshine. July can rightly be called the month of butterflies. The spring azure, the American copper, the white cabbage, the yellow sulphur, the dusky meadow browns, and the queenly swallowtails, striped and belted with gay colors, have been with us for some time, a few of them since early March. As spring passes into summer the butterfly population reaches its maximum, for now there appear on the scene the dappled band of fritillaries, variegated by odd dashes and spots of burnished silver, the anglewings with their peacock eyes, the banded and spotted purples, and many others with names redolent of romance and far away places: the painted lady, the red admiral, the wanderer, the gray comma, the silver-spotted hesperid, the tawny emperor, the hoary elfin.

And the moths, no less numerous but of a more somber hue, as befits creatures of the night, flock to lighted windows and visit the blossoms of the bladder campion and bouncing Bet. Their larger brothers and sisters, the hawkmoths, prefer the yellow flowers of the evening primrose.

Few of us have had the interest to observe the butterflies except in a most cursory manner. But if you study them closely, you will find that some species visit flowers indiscriminately, being rather catholic in their tastes, while others are very selective and visit the blossoms of only a few plant species, perhaps only one or two. But as butterflies are more or less intimately associated with their food plants, we look for certain species in fields, others in meadows, still others in waste places and along roadsides. Or we watch for them along the woodland border, in the woodlands, or in the marsh and swamp. There is a certain amount of overlapping, since butterflies can easily pass from one habitat to another, but usually they stay where they belong.

Butterflies also have their own distinctive habits and behavior patterns. Watch an American copper on a hot, sunny day and you will see the little butterfly dart at every passing object. Even the male pearl crescent chases every shadow. Three or four buckeyes often rise into the air, where they buffet each other about, rising and falling, as they engage in their aerial pugilistics. What a contrast to the lazy, easygoing satyrs and wood nymphs, the purposeful fritillaries, the bustling skippers, the sedate monarch, the vacillating blues that never seem able to make up their minds what to do.

Some butterflies are always hungry and seem perpetually on the move, flying from one blossom to another; others, less greedy, spend long hours sunning themselves. A few species, such as the little azure, the pearl crescent, the tiger swallowtail, and the sulphurs are as fond of water as the sugared sweets of flowers and frequently gather at a roadside puddle, the sulphurs sometimes congregating by the hundreds. We often see the sulphurs suddenly rise into the air, flutter about a bit, then settle down again.

It may surprise you to learn that some butterfly species are extremely pugnacious and will chase or drive away from their territory not only other butterflies but other insects and, what is even more amazing, birds, dogs, and even people. Butterflies have their favorite perches, too, and their own individual resting positions, whether on a leaf, a twig, or the ground. It is apparent to the most casual observer that they have their own flight patterns. Watch a monarch, and see how effortlessly it sails through the air. Or watch a grass nymph, and observe how weakly it flutters above the grasses, into which it quickly drops if frightened; or keep your eye on a skipper as it darts erratically about. Just as we can identify many birds at a distance from their manner of flight, we can do the same with the butterflies.

Other insects also have their own flight patterns. Compare a dragonfly, a bee, and a grasshopper, and distinctions are obvious. Contrary to what one might expect, large wings are not essential to good flight. The dobson flies and golden-eyes have comparatively large wings, but both are poor flyers. A butterfly certainly cannot match the speed or aerial gymnastics of the dragonflies or the male botfly, one of the fastest insects. And try to catch a housefly, which has only one pair of wings.

A housefly's wings, or the wings of any insect, may seem delicate structures, but they are actually tough and strong. Look at a wing of the housefly with your lens and it will appear as a piece of parchment divided into a number of areas by structures that appear to be ribs. A wing is a saclike fold of the body wall, although its saclike structure is not apparent, since the two walls extend outward over the same area and are so thin and so close together that they appear as a single membrane. The dual nature of the wing may be seen, however, where the walls remain separated.

The manner in which the wing veins are arranged is peculiar to the housefly, and in no other species of insect will you find them arranged precisely the same way. Veins are known by certain terms according to their position: costa, subcosta, radius, media, cubitus, and anal, as shown in Figure 46A. The areas into which the wing is divided are called discal, costal, etc., or may merely be

FIGURE 46

*Wings
and Wing Scales*

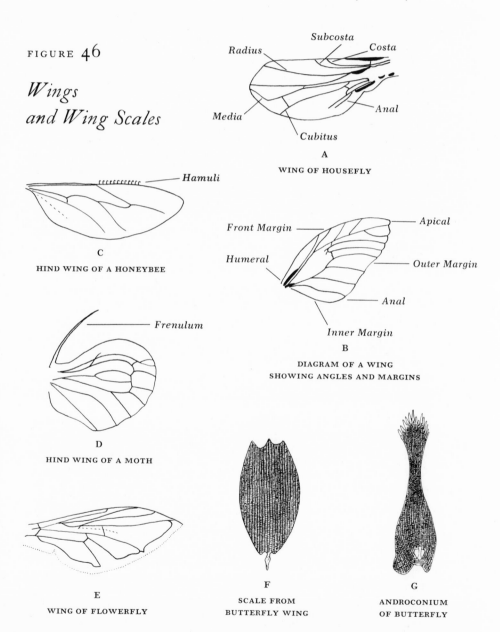

A

WING OF HOUSEFLY

C

HIND WING OF A HONEYBEE

B

**DIAGRAM OF A WING
SHOWING ANGLES AND MARGINS**

D

HIND WING OF A MOTH

E

WING OF FLOWERFLY

F

**SCALE FROM
BUTTERFLY WING**

G

**ANDROCONIUM
OF BUTTERFLY**

indicated by the letters R_1, R_2, R_3, etc. The arrangement of the veins is known as venation, or neuration, and serves as a means of classifying insects. The wings of insects present such countless differences that an expert can usually refer a detached wing to its proper genus, and often to its species.

With few exceptions, wings are present in all adult insects and are more or less triangular in shape with three margins (Fig. 46B): front (costal), outer (apical), and inner (anal); three angles: humeral (at the base of the costal), apical (at the apex of the wing), and anal (between the outer and inner margins). Typically there are two pairs of wings, but in true flies the second pair has been replaced by a pair of knobbed, threadlike organs called halteres. They are also known as balancers, because at one time it was believed they functioned in the same manner as the weighted pole carried by tightrope walkers.

Recent investigations, however, have shown that halteres act in a different manner. They vibrate very rapidly during flight. Their frequency is about the same as the wing beat, but they are usually in antiphase. A single muscle provides the upbeat of the muscle, but there is no antagonistic muscle, the downbeat being brought about by the elasticity of the hinge. Furthermore, the two halteres do not move in the same plane, because each is angled posteriorly. They function like a gyroscope and in insect flight are the counterpart of the bank-and-turn indicator of an airplane.

Much as we would like to discuss the mechanics of insect flight, lack of space prevents us. But as the oarsmen of a boat can obtain greater efficiency if they all pull their oars in unison, so an insect can make greater use of its wings if the fore and hind wings act together. The synchronous action of the two pairs is brought about in some species by the fore wings overlapping the hind ones, but in other species there are structures that fasten the two wings together on each side. Examine the outer costal margin of a honeybee's hind wing, and you will see a row of hooks, hamuli, which fasten into a fold on the inner margin of the fore wing (Fig. 46c). Or look at the hind wing of almost any moth and you will find a strong spinelike organ or a bunch of bristles, the frenulum (Fig. 46D), which in some forms engages a membranous loop on the fore wing. In one family of moths, the Hepialidae, the fore and hind wings are held together by a slender, fingerlike structure, the jugum, and in several groups of insects by a structure called the fibula.

The front wings of many insects have been modified and are more useful for protection than for flight. In the grasshoppers they are leathery and called tegmina; in the beetles they are usually horny and are known as elytra; and in the true bugs the base of the wing is thickened and the apex of the wing membranous, forming what is termed a hemelytron. The brilliantly colored flowerflies we have mentioned so often can be identified by a false vein in the wing shown in Figure 46E.

In some insects—caddis flies, for instance—the wings are densely clothed with long, silky hairs; in others, such as butterflies and moths, they are covered with scales. Handle a butterfly or moth and your fingers become covered with a dust which, examined with a microscope, is seen to consist of minute scales (Fig. 46F). If you examine a section of the wing in the same way, you will probably find that the scales are arranged more or less regularly, like the scales of a fish or the shingles on a house; however, in some moths they are merely scattered over the surface of the wing. Scales are not confined to the wings but occur on the body, legs, and other appendages. They are also found on other insects, such as silverfish and carpet beetles. Scales are merely modified setae or hairs that, instead of growing long and slender, grow very wide in comparison with their length. One striking feature of the scales on butterflies and moths is that the upper exposed surface may be marked with very fine longitudinal ridges as well as with a series of transverse ridges between them. The primary function of the scales is protection; a secondary one is ornamentation, for the beautiful colors and markings of these insects are due entirely to the scales and are destroyed when the scales are removed.

On the wings of certain male butterflies, some of the scales are modified as scent organs that secrete a fluid with an odor supposed to attract the females. Called androconia (Fig. 46G), these scent organs come in a variety of forms. They usually occur in patches on the upper surface of the fore wings and are generally concealed by other scales.

We speak of hairy caterpillars, but in insects hairs are called setae. A seta has its origin in a single hypodermal cell, is actually an extension of the epidermal layer of the cuticula, and is hollow. Most setae are bristlelike, but there are numerous modifications; some are stout and firm, others have lateral extensions, and still others are flat, wide, and comparatively short. And the setae are not always simple in form but may be toothed, branched, or otherwise modified, as they are in various bees (Fig. 47A). Such hairs serve to hold pollen.

Generally, hairs are a sort of clothing and function primarily as a protection for the body and appendages, probably being more effective in this respect than odors or repellent fluids. Many birds will eat ill-flavored insects, but they usually leave hairy caterpillars alone, though the fall webworm, tent caterpillar (Fig. 47B), woolly bear (Fig. 47C), and gypsy moth are eaten by various birds. Such instances are exceptional. Hairs are also effective in protecting some hibernating caterpillars, like the woolly bears, from sudden changes in temperature.

In some insects hairs are glandular; that is, they serve as an outlet for the secretions of glands with which they are associated. Certain hairs on brown-tail moth caterpillars emit a fluid which produces an inflammation of the skin in humans similar to that produced by poison ivy. The caterpillar of the io moth (Fig. 47D) is armed with venomous spines which are not only sharp but brittle and easily broken. Venomous spines also occur on the saddleback caterpillar (Fig. 47E) and the spiny oak slug, so if you intend to examine these insects with your lens, gloves are advised.

We can inspect the glandular hairs of the housefly without fear, however. Look between the claws on its feet (Fig. 47F). The cushionlike structures (pulvilli) have hollow hairs, the tenent hairs, through which is exuded a sticky fluid that enables the fly to walk on smooth surfaces, such as up a window pane or across the ceiling. But unless the hair tips are free from dust, they do not function well; hence the fly always seems to be busy cleaning its feet by rubbing them against each other and its body. Other insects possess tenent hairs too. While you have the housefly in your hand, look at the dense growth of hairs on its legs and body. It is in such hairs that the microorganisms the housefly has been charged with carrying are lodged.

C
WOOLLY BEAR

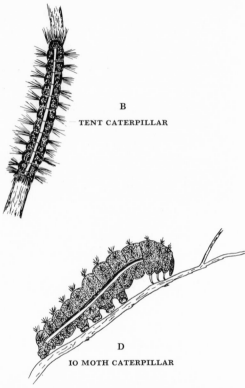

B
TENT CATERPILLAR

D
IO MOTH CATERPILLAR

A
HAIRS OF VARIOUS BEES

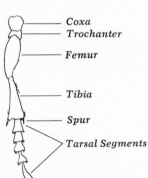

Coxa
Trochanter

Femur

Tibia

Spur

Tarsal Segments

Claws

F
LEG OF HOUSEFLY

E
SADDLEBACK CATERPILLAR

FIGURE 47

The Hairs of Bees,
Flies, and Caterpillars

Finally, there are hairs that are a part of a sense organ. They may be tactile (either over the entire integument or locally), olfactory, or auditory; or they may function in some other sensory manner. We shall return to them shortly.

Caterpillars are as abundant during July as their winged parents. We find not only the monarch caterpillars on milkweed plants, but also the gay harlequin caterpillars clothed with tufts of orange, black, and white. They feed as unconcernedly as the monarch caterpillars, since the birds don't like the hairy caterpillars; but doubtless they get added protection from the acrid nature of their food. Measuring worms are fairly common on trees and shrubs but often escape detection, for they cling to a twig with their prolegs and with their bodies extended outward look like twigs (Fig. 48A). If you wonder how they can remain in such a position for any length of time without becoming exhausted, look at them through a reading glass. They are supported by a strand of silk extending from the mouth to the twig. The thread serves as a sort of guy rope and is under considerable tension, for if you cut it the caterpillar falls back with a sudden jerk.

Our common walking stick or stick insect also resembles a twig (Fig. 48B). Both this insect and the measuring worms illustrate protective resemblance. So, too, do Nerice bidentata (Fig. 48C) and the larvae of certain sawflies (Fig. 48D). In many instances color adds to the deception, helping the caterpillars to blend with their background. There are countless other examples, such as the treehoppers that resemble spines (Fig. 36F), and the caterpillars of the viceroy and giant swallowtail butterflies that look like the excrement of birds. When disturbed, certain weevils have the habit of dropping to the ground, where they remain immovable and simulate bits of soil or little pebbles so well they can be distinguished only with difficulty.

Sometimes I think that July might rightly be called the month of beetles, since these insects are even more numerous than butterflies, though not quite so conspicuous or showy. Countless potato beetles, Mexican bean beetles, cucumber beetles, flea beetles, and weevils occur in almost any vegetable garden; flatheaded and roundheaded borers in the orchard; elm leaf beetles and willow beetles along the roadside; and bark beetles, timber beetles, borers, and weevils in the woods.

Beetles come in many shapes, sizes, and colors. Tortoise beetles —small, bright, golden-green or iridescent insects with the prothorax and wing covers broadly expanded to form an approximately circular or oval outline suggesting the shell of a tortoise—are beautiful to the naked eye; beneath the lens their beauty is breathtaking.

Few insects have the matchless beauty of the gold bug, commonly found on the bindweed, clinging like a drop of molten gold

FIGURE 48

Examples
of Protective Resemblance

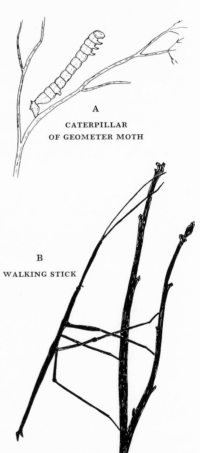

A

CATERPILLAR
OF GEOMETER MOTH

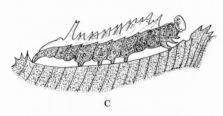

C

NERICE FEEDING ON ELM LEAF

B

WALKING STICK

D

SAWFLY LARVA FEEDING

to the leaves. The slender, long-legged tiger beetles are also brightly colored and may be seen on sunny days along dusty country roads, well-beaten paths, and the shores of streams. They are predaceous and capture their prey by chasing it; the larvae are also predaceous but have a different approach: they dig a burrow (Fig. 49A) and lie in wait in the opening. The larva deserves to be examined with the lens, for it is a most uncouth creature with a large head and prominent rapacious jaws, with which it seizes any insect that ventures within striking distance. You will observe that it has a hump with forward-pointing hooks to anchor itself in its burrow when wrestling with an unusually large victim. The best way to get hold of the larva is to thrust a straw down the burrow and withdraw it slowly; .the insect will be found savagely clinging to the end of the straw as if angry at the intrusion.

The ant lion has improved upon the tiger beetle's method of getting food, for it excavates a conical pitfall (Fig. 49B) in sandy or crumbly soil. The manner in which the ant lion fashions its trap is most interesting and may be observed at close range by transferring the insect to a shallow box or dish containing some sand. Look for it near the foundation of a house, or in any sheltered spot where the soil is sandy.

The ant lion is a queer-looking insect, and if its mouthparts are examined with the lens they will be found similar to those of the aphislion. The little animal can undergo long fasts, since its food supply is uncertain and many days, even weeks, may elapse between meals. Hence the length of an ant lion's life is variable, but when fully matured, it spins a loose globular cocoon (Fig. 49c) of silk and sand. The adult is a delicate, gauzy-winged insect not unlike a dragonfly.

Most of us remember the little rhyme which goes:

> *Lady bird, lady bird! Fly away home,*
> *Your house is on fire,*
> *Your children do roam,*
> *Except little Nan, who sits in a pan*
> *Weaving gold laces*
> *As fast as she can.*

A

LARVA OF TIGER BEETLE
IN BURROW

B

ANT LION IN PIT

D

LARVA
OF LADYBIRD BEETLE

C

COCOON
OF ANT LION

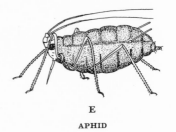

E

APHID

FIGURE 49

Aphids, Ladybugs,
Tiger Beetles, and Ant Lions

The ladybird, of course, has no home and never did have one. The "children" do roam, however—in search of aphids and scale insects. All, that is, except "little Nan." She, alas, cannot "roam" because she is the yellow pupa and is securely tied to the plant by the handle of the "pan."

Both the adult beetles and the larvae are predaceous to a high degree and can be numbered among our most beneficial insects because of the large numbers of aphids and scale insects they consume. It seems hardly necessary to describe the adults, which are more or less hemispherical and generally red or yellow with black spots, or else black with white, red, or yellow spots, though some species are quite variable. The larvae, however, are another matter; they are common enough, but ordinarily do not come to our attention. Viewed with the lens (Fig. 49D), they are fairly long and velvety and are covered with warts or spines. Their six queer short legs seem more suited for clasping a twig than for walking. Most of them are black and prettily marked with blue, orange, or yellow, but under the lens they appear as some sort of grotesque ogre— which they must certainly seem to the aphids and scale insects on which they prey. Several species are herbaceous; a common species is the squash ladybird.

To find ladybird larvae one need only locate an aphid colony, and since aphids are abundant and occur on almost every kind of plant, this should not be difficult. Often called plant lice, a name that is rather self-evident, aphids are flask-shaped (Fig. 49E) in outline, with two large eyes set rather far apart. They are usually green, though some species are colored differently; a few have the most bizarre and striking ornamentations.

Aphids are visible to the naked eye, but they can be observed better with a reading glass or a hand lens. If a colony is examined, it will be found that they are in all stages of development and are engaged in various activities. Some have their beaks in the plant tissues, their hind legs high in the air and their antennae curved backward as they suck the juices. Others have their beaks tucked under their bodies and walk slowly about, stiff-legged, perhaps looking for a likely spot in which to thrust their beaks. Still others just sit quietly and gaze out at the world with their large eyes. Sometimes they occur in such numbers that those moving about must climb over the backs of the others, and the smaller ones are so tightly pressed between the larger ones that it would seem as if they would be squeezed to death. These little insects may appear innocent enough, but they are actually very destructive to plant life. If it were not for their enemies that keep them under control, they could become a real menace.

Aphids appear defenseless, but some of them can secrete a waxy substance when confronted by an enemy and thus effect their escape. Most aphids also secrete a sweetish liquid called honeydew, which is much desired by bees, wasps, and ants. The bees and wasps take it where they can find it, but some species of ants "adopt" the aphids and take care of them as "cows" in return for a steady supply. One species of aphid, the corn-root aphid, has become largely dependent on a species of ant, the cornfield ant. The ants store the aphid eggs in their underground nests and care for them throughout the winter; then in the spring when the stem mothers hatch, the ants transfer them to the roots of various weeds on which they feed, and later, when corn plants become available, move them to the roots of the corn.

As you watch the aphids through your magnifying glass, you will probably see the ants stroking them with their antennae until they respond to such caresses with a glistening drop of honeydew, which is immediately snatched up by the ants. Ants, of course, use their antennae for other purposes than stroking aphids. You have doubtless read that they communicate with one another by means of their antennae, and perhaps you have actually seen them do so. The antennae are not the simple structures they appear, but are highly developed organs.

The lens show that the ant's antennae consist of twelve segments linearly joined together. The antennae of insects are commonly regarded as organs of touch, but in the ant certain segments have been modified for other purposes. Thus, by means of segment number 12 (counting from the point of attachment) the ant can detect the odor of its own nest and distinguish its nest from others; by number 11 it can detect the odor of any descendant of the same queen and can recognize members of its community, whether it is in the nest or out on a foraging expedition; by number 10 it can smell the odor of its own feet, so that it can retrace its steps when on the trail; and segments 8 and 9 tell it how to care for the young of the colony. If an ant is deprived of these segments, it has no further caste as a social insect and loses its standing in the community.

Since the ant's antennae are essential to its welfare, they must be kept clean to function properly. The ant has all the necessary toilet articles for removing dirt and other debris from its body. Look closely at the leg: you will see a curved, movable, comblike spur on the distal end of the tibia, and opposite it, on the base of the metatarsus, a concavity tipped with hairs. This is the antennae cleaner (Fig. 50A). In cleaning its antennae, the ant lifts its leg over the antenna and then draws it through the space between the spur and the hairs, which effectively act as a brush. A dirty brush is of little value, so after cleaning the antenna the ant cleans its brush by passing it through the mouth.

While you have an ant in your hand, observe its general

anatomy: the head with its ever-moving antennae, the large jaws armed with teeth like those of a saw, which are used for all sorts of activities; the slender thorax; the fairly long and efficient legs that carry it swiftly over the ground and in some cases propel it into the air in a series of leaps; and the thin pedicel, a characteristic of the ant as well as the wasp, with which the abdomen is attached to the thorax.

The habits and behavior of insects are determined by instinct, in some instances perhaps by a certain degree of intelligence, and by their reaction to various stimuli such as light, chemicals, temperature, and so on. And since they respond to such stimuli, it follows that they must have receptors capable of receiving and translating sensory impressions and stimuli into impulses that bring about corresponding responses.

Such receptors are known as sense organs and in insects are distinguished by the kind of stimulus that acts on each. Hence the following sense organs are recognized: (mechanical) touch and hearing; (chemical) taste and smell; and (light) the compound eyes and ocelli. The sense organs of touch, hearing, taste, and smell are basically modifications of the armorlike integument or external covering, and may take the form of peglike projections [*Sensillum cœloconicum* (Fig. 50B), *S. basiconicum* (Fig. 50c), and *S. styloconicum* (Fig. 50D)] that are olfactory in function; bristles or fine hairs [*S. chœticum* (Fig. 50E) and *S. trichodeum* (Fig. 50F)] that respond to touch and flask-shaped cavities [*S. ampullaceum* (Fig. 50G)] that are auditory. Sense organs may occur on almost any part of the insect, but they are most numerous and varied upon the head and its appendages, particularly the antennae.[1]

The antennae are always a single pair and are situated near the compound eyes, in some instances between them. In some insects the segments are all alike; in others they differ in form and size. Various terms are applied to the segments of an antenna, as shown in Figure 50H. The first or proximal segment is called the

---

[1] Essentially organs of touch, the antennae are modified in some insects to function as auditory organs and in others as organs of smell. Indeed, they may even be adapted for other than sensory purposes; as for respiration (the aquatic beetle Hydrophilus) and for mating (the male beetle Meloe).

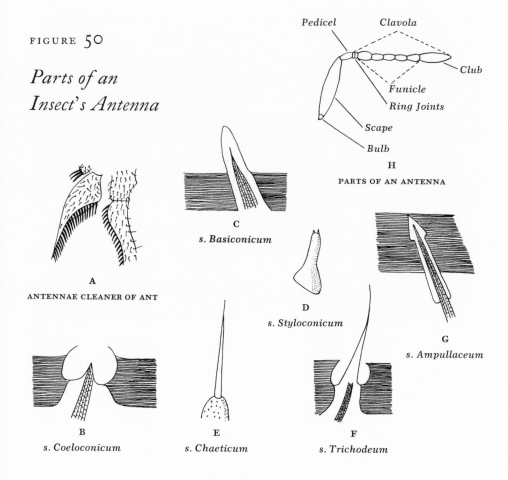

FIGURE 50

Parts of an Insect's Antenna

Pedicel

Clavola

Club

Funicle

Ring Joints

Scape

Bulb

H

PARTS OF AN ANTENNA

A

ANTENNAE CLEANER OF ANT

C

s. Basiconicum

D

s. Styloconicum

G

s. Ampullaceum

B

s. Coeloconicum

E

s. Chaeticum

F

s. Trichodeum

scape, the second the pedicel. The remaining segments are collectively known as the clavola. In some species the first one or two segments of the clavola are much smaller than the others and are called ring joints; in others the distal segments are more or less enlarged and are termed the club. The part of the clavola between the club and the ring joints, or when the latter are not specialized, between the club and the pedicel, is known as the funicle.

There are various forms of antennae. The filiform (Fig. 51A), in which the segments are of nearly uniform thickness, giving the antennae a threadlike appearance, is the more generalized. We find this form on stinkbugs (Fig. 51B)—fairly common insects abundant on various plants. In the setaceous (Fig. 51C) or bristlelike form, the segments are successively smaller, the entire organ taper-

FIGURE 51

*The Stinkbug,
June Bug, Termite,
and Types of Antennae*

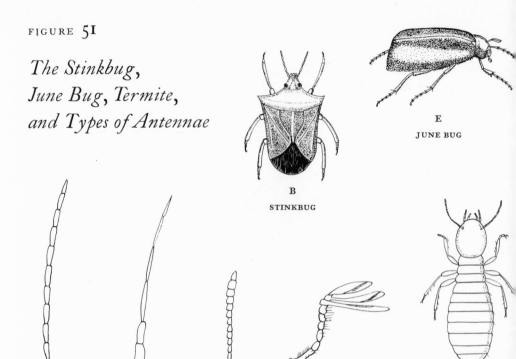

E
JUNE BUG

B
STINKBUG

A
FILIFORM

C
SETACEOUS

D
CLAVATE

F
LAMELLATE

I
WORKER TERMITE

ing to a point (dragonflies). Where the reverse holds true—that is, where the segments gradually become broader so that the antenna assumes the form of a club, as in butterflies—the antenna is said to be clavate or club-shaped (Fig. 51D). In some species, such as the June bug (Fig. 51E), the segments that compose the tip or knob are extended on one side into broad plates forming a somewhat lamellated (plated) structure; hence the term lamellate (Fig. 51F) for this type of antenna. The term serrate (Fig. 51G) or sawlike is self-explanatory. In this form the segments are triangular and project like the teeth of a saw (buprestid beetles). Other forms are the moniliform (Fig. 51H), in which the segments are more or less spherical, suggesting a string of beads, as in termites (Fig. 51I); comblike or pectinate (Fig. 51J), in which the segments have long projections on one side, like the teeth of a comb; the capitate (Fig. 51K), in which the terminal segment or segments form a large knob (engraver beetles); and the geniculate (Fig. 51L), in which the

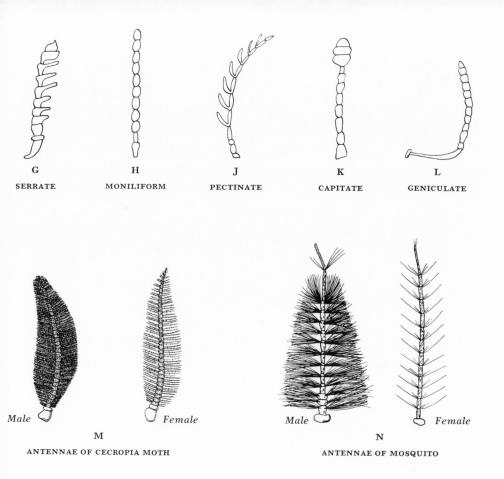

G	H	J	K	L
SERRATE	MONILIFORM	PECTINATE	CAPITATE	GENICULATE

Male   *Female*   *Male*   *Female*

M   N

ANTENNAE OF CECROPIA MOTH   ANTENNAE OF MOSQUITO

antenna is bent abruptly at an angle, like a bent knee (bumble-bees).

By examining the antennae of any insect with your lens and comparing it with the illustrations, you can easily recognize the type of antennae on your specimen. On comparing the antennae of the male and female cecropia moth (Fig. 51M) and the male and female mosquito (Fig. 51N), you will find that the antennae of the male moth are larger and more feathered than those of the female and that the antennae of the male mosquito have a larger number of long, slender fibrillae than those of the female. The reason for such differences is that the antennae of the male moth function as organs of smell, and the antennae of the male mosquito function as organs of hearing, the delicate fibrillae of various lengths being sent into sympathetic vibrations by the note of the female. The object in both instances is to bring the sexes together.

In many ways the scale insects, on which ladybird beetles also feed, are as interesting as the aphids. They present some curious features: the males entirely lack a mouth and have only a single pair of wings. Indeed, in a few species they are entirely wingless or have only vestigial wings, their hind wings having been replaced by a pair of club-shaped halteres. Unlike the males, the females are always wingless, as are the young; and in some species the females are also legless. The females have either a scale-like or gall-like form, or they may be grublike and clothed with wax. The waxy covering may be in the form of powder, large tufts, plates, a continuous layer, or a thin scale beneath which the insect lives.

Although scale insects are visible to the naked eye, few details can be seen without the lens. A widely distributed species is the oystershell scale (Fig. 52A). It is usually found on fruit trees and various shrubs, but it also occurs on other plants; lilac branches are sometimes covered with it. The scales, curved like oyster shells, are about ⅛ inch long, and their brownish color matches the dark bark on which they are found. Should you examine the scales in winter, you will find upward of a hundred white eggs beneath each scale, together with the dead body of the female.

Sometime if you find pine needles that appear to be covered with a white powder (Fig. 52B), look at them with your lens and you will see that they are covered with long, narrow, snowy-white scales. This scale insect is known as the pine-leaf scale. The stems of rose bushes often present a whitish scurvy appearance, evidence of an attack by the rose scale. White scales that are irregularly oval with a yellowish point and about ⅒ inch long on apple branches reveal the presence of the scurvy scale. Years ago when I knew little about insects, I came across a maple tree whose branches seemed to be covered with bits of cotton. I was curious to know how they got there, and when I investigated found that they were actually tufts of a cottony material protruding from oval brown scales. The cottony maple scale (Fig. 52C) is common not only on maples but on Osage orange and grape; during the summer it is not unusual to find the twigs of these plants festooned with cottony tufts.

Of all our scale insects the San Jose scale (Fig. 52D) is the best known, principally because of the tremendous damage it once inflicted on orchard trees. It infests many varieties of fruit trees and garden shrubs, and if you live in a place where the winters are not too cold, you should have little trouble finding it. The female is yellowish, circular, legless, and slightly smaller than the head of a pin. She is covered with a dark-gray, circular, waxy scale, $\frac{1}{16}$ inch in diameter, slightly elevated in the center into a nipple formed by cast-off skins, and surrounded by a ring that varies from pale yellow to reddish-yellow. Smaller black scales, somewhat elongated in form, are the males.

To return to our discussion of beetles, one of the ground beetles or carabidae has the unique habit of setting up a smoke screen when confronted by an enemy. This insect, called the bombardier, has a pair of anal glands that secrete an evil-smelling fluid which volatilizes explosively, with a noise like that of a tiny popgun, on contact with the air and forms a spray that looks like smoke. By the time a would-be captor has recovered from astonishment at the noise and smoke in its face, the bombardier is at a safe distance. The bombardier can be found beneath stones and other objects lying on the ground and can be observed in action with a reading glass. The head, prothorax, and legs are reddish-yellow and the wing covers dark-blue, blackish, or greenish-blue.

The bombardier beetle is not unique in being able to eject a

FIGURE 52

Scale Insects

A	B	C	D
OYSTERSHELL SCALE	PINE-LEAF SCALE	COTTONY MAPLE SCALE	SAN JOSE SCALE

repellent fluid, although it is the only one I know of that does so and makes a noise at the same time. The "molasses" of the grasshopper is said to be ill-tasting to birds and other animals, and on the Pacific coast the pinacate bug defends itself by elevating the hind end of the body and discharging an evil-smelling, oily fluid, in the manner of the skunk. The larva of the American sawfly (Fig. 53A) is able to squirt jets of a watery fluid, and stinkbugs protect themselves by giving off a secretion with an unpleasant odor. All these insects will perform for you if molested, and with a hand lens you can locate the openings of the glands that secrete the repellent fluids.

Many insects have hypodermal glands that open into saclike invaginations of the body wall which can be evaginated and the secretions ejected when desired. Called osmeteria, they occur in the caterpillars of the swallowtail butterflies. When the caterpillars are disturbed, the osmeteria are thrust out and give off a liquid with an indescribably repellent odor. Find one of these caterpillars, squeeze it, and with the lens observe what happens (Fig. 53B).

Even the blood of insects may serve as a repellent. The blood of ladybird beetles, fireflies, and oil beetles contains cantharidin, an extremely caustic substance that is an almost perfect protection against birds, reptiles, and predaceous insects. In oil beetles the blood issues as a yellow fluid from a pore at the end of the femur when the insects are handled. Our most common species is the buttercup oil beetle, found on the leaves of various buttercups.

Ground beetles are so named because they commonly stay on the ground, lurking by day under stones and other cover and coming out at night to roam in search of prey. Most of them are brown or black, but some are brilliantly colored. The searcher, one of the larger and more beautiful of the ground beetles, has green or violet wing covers, on which, as the lens will show, are a number of fine parallel grooves or striations. Another species, the fiery hunter, may be easily recognized by the rows of reddish or copper-colored pits on the wing covers. It is these fine lines and pits that produce the brilliant colors.

In insects (and in other animals too, for that matter), colors are classified as structural and pigmental. Structural colors are due to structural peculiarities, such as the grooves and pits on the wings of the searcher and the fiery hunter, and the striae or grooves on the scales of a butterfly. These grooves and pits break up light into its component parts or wavelengths in much the same way that a prism forms a spectrum, the colors produced being determined by the distance between the grooves or striae. The iridescence of a fly's

FIGURE 53

Defensive Weapons and Protective and Warning Coloration

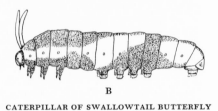

B

CATERPILLAR OF SWALLOWTAIL BUTTERFLY

A

LARVA
OF AMERICAN SAWFLY

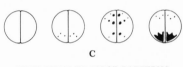

C

VARIATIONS IN COLOR PATTERNS
OF ELYTRA OF MEXICAN BEAN BEETLE

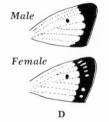

Male

Female

D

FRONT WINGS
OF COMMON
SULPHUR BUTTERFLY

E

CATOCALA MOTH
ON BIRCH

F

CATERPILLAR
OF TIGER SWALLOWTAIL

G

ALAUS,
EYED ELATER

wing is created in the same way, the light that falls on it being separated by the two parchmentlike membranes or lamellae, and the result-
ing colors being determined by the distance between the lamellae. If
you have studied physics, you know that color produced in this
manner is known as interference. Sometimes the interference colors
of butterfly scales may be due not only to surface markings but also
to the lamination of the scales and to the overlapping of two or
more scales. In the case of the fiery hunter beetle, the pits alone do
not produce any color; it is only when they are combined with a
reflecting or refracting surface and a pigment layer that iridescence
results. The silver-white color of certain insects is a result of the
total reflection of light by scales, by air-filled sacs or pockets, by air-
filled tracheae, or by a film of air incorporated in the hairs of the body
that many aquatic insects, such as the diving beetles, carry with them.

Pigmental colors are produced through the absorption and reflection of the various wavelengths that compose light by chemical
substances called pigments. The green color of many caterpillars
and grasshoppers is due largely to the ingestion of chlorophyll,
the green coloring matter found in the leaves they eat. Similarly,
insects that feed on the blood of higher animals become red because
of the ingested hemoglobin. Pigments may be taken directly from
the food, may be manufactured indirectly from the food, or may be
excretory products. The black and brown colors of insects are due
to by-products of metabolism; they are nitrogenous substances
known as melanins, and are diffused in the outer layer of the
cuticula.

There are other pigments in leaves besides chlorophyll. The red
and yellow colors of insects come from carotene and xanthophyll
ingested with the leaves and deposited in the cuticula and hypo-
dermis. It is interesting to note that an insect which feeds on the
Colorado potato beetle obtains its yellow from the potato beetle,
which in turn gets its color from the potato leaves. A substance
known as anthocyanin produces the blue, red, and purple colors in
flowers, fruits, leaves, and stems. It is also responsible for the same
colors in many insects, though we are doubtful about the blues,
since blue, violet, and green are generally structural colors.

Certain other substances produce the pink, purple, and green hues in some insects, while in others the red and yellow pigments are excretory derivatives of uric acid. Dull yellows and browns are often derived from the tannin found in leaves. The subject is extensive and is further complicated by the effect that such external factors as temperature, moisture, and light have on the formation of pigments and color patterns. For if you were to compare two specimens of the same species, you would probably find that they are not colored exactly alike. Collect a hundred Colorado potato beetles, and no two would have exactly the same pattern on the pronotum. The wing covers of the Mexican bean beetle show countless variations, a few of which I have illustrated in Figure 53c. The ilia underwing may be found in more than fifty varieties, each of which might have its own name, were it not that they run into each other.

Sometimes the normal color of an insect is replaced by another; the red in the red admiral butterfly is often replaced by yellow. Caterpillars that are hatched from the same batch of eggs may show different colors if separated and fed differently; indeed, they do not have to be separated, for sometimes caterpillars of the same brood feeding on the same plant under the same conditions may show variations in marking. Later in the summer, locate a nest of the fall webworm and your lens will show considerable variation in the markings of the caterpillars.

Albinism, which often occurs in higher animals, sometimes occurs in insects; the common yellow roadside or sulphur butterfly is frequently albinic. Melanism is also sometimes observed.

Many insects that have several broods a year show not only differences in color or color patterns, but sometimes differences in form. We have already mentioned the spring azure and the pearl crescent. The question mark or violet tip has two forms that differ not only in coloration but in the shape of the wings; and the ajax swallowtail has three forms, each progressively larger, with longer tails on the hind wings. And there are others. Seasonal increase in size is probably due to increased metabolism as a result of progressively higher temperatures, since it is well known that warmth stimulates growth. Temperature also has an effect on pigment formation, as Edwards demonstrated many years ago (see page 142).

Finally, in many insects the sexes may be distinguished by differences in color or color patterns, as shown in Figure 53D.

Color is doubtless an advantage to many insects when their color or color patterns blend in with their surroundings. The catocala moths (Fig. 53E) are a case in point. They generally have brilliantly colored hind wings of red, orange, and black. Their front wings, however, vary in color from white to gray and brown. In flight the moths are readily seen, but when they alight or are at rest, the fore wings cover the hind wings, and the moths become inconspicuous. Indeed, they look so much like the bark of the trees on which they commonly rest that they are barely discernible. Other insects also illustrate protective coloration; have you ever noticed, for instance, how much grasshoppers resemble the soil in color? Or how some caterpillars are not only green but have oblique lateral stripes that divide the solid green and make the resemblance to the leaves on which they feed even more pronounced?

Then there are some insects which are vividly colored or have hideous markings that make them quite conspicuous. The caterpillar of the tiger swallowtail (Fig. 53F) and the eyed elater (Fig. 53G) are two common examples. In these insects the colors and markings presumably frighten potential enemies.

Although mud daubers appear in June, when they may be seen flitting about the flowers or sunning themselves on a fence rail, it is not until July that they become seriously engaged in building their nests. A common species is a black or brown wasp (Fig. 54A) with yellow spots and yellow legs. I often see one at the edge of a puddle, pond, or stream, running around and digging here and there until she finds mud of the proper consistency. Then she plunges her head down into it and extends her abdomen in the air. When she has cut out a pellet of mud about the size of a pea, using her jaws for the purpose, she mixes it with her saliva and then flies to the spot she has selected for her nest. She attaches the pellet to the support, if she is just beginning construction, or to the mass of mud she has previously gathered, if her nest is underway. The site of the nest may be the side of a building, the rafter of a barn, or the wall or beam of an unfinished attic. The name mud dauber is rather an uncomplimentary term, for the nest, despite its somewhat slovenly

FIGURE 54

The Mud Dauber
and Nests of Various Wasps

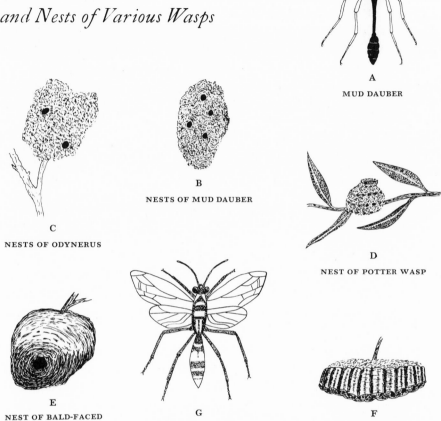

A
MUD DAUBER

B
NESTS OF MUD DAUBER

C
NESTS OF ODYNERUS

D
NEST OF POTTER WASP

E
NEST OF BALD-FACED
HORNET

G
PAPER WASP POLISTES

F
NEST OF PAPER WASP

external appearance, is cleverly and skillfully constructed (Fig. 54B).

Wasps are irritable and nervous creatures, and it is advisable to leave them alone. The mud dauber, however, doesn't seem to mind if you watch her while she builds, so if you station yourself at a respectable distance in order not to interfere with her, you can watch her through a reading glass and observe how she plasters the mud and fashions it into a nursery for her young.

Another capable mason, Odynerus, makes a nest about the size of a hen's egg and usually attaches it to the twig of a bush (Fig. 54c). The most advanced example of wasp masonry is found in the nest of the jug builder or potter wasp. It is saddled on the twig of a tree or bush and is an exquisite little object (Fig. 54D) even to the naked eye, and more so when viewed with the lens. It is about half an inch in diameter, with a delicate liplike margin around the small opening, which is sealed when the wasp has finished her labors, and is worthy of the skill of a master craftsman.

Other wasps, such as yellow jackets, paper wasps, and bald-faced hornets, are also engaged in building nests, adding to them, and rearing young. Yellow jackets, which are brightly colored insects, usually construct their paper nests in holes in the ground, enlarging them as the need arises. Sometimes, however, they build them in a stump or under some object lying on the ground. These wasps can get very nasty, and it is wise to leave them alone. But when cold weather sets in and the wasps have left, the nest can be examined.

The extremely large paper nests (Fig. 54E) we see suspended from the branches of trees are made by bald-faced hornets, black wasps with a white face. The nests are similar to those of the yellow jacket except that the paper has a different color and is of stronger quality. In fact, these nests are so strong that they endure well into winter, in spite of being buffeted by rain, sleet, and wind.

Nests like the one shown in Figure 54F are the most common of all wasps' nests and are found hanging beneath the eaves of buildings, in attics, in barns and sheds, and in garages. They are made by the paper wasp Polistes (Fig. 54G), uniformly brown insects common in our gardens. The nest consists of a single naked comb and resembles an open umbrella without a handle and hung by its tip. Look at an active community through your reading glass, and you will find that nearly every open cell contains either a white egg or a chubby little soft-bodied grub. The closed cells contain pupae in various stages of transforming. The grubs, which hang head down, are held in place by a sticky disk, and later by their enlarged heads that completely fill the openings of the cells. They are constantly fed by the workers, first on nectar and fruit juices, later on more substantial food such as the softer parts of caterpillars, flies, bees, and other insects.

During the month of July, toads, leopard frogs, tree frogs, and wood frogs complete their metamorphosis and hop out of the water. I cannot understand why so many people dislike the toad, for he is anything but what they think he is: cold, slimy, repulsive—the symbol of evil. He can, of course, be both cold and slimy under certain conditions. But if he is not exactly beautiful, neither is he entirely unattractive when you look at him dispassionately. As for being a symbol of evil, he is, on the contrary, one of our benefactors. For it has been found that 88 per cent of his food consists of insects and other small creatures considered garden pests. For those of you who are statistically minded, it has been estimated that he eats in three months some ten thousand injurious insects. Of this number 16 per cent are cutworms, 9 per cent are caterpillars, and 19 per cent are weevils and other injurious beetles. In terms of dollars and cents, a toad was once valued at $19.44 for a single season because of the cutworms alone that he devours. I wonder what a toad is worth today. And if you want to look at his skin with your lens or reading glass, warts and all, it is as if you were looking at a section of rugged terrain with its hills and valleys from a balloon high in the sky. The warts will not harm you.

As soon as the tree frog has left the pond and taken up his abode in a tree or shrub, we begin to hear his long, reedy tremulo. But try to find him; he remains as invisible as Perseus in his charmed helmet, because he blends with his surroundings. He can pass for a green leaf, a plant stem, or a small excrescence on the gray trunk of a birch or lichen-covered oak, for he has an extensive wardrobe and can change his suits almost at will. In a way he resembles the chameleon, but unlike that lizard his color changes are due more to environmental conditions than to emotional disturbances.

The tree frog is a squat little creature, somewhat clumsy looking, but he is actually a skilled acrobat, leaping through the air as easily as the man on the flying trapeze. I have often seen him jump into the air after a flying insect, quite unconcerned about where he is going to land. Just when you think he is going to fall to the ground, he grabs a twig or other support, perhaps with one foot, pulls himself up, blinks his eyes, and settles comfortably.

Wood frogs are essentially creatures of the woods. As soon as

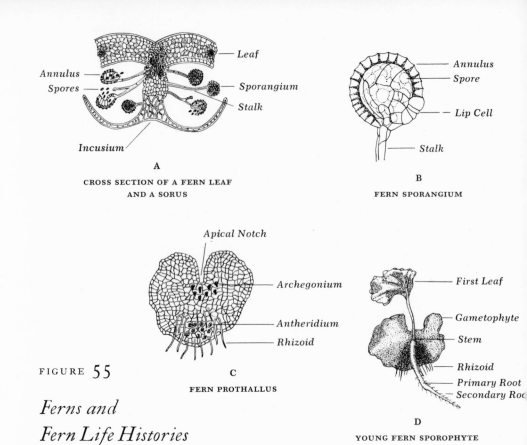

Annulus
Spores

Leaf

Sporangium

Stalk

Incusium

A

**CROSS SECTION OF A FERN LEAF
AND A SORUS**

Annulus
Spore

Lip Cell

Stalk

B

FERN SPORANGIUM

Apical Notch

Archegonium

Antheridium

Rhizoid

FIGURE 55

C

FERN PROTHALLUS

First Leaf

Gametophyte

Stem

Rhizoid

Primary Root

Secondary Roc

D

YOUNG FERN SPOROPHYTE

*Ferns and
Fern Life Histories*

they come out of the water they make for the woodlands, where we henceforth find them, among the mosses and ferns that are now at their maximum of luxuriant growth.

Most of us know ferns as plants only a few feet tall, or at most 5 or 6 feet; but go into the tropical rain forests and you will find them as tall as trees—40 feet or more. It is not surprising they should be so tall, for their remote ancestors were as tall, if not taller. Though evolutionary processes bring about many changes, the ferns of today are remarkably like those that lived fifty million years ago; fossil ferns have been discovered that are practically identical with living species. Look at some of them and you will be amazed at their resemblance to the maidenhair, hay-scented, ostrich, and cinnamon ferns.

Compared with man's life span, fifty million years is an incomprehensible length of time; yet when we look at a fern today, we

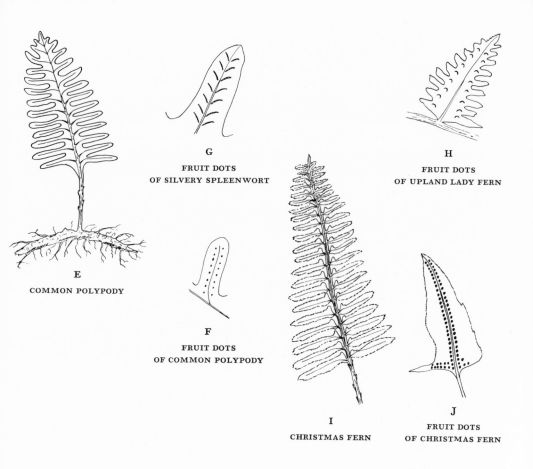

G

FRUIT DOTS
OF SILVERY SPLEENWORT

H

FRUIT DOTS
OF UPLAND LADY FERN

E

COMMON POLYPODY

F

FRUIT DOTS
OF COMMON POLYPODY

I

CHRISTMAS FERN

J

FRUIT DOTS
OF CHRISTMAS FERN

can quickly bridge the gap and visualize the dense forests of treelike ferns that grew in the marshes of the youthful sea-born continents. The ferns and their relatives were the dominant flora; indeed, they constituted the only form of vegetation. If they have given way to other forms since then, and now occupy a position secondary to the flowering plants, they are by no means completely overshadowed by them. Go wherever you wish—stroll down a country lane; follow a woodland trail; pause by the edge of a pond, or where the marsh forms a transition zone; or climb a rocky cliff; everywhere you will find the tribe of ferns well represented.

To the untutored eye all ferns may seem alike, but there are differences—in the form of the leaf, in the way they grow, in the places where they grow. Who can fail to recognize the sensitive fern with its large green fronds, often as much as 3 feet in length, deeply lobed, and the segments at the base spreading into narrowing wings at each side of the main stem? Or the cinnamon fern, rising from the ground in graceful vase-shaped clusters; or the royal fern with its oval or elliptical leaflets; or the bracken with its fronds three-divided? These are recognizable features, to be sure, and we must admit that there are many ferns not easily identified except by an expert. To those who do not know ferns too well, the fruit dots or spore cases provide an easy way to learn about these neglected plants, for they differ widely in shape, size, and position and are characteristic for the species.

The spore cases usually appear in July as small brown or orange spots on the lower surface of the leaf. When viewed with a lens, they are small globular objects set on the ends of stalks (Fig. 55A). Under the microscope a spore case (sporangium) is seen to be surrounded by a jointed ring, or annulus (Fig. 55B). When the spores within the sporangium are mature, the ring straightens hygroscopically, breaks the thin walls, and frees the spores, which are scattered by the air currents. Those that fall on favorable soil germinate quickly and develop into small green heart-shaped structures of delicate and evanescent tissue. Called prothalli (Fig. 55C), these structures bear fine hairs (rhizoids) that function as roots. Among the roots there are small compact bodies (antheridia) in which sperms are produced, and just below the cleft, small bottle-shaped bodies (archegonia), each of which contains an egg. On access to water the antheridia swell and expel the sperms, which swim about in the water until one of them enters an archegonium and fertilizes the egg.

As soon as the egg is fertilized it begins to divide, and by successive divisions a number of cells are formed. Some of these cells gradually develop into a mass, called a foot, which penetrates the prothallus and absorbs nourishment from it; others develop into a little root that penetrates the soil; and still others become a small leaf that curls up around the edge of the prothallus until it becomes exposed to air and light, when it begins to manufacture food (Fig. 55D). Finally, yet other cells develop into a stem that grows slightly downward and then horizontally beneath the surface of the ground, forming a rhizome. From the rhizome additional leaves and roots develop which at once begin to perform their functions of manufacturing foods and absorbing the necessary elements for their manufacture. And as the plant begins to grow, the primary leaf and any others that may have developed at the same time, as well as the rhizoids, the foot, and the prothallus, begin to wither and die, leaving in their place the familiar leafy fern plant we know so well.

The leaf of a fern is called a frond, and in most ferns the fronds are cut or divided into leaflets (pinnae) which, in some species, are further subdivided into still smaller leaflets (pinnules). The spore cases or sporangia are usually clustered in groups, called sori. The sori, which commonly occur on the pinnae, may be linear, oblong, kidney-shaped, or curved, and thus serve as diagnostic characters. For instance, in the common polypody (Fig. 55E) they are rather large, yellow-brown, round, and located midway between the midvein and the margin of the pinnae (Fig. 55F).

By examining the sori with your hand lens, you can identify most of the ferns, though you will have to get a book on the subject for reference. In the silvery spleenwort the sori are oblong, slightly curving, and arranged in a double row at an angle to the midveins of the pinnules (Fig. 55G); in the upland lady fern they are horseshoe-shaped and curve away from the midveins (Fig. 55H); in the Christmas fern (Fig. 55I) they are round and arranged in two lengthwise rows near the midveins (Fig. 55J); in the marsh fern they are kidney-shaped in two parallel rows; in the ostrich fern they look like tiny gun cases; and in the Virginia chain fern they are set in lines like the links of a chain. Once you have discovered how interesting the study of ferns can be, I think you will find that knowing them better might prove a fresh and stimulating experience.

Like butterflies, wild flowers seem to reach their maximum abundance in July. The summer species, more robust and showier than those of spring, decorate the landscape with colors, tints, and hues that man in his most vivid imagination could not hope to duplicate. One of the showiest of these summer wild flowers is indisputably the butterfly weed that, from a distance, seems to fire the pasture with creeping flames of orange-red. This is the handsomest member of the milkweed group, and above its flat-topped flower clusters, swallowtails, cabbage whites, roadside yellows, fritillaries, coppers, blues, painted ladies, coral hairstreaks, pearl crescents, monarchs, and many other butterflies float, alight, sip, and sail away. Here indeed is a butterfly flower.

In wet meadows and along the borders of swamps, a purple mist seems to rise from the countless wands of the purple loose-strife. The long-petaled flowers are trimorphic; that is, the stamens and styles are three different lengths, and since six different kinds of yellow and green pollen are produced on the two sets of three stamens, one would expect that there would be eighteen ways in which the insects could transfer the pollen. But only pollen brought from the shortest stamens to the shortest style, from the middle-length stamens to the middle-length style, and from the longest stamen to the longest style can effectually fertilize the flower. Examine the pickerel weed and you will find it similarly trimorphic.

The showy flower spike of the pickerel weed, crowded with ephemeral violet-blue flowers above the rich glossy spathe, invests this water-loving plant with a unique charm. Each flower, marked with a distinct yellow-green spot, lasts only a day, but it is followed by a succession of blooms on the gradually lengthening spike, so that the plant blooms well into fall. The flower cup is funnel-formed and six-divided, the upper three divisions united and the three lower ones spread apart. Three of the six stamens are long and protruding, the other three short and often abortive; and the blue anthers are placed in such a way that it is impossible for an insect to enter the flower without brushing against them.

In his poem "The Humble Bee" Emerson speaks of the chicory as matching the sky. The flowers, which are an inch to an inch and a half across, are a beautiful blue color, but as the plant grows along roadsides, in waste places, and in similar situations, we regard it as a weed and remain unaware of its distinctive beauty. The flowers, similar in form to the dandelion, close in rainy or cloudy weather and open only in the sunshine. There are few florets in a single head, but these are highly developed, with gracefully curving, branching styles. Some of us may associate the plant with coffee, for the root has been used as a substitute or adulterant for this beverage; chicory roots have also been boiled and served in place of carrots, but they never became a popular substitute.

Few plants could be more innocent-looking than the tiny

sundew of swamps and bogs, with its raceme of buds and solitary little blossom that opens only in the sunshine, but it can be a deadly trap to fungus gnats and other small insects. These tiny creatures, attracted to the jewels that dot the leaves and sparkle in the sunshine, find to their dismay that the jewels are a viscid substance (secreted by glands in the leaves) more sticky than honey and which clings to their feet like glue. Then the pretty reddish hairs that cover the leaves reach out like tentacles and clasp them in a slowly closing, tight embrace; should an insect struggle in a vain attempt to escape, the hairs only move the faster. No torture implement of the Inquisition was more cleverly designed than the leaves of the sundew, for once an insect becomes imprisoned, the leaf rolls inward and forms a temporary stomach for a complex fluid, similar to the gastric juice in the stomach of animals, which is secreted to digest the hapless victim. You can watch the operation of the trap by placing a small insect on the leaves.

Collect some of the golden-yellow flowers you find growing in the nearby field or along the roadside, hang them in your window, and you will avert the evil eye and the spells of the spirits of darkness. You may not believe in this superstition, but people did not long ago, and maybe they still do in some places. Were it not that the brown petals of the withered flowers remain on the flower stems, St.-John's-wort might be an attractive plant, but it always seems to have an unkempt, untidy look. There are a number of species and some related species, such as St.-Andrew's-cross and St.-Peter's-wort. They are all much alike, differing only in minor details.

With its orange-yellow flowers spotted with reddish-brown and hanging like jewels from a lady's ear, the jewelweed (Fig. 56A) is known to all who know the outdoors. The flowers are irregular and perfect, with three sepals and petals, but these are not easily distinguishable. One of the sepals is large and sac-shaped and, contracted into a slender incurved spur, is admirably adapted to the long bill of the hummingbird; but you have to examine the flower closely to see how well it serves its purpose. Of course, insects visit the jewelweed too, particularly the long-tongued bumblebees.

On a dewy morning or after a shower, stop by the jewelweed

where it grows along a woodland brook or in the moist ground by the edge of a pond, and you will find the notched edges of the drooping leaves hung with dewdrops that sparkle like jewels in the

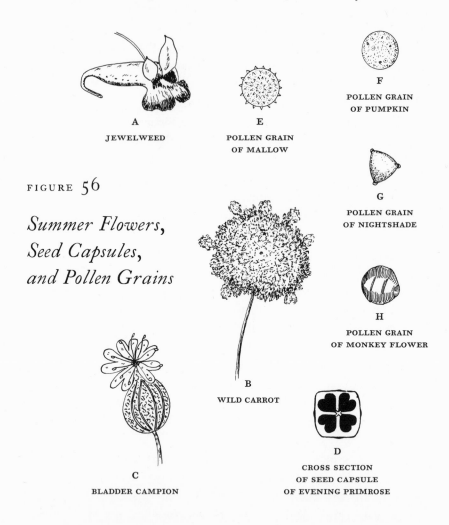

A
JEWELWEED

E
POLLEN GRAIN
OF MALLOW

F
POLLEN GRAIN
OF PUMPKIN

G
POLLEN GRAIN
OF NIGHTSHADE

H
POLLEN GRAIN
OF MONKEY FLOWER

FIGURE 56

*Summer Flowers,
Seed Capsules,
and Pollen Grains*

B
WILD CARROT

C
BLADDER CAMPION

D
CROSS SECTION
OF SEED CAPSULE
OF EVENING PRIMROSE

sunshine. Could another name for the plant be more fitting? Still, the jewelweed does have another name—touch-me-not. Touch the fruiting capsules when ripe; they open with startling suddenness and expel the tiny seeds to a distance of as much as 4 feet. If you have never done it, I guarantee you will jump at the unexpected volley from the miniature machine gun.

Typical of the summer wild flowers, the meadowsweet bears pink and white flowers that beneath the lens look like miniature apple blossoms. They grow in beautiful pyramidal clusters which are always crowded with innumerable small bees, flowerflies, and beetles gathering the accessible nectar. A cousin, the steeplebush, has smaller flowers; they are deep rosy pink and set on steeplelike spikes; unfortunately, the succession of blooms is slow and downward, so that the top is invariably in a half-withered condition. Yet the bright spires attract our attention no less than the eyes of the insects. Why are the undersides of the leaves so woolly? Not to guard the plant from crawling insects that could only benefit it, but to protect it against the vapors that rise from the damp ground in which it grows, for they would clog the pores of the leaves and prevent them from getting rid of excess water that would otherwise drown the plant.

About our dooryards and in our gardens in the Northeast, the dayflower often takes possession of the soil to the exclusion of other plants; elsewhere it may be found on river banks and other wet, shady places. There are several species essentially Southern in range; our Northeast species is a naturalized Asiatic which has extended its range to Texas. The flowers open only in the morning and close by noon. They are rather odd from a botanical point of view: the three sepals are unequal, and so, too, are the three petals, two of which are rounded, showy, and blue, while the other is inconspicuous and somewhat whitish. Of the six stamens, three are fertile, and one of these is bent inward; the other three are sterile and smaller, with cross-shaped anthers. It is altogether an unusual flower, one deserving a little study with the lens.

During the month of July it seems we can't stir far from the house without finding the spreading dogbane. This plant seems to be everywhere; it isn't, of course, but we do find it in fields and thickets, beside roads, lanes, and stone walls, its delicate and beautiful white-pink and rose-veined bells suggesting pink lilies of the valley. The little veins show where the nectar is, and butterflies, bees, flies, and beetles come to feast. But some of these pay for having trespassed on what is, in a sense, the butterflies' preserve, since butterflies are best fitted to serve the dogbane; many bees, flies, and other insects remain prisoner in the flower's trap of horny teeth until death from starvation at last releases them.

In July the dogbane beetle adopts the plant and remains inseparable from it; I have never found the insect on any other plant. Resplendent in metallic green and red of incomparable luster, the beetle is a beautiful insect, and when they are numerous on the plant the dogbane appears studded with jewels. To prevent capture, the beetle draws its legs up beneath its body and drops to the ground, where it becomes lost to view in the grass with which its colors blend so well.

One must examine the lacy umbels of the wild carrot (Fig. 56B) with the lens to appreciate their delicate structure and perfection of detail—the naked eye cannot do it. What appears to the casual observer as a single cluster is actually a number of small clusters. Notice that the small white flowers are disposed in a radiating pattern like a handmade piece of lace. In the very center of the cluster, see the tiny purple floret which is not a part of any of the smaller clusters but is set upon its own isolated stalk. No one has yet been able to account for it. Over sixty different kinds of insects may be taken from the flat-topped clusters of the wild carrot, because the flowers secrete an abundant amount of nectar which is easy to reach. Even the shortest-tongued insects can sip the nectar in less time than it takes them to sip the nectar from the tubular florets of the Compositae. Here is the reason, too, for the plant's wide distribution: it may be found growing in fields and waste places almost everywhere. When the flowers have served their purpose, the entire cluster dries and curls up to resemble a bird's nest.

The wild carrot has many relatives—the snakeroots, pennyworts, water hemlocks, sweet cicely, water and meadow parsnips, and the garden carrot, celery, coriander, caraway, and fennel, many of which are escapes. Like most families there is a black sheep; though here there are several—cowbane and poison hemlock, among others, poisonous to both man and animals. It was from the roots of the hemlock that the drink given to Socrates was made, or so it is believed.

We are all familiar with the bouncing Bet, at least by sight, for it is common along roadsides, banks, and waste places. An escape from colonial gardens, the pink or white blossoms, scallop-tipped, with an old-fashioned spicy odor, remotely remind us of the pinks. Butterflies, which delight in bright colors, have little interest in the flowers, but the night-flying moths can see them in the darkness and are attracted to them by their fragrance, which becomes more pronounced after sunset. Colonial housewives bruised the leaves in water and made a cleansing soaplike lather that served them well. Hence the plant is also known as soapwort.

I had known the bladder campion for years before I took a good look at it. And what a revelation when I saw, with my lens, its greatly inflated flower cup (Fig. 56c); pale-green and beautifully veined, it is not unlike a citrus melon. The white flowers gleam in the darkness and guide the moths that are no less seduced by the strong perfume which scents the night air.

But unlike these two members of the pink family, their relatives the evening lychnis and the night-flowering catchfly open their blossoms only at dusk and close them by morning. The flowers of both are white and strongly scented and the stems of both plants are covered with sticky hairs to catch flies, no doubt, but the hairs are more effective against the ants that would pilfer the nectar without doing either plant any service. The beautifully marked calyx of the catchfly resembles spun glass when seen with the lens.

By day the evening primrose has little to commend it, for when we see it by the roadside or in a thicket or fence corner, its wilted, faded flowers and hairy capsules, crowded among the willowlike leaves at the top, give it a rather bedraggled appearance. But when the sun has set in the western sky and twilight begins to creep over the landscape, developing buds begin to open, and pure-yellow, lemon-scented flowers appear that gleam like miniature moons when other flowers have melted into the deepening darkness. Bumblebees and honeybees may visit the flowers in the morning before they have had a chance to close, but it is the night-flying moths on which the plant depends for cross-pollination. It is not without reason that the flowers are yellow, that their fragrance becomes stronger with the night, that the nectar wells are located in tubes so deep that none but the moths' long tongues can drain the last drop, or that the golden pollen is loosely connected by cobwebby threads on eight prominent and spreading stamens. The Isabella tiger moth is perhaps the plant's chief benefactor, but there are others; when morning comes we may find a little rose-pink moth, its wings bordered with yellow, asleep in a wilted blossom, and since the flowers turn pink when faded, the moth is safe from prying eyes. Later in the summer when enough seed has been set, the primrose changes its habit, and the flowers remain open all day. Cut open one of the brown seed capsules (Fig. 56D) and you can see the small brown seeds neatly arranged in eight parallel rows.

Nature, if anything, is always practical, and she so designed the fireweed to grow in burned-over places where the spikes of the beautiful brilliant-magenta flowers, on stems as much as 8 feet tall, soon cover the ugliness of the blackened ground. Beginning at the bottom of the long spike, the flowers open in slow succession throughout the summer, leaving behind slender, velvety, gracefully curved, purple-tinged seed pods that are as effective as the flowers in covering the ravages of fire.

These are but some of the wild flowers that we might examine with profit as we wander about the countryside. There are many more, to be sure: along the roadside the common persicaria, its little pink flowers opening without method anywhere on the spike, and its kin the knotgrass, lady's thumb, and buckwheat. Here, too, sweet clovers, with leaves that go to sleep at night, scent the air, their papilionaceous blossoms an open invitation to all sorts of pollenizing insects. Honeysuckles also grow by the wayside and in woodland thickets, where wood lilies likewise gleam from the beds of feathery gold-tinged ferns. In fields and meadows lilies nod in the passing breeze, and in quiet ponds water lilies spread their chalices to welcome bees, flowerflies, beetles, the "skippers," and other winged creatures.

Wild roses are common everywhere: the meadow rose on rocky ground, the pasture rose in pastures and similar places, the sweetbrier in fields and woodland thickets, where the parasitic Indian pipe and pinesap find the soil most congenial. There are the vetches too, with their blue or purple beanlike blossoms and many small leaflets, spreading over the herbage of fields and waysides, together with their numerous relatives: the wild bean, hog peanut, ground-

FIGURE 57

*Mouthparts of a Flower Beetle
and Adaptive Modifications*

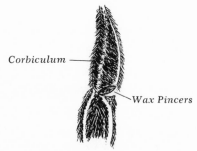

Corbiculum

Wax Pincers

B

PORTION
OF LEFT HIND LEG
OF HONEYBEE

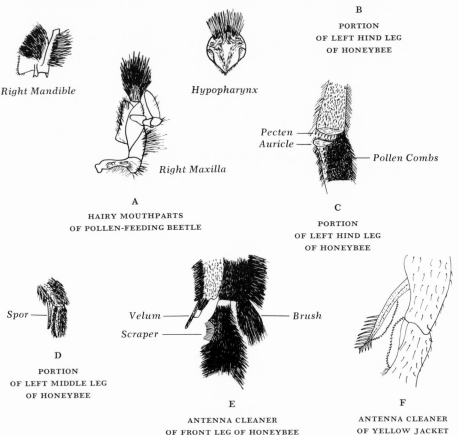

Right Mandible

Hypopharynx

Right Maxilla

A

HAIRY MOUTHPARTS
OF POLLEN-FEEDING BEETLE

Pecten
Auricle

Pollen Combs

C

PORTION
OF LEFT HIND LEG
OF HONEYBEE

Spor

Velum

Scraper

Brush

D

PORTION
OF LEFT MIDDLE LEG
OF HONEYBEE

E

ANTENNA CLEANER
OF FRONT LEG OF HONEYBEE

F

ANTENNA CLEANER
OF YELLOW JACKET

nut, wild senna, partridge pea. And all have their own kind of
pollen grains, which vary as much as the eggs of insects. Place
pollen grains from different flowers on the stage of the microscope,
and you will have them in all shapes, colors, and sizes. I have
shown a few of them in Figures 56E, F, G, H.

The insects that visit flowers for their pollen and nectar and that
are effective as pollenizers are variously modified in some way to

serve such a purpose. The adaptations are not as manifold or as exquisite as those of the plants, since the plants had to adjust themselves to the insects as a matter of necessity. Yet the insects have to some extent been influenced by floral structure, and many have been uniquely modified structually and physiologically to accommodate them to the flowers they visit.

Many pollen-gathering insects are clothed with hairs that are usually dense and often twisted, branched, or barbed. Examine with your lens any insects like mining bees, bee flies, flowerflies, or pollen-feeding flower beetles (Fig. 57A), and you will see that hairs occur not only on the body but also on the mouthparts. A few bees are provided with special structures for carrying pollen known as pollen baskets or corbicula. In the honeybee the pollen baskets (Fig. 57B) are on the outer surface of each tibia. When you examine them with the lens you can see that they are merely a sort of receptacle.

Station yourself near a flower visited by honeybees and, with a reading glass, watch a bee as it climbs over the flower. Look closely and observe how the flexible branching hairs of the body and legs entangle the pollen grains. With special structures called pollen combs (Fig. 57C), located on the inner surface of the hind tarsus, the bee combs the pollen grains out of the hairs and transfers them to the baskets, filling them to capacity and pressing the pollen down into the baskets by means of the auricle. On the bee's return to the hive, it thrusts its hind legs into a cell and then, using a spur located at the apex of the middle tibia (Fig. 57D), pries off the pollen, slipping the spur in at the upper end of the corbiculum and then pushing it along the tibia under the mass of pollen.

By the very nature of the bee's activities, its antennae become covered with pollen grains and various kinds of debris and must frequently be cleaned. On the tarsus of the front leg there is a concavity or semicircular scraper that the insect uses for such a purpose (Fig. 57E). When cleaning its antennae, the bee raises its leg and passes it over one of them, which then slips into the scraper. The bee now bends its leg at the joint where the tibia and tarsus meet, and as the leg is bent an appendage called the velum falls into place to complete a circular comb, through which the antenna is drawn. The comb, in turn, is cleaned by means of a brush of hairs on the front margin of the tibia. The yellow jacket also has an antenna cleaner; it is located on the front leg and is really a comb in every sense of the word, and quite ornate, too (Fig. 57F).

As we have already noted, insects respond to the conditions of their environment and to various stimuli by means of sense organs, such as touch, taste, and so on. Not the least of these sense organs are those of sight. Insects have two kinds of eyes (though both may not occur on every species): simple eyes called ocelli, and compound or faceted eyes.

The simple eye is constructed on much the same plan as the human eye, but its capacity for producing images is extremely limited. The form of the lens is fixed, as well as the distance between the lens and retina; hence there is no power of accommodation, and most external objects are out of focus. Actually, it is probable that all a simple eye can do is to distinguish between light and darkness, though it may possibly form a crude image at close range.

The compound eye is quite different. Externally the convex surface of this eye is covered by a modified transparent section of the cuticula called the cornea, which is divided into hexagonal areas or facets. Each facet is only the external part of a long, slender visual rod known as an ommatidium. A compound eye is made up of a number of similar ommatidia lying side by side, but separated from one another by a layer of dark pigment cells. Although there is more to it than this, a fuller discussion will serve no useful purpose; it is enough to say that the compound eye forms a mosaic image, that it can detect movement, and that it is sensitive to the various wave lengths which compose the spectrum, for honeybees seem to prefer blue flowers, the white butterflies white flowers, and the yellow butterflies yellow flowers.

If you examine the compound eye of almost any insect with your lens, you can readily see the facets. But should you try to count them, you are likely to stop before long; for in the common housefly there are 4,000 of them in one eye alone. And this is really a small number compared with the 17,000 in the eye of a swallowtail butterfly, the 25,000 in the eye of the beetle Mordella, and the 27,000 in the eye of a sphingid moth. Ants have a relatively small number: less than 400.

For the most part, the trees have ceased their flowering except in the South, where the magnolia may still be in bloom. The same can be said for the shrubs, though the sweet pepper bush and buttonbush are just coming into their own, the former with fragrant white, narrow, upright clustered spikes, the latter with white, fragrant, spherical heads. Both are water-loving plants, and we find them growing along the banks of streams and ponds, in wet woodland thickets, in swamps, and on low ground almost everywhere. Since both are fragrant, the pepper bush with a spicy odor, the buttonbush with a fragrance that suggests the jasmine, it is unnecessary to say that both play host to a multitude of nectar-seeking insects. To appreciate its perfection of detail, we must examine the buttonbush's little "cushions of pins," as someone aptly called them, with a magnifying glass. If we begin to count the minute florets, we might count a hundred or so before we become confused and tire of the exercise. Examine a floret carefully, and you will discover that its long tube is primarily adapted for the long-tongued insects; hence butterflies are the most abundant visitors. But there are many others: honeybees, bumblebees, mining and carpenter bees, flies, wasps, and beetles.

Should you wonder why a bee should be called a carpenter, split a few dead twigs from the sumac, elder, or bramble. In at least one of them you should find a tunnel down the center partitioned off as shown in Figure 58A. The partitioned tunnel is an insect apartment house built by the little carpenter bee, a small, beautiful insect with a metallic-blue body.

A reading glass or hand lens can show details of construction much better than the naked eye. Each partition is made of small bits of pith glued together, and each forms the roof of one cell and the floor of the one above it. Each compartment, except the topmost one, which is occupied by the mother bee herself until her family has grown up, contains some beebread (a paste made from honey and nectar) and an egg—the beebread being food for the grub when it hatches from the egg. The egg in the lowest compartment hatches first, and the grub is the first to develop into an adult bee. Is this bee imprisoned at the bottom of the tunnel, or does it make an exit hole in the side of the twig? The answer to both parts of the question is no; the bee simply tears down the roof of its cell.

But then it has to wait patiently for the occupant of the cell above to transform into an adult bee. This bee in turn tears down the roof of its cell, and so the procedure is repeated by each successive bee

FIGURE 58

Nests of Various Bees

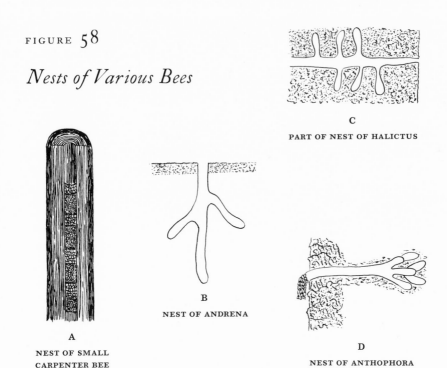

C

PART OF NEST OF HALICTUS

B

NEST OF ANDRENA

A

NEST OF SMALL
CARPENTER BEE

D

NEST OF ANTHOPHORA

until all have transformed. When this stage has been reached, the remains of the partitions are pushed down to the bottom of the tunnel, and the bees are led forth by the mother and fly out into the sunshine.

They do not, however, abandon the apartment house. After a while they all return and remove the remains of the partitions and other refuse from the tunnel, the mother bee and the young working together. Then the tunnel is used again by one of the bees. Should the brood be a late one—that is, mature in the fall—the bees make use of the apartment house as a winter home. If you want to know what goes on in these apartment houses, open them at different times of the year.

There is a larger species of carpenter bee, a large black insect that resembles a bumblebee in size and appearance and is called the large carpenter bee. It builds an apartment house similar to that of the small carpenter bee, except that it tunnels into solid wood, such as a beam or timber.

Should you also wonder why certain bees are called mining bees, it is because they excavate tunnels in the ground. These are more difficult to find than the apartment house of the little carpenter bee. There are several species of mining bees, and each excavates its own kind of tunnel. I have pictured several of them in Figures 58B,C,D. Mining bees are gregarious; as many as a hundred burrows may be found within a given area.

When the grasses of spring begin to fade and lose their beauty, others appear and take their place. Among these are the fescues, which for a short time are predominant in the fields. The meadow fescue we have already mentioned; there is also the red fescue common along the wayside, and the sheep's fescue and slender fescue that seem to prefer dry locations. In the roadside thicket we find the hispid panic grass, and where the meadow meets the woodland is the fringed bromegrass, its stems rising above other grasses and its panicles large and conspicuous. With its squirrel tails of bearded spikes, the squirreltail grass blossoms in our gardens and dooryards and rapidly becomes a weed if we are not careful; it is better confined to waste places where it can spread its bristly flowering heads unrestricted. Later in the month other grasses appear on the scene, but these are the grasses of August, when summer is at its height.

As the days go by, mushrooms become more numerous: the meadow mushroom, which is the one cultivated to serve as food; the tall, stately destroying angel (Fig. 59), its shimmering satiny whiteness symbolic of innocence, but one of the most poisonous known; the equally deadly fly amanita, with its danger signals of warts and concentric rings; the beautiful Jack-o'-lantern; the dainty coral mushroom that borders the woodland path; and the green and red Russulas that vie with the flowers in the brilliancy of their coloring. There are other forms, but they are just beginning to appear, and rightly belong to August. Yet when we see them we think of August and realize that summer is pushing on. There are other signs, too: the woodchuck is beginning to stuff himself with all sorts of juicy morsels and tender tidbits; the leaves, so green and

FIGURE 59

The Destroying Angel Mushroom

fresh and tender a few short weeks ago, are becoming yellow and brown, torn and chewed, mute testimony that the harsh summer sun and countless caterpillars have taken their toll; fruits and berries are beginning to take form and show faint signs of coloring that will shortly become more brilliant; and some birds are beginning to get restless. One of these days we will tear another month from the calendar, which is merely a man-made device to inform us that time hurries on.

AUGUST

"The quiet August noon has come;
A slumberous silence fills the sky."

WILLIAM CULLEN BRYANT

"A Summer Ramble"

AUGUST

UGUST IS USUALLY HOT AND SULTRY, AND there are days when hardly a leaf moves and the sun is almost obscured by the copper sky. This is the kind of weather when many of us seek relief at the seashore or in the cool mountains, though some of us prefer to stay home and get the maximum from our investment in an air conditioner. Sometimes black clouds suddenly take form and push the thermometer down a degree or two. But after the rain has fallen and the sun shines again, the weather becomes even more intolerable, for now the humidity increases and adds to our discomfort.

The mammals, feeling the heat, are less active than at other times of the year and move about only to get something to eat. The birds hide in cooling trees and are even less vocal now than in July. But the countryside is not altogether silent, for the meadow grasshoppers sing all day, the dog-day cicada drones its monotonous tune, and the katydid calls its familiar "Katy" from a nearby thicket. And as the sun begins to sink in the sky, the field crickets and tree crickets begin to tune their instruments and throughout the night fill the air with their orchestral music.

Many different kinds of insects are able to produce some kind of sound, though I doubt if any are more universally known than the crickets. Of our own species of crickets, the one we know best is the black field cricket. If you have never seen one of them chirp (only the males do), find one and you will observe that he raises his wing covers to an angle of 45 degrees and rubs them together. There is more to it than this, however, for if you examine one of the wing covers (Fig. 60A) with your lens, you will see that the veins form a peculiar scroll pattern which serves as a framework for making a sounding board of the wing membrane by stretching it out like a drumhead. Note also that near the base of the wing there is a heavy crossvein covered with transverse ridges, called a file, and on the inner edge of the same wing near the base, a hardened area known as the scraper. The cricket sounds his notes by drawing the scraper of the under wing cover against the file of the overlapping one. We can produce a similar sound by running a file along the edge of a tin can.

As the wing covers are excellent sounding boards (tympani) and quiver when notes are made, they set the surrounding air to vibrating, thus giving rise to sound waves that travel a considerable distance. It is interesting to note that the cricket can use his wing covers alternately; that is, he can use one wing cover as a scraper and the other as a file and then reverse them, thus reducing their wear and tear.

At one time it was believed that the male crickets chirp to attract the females, but this view was discarded when it was found that the females do not always respond. Actually, their chirps, as we hear them, have no meaning to crickets. We now know that when crickets chirp, they also produce notes too highly pitched to be audible to us. It is these ultrasonic notes they use to communicate with one another, the chirps being merely incidental. It has also been established that the field cricket has at least three basic sound signals: a calling note, an aggressive chirp, and a courtship song. On the tibia of each front leg is a small white, disklike spot; it is the auditory organ or ear (Fig. 60B).

Male katydids make sounds in much the same way as crickets, though they are left-handed musicians, having a file on the left wing only (Fig. 60c). It consists of about fifty-five teeth. When the insect chirps, it spreads its wing covers, then closes them gradually, and as they close the scraper clicks across the teeth, making from twenty to thirty sharp ticking sounds in rapid succession.

A fairly common species of grasshopper (Stenobothrus) makes a sound by rubbing its pegged hind legs against the elevated veins of the front wings. The two wings and legs constitute a pair of violinlike organs, the thickened veins corresponding to the strings, the mem-

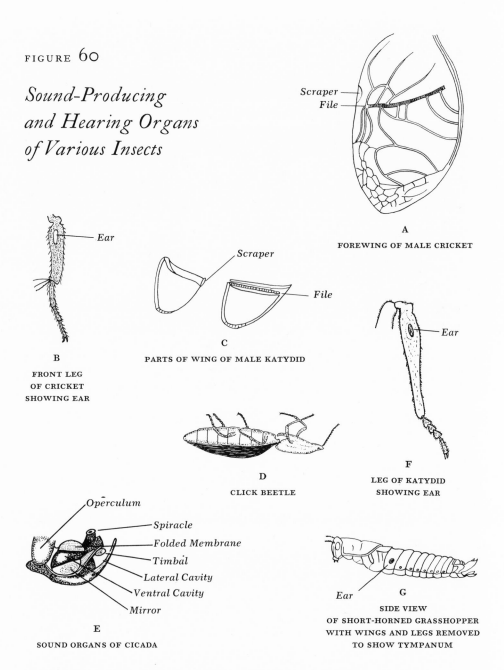

FIGURE 60

Sound-Producing and Hearing Organs of Various Insects

Scraper
File

A
FOREWING OF MALE CRICKET

Ear

B
FRONT LEG
OF CRICKET
SHOWING EAR

Scraper
File

C
PARTS OF WING OF MALE KATYDID

Ear

F
LEG OF KATYDID
SHOWING EAR

D
CLICK BEETLE

Operculum
Spiracle
Folded Membrane
Timbal
Lateral Cavity
Ventral Cavity
Mirror

E
SOUND ORGANS OF CICADA

Ear

G
SIDE VIEW
OF SHORT-HORNED GRASSHOPPER
WITH WINGS AND LEGS REMOVED
TO SHOW TYMPANUM

brane of the wings to the resonating chamber, and the series of pegs to the bow. When playing, the grasshopper assumes a nearly horizontal position; then, raising both hind legs at the same time, rubs them against the wings. The Carolina locust that occurs on road-sides, in waste places, and on well-beaten paths (though we usually don't see it except when in flight, for its color matches the ground) makes a crackling sound as it flies by rubbing the hind wings against the fore wings.

Sounds produced by an insect rubbing one part of its body against another part is known as stridulation. We find rasping organs, varying greatly in form and in their location on the body, in a great many insects. The water boatman and the back swimmer make a clicking sound by rubbing their front legs against their probiscides; the sphingid moths use their palpi. The powder-post beetles rub their front legs against a projection at the posterior angle of the prothorax. We can see how these insects do it by watching them with a reading glass, though powder-post beetles may be rather difficult to find, hidden away as they are in furniture, construction timber, and the like. Click beetles (Fig. 60D) are much easier to find. When placed on their backs, they spring into the air, turn over, and land on their feet; hence they are also known as skipjacks, spring beetles, and snapping bugs. They have a spine on the lower surface of the body that normally rests in a groove on the mesothorax.[1] The connection between the prothorax and meso-thorax is more flexible than in most insects, and as the prothorax is flexed upward, the spine slips over a sharp edge on the anterior margin of the mesothorax and produces a clicking sound.

Many insects set up sound waves by vibrating their wings; oth-ers, such as the deathwatch beetles, make a faint ticking sound by bumping their heads against the sides of their burrows; and certain wood-boring insects make faint clicking notes as they gnaw through the wood. Such sounds are incidental and are not produced by specific sound-producing organs. In flies and certain other insects, sound is created by a membrane in the spiracles that vibrates as air is being taken in and expelled. The queen bee, the blowfly, and the May or June beetle make sounds in this way and are called drum-mers. The classical drummer of the insect world is the cicada, whose sound organs are the most complicated found anywhere in the animal kingdom.

---

[1] The thorax is divided into three segments: prothorax, mesothorax, and metathorax.

These sound organs vary somewhat in the different species of cicada, but they are essentially as follows: on the underside of the third thoracic segment of the male, there are two large plates called opercula, which can easily be seen with the lens and which can be lifted slightly. Each operculum serves as a lid covering a pair of cavities that contain the sound organs; it functions as a protective covering. The cavities are known as the ventral cavity and the lateral cavity. The lateral cavity is provided with a single membrane called the timbal, the ventral cavity with two membranes called the folded membrane and the mirror. Within the body of the insect is a large air chamber that communicates with the outside through a pair of spiracles (Fig. 60E). The cicadas produce sound waves by setting the timbal and its attached muscles into motion. The air in the air chamber, acting as a resonator, transmits the sound waves created by the vibrating timbal to the folded membrane and the mirror, which amplify them. The timbal, however, is the primary source of the sound, for if it is removed the insect becomes mute.

Sounds produced intentionally by insects are of no value unless they can be heard by the other insects that are supposed to hear them; hence most insects have some sort of auditory organ. The antennae of the male mosquito function as a hearing organ, and it is believed they serve a similar purpose in ants, and possibly butterflies and moths. The hairs of certain caterpillars are able to pick up sound waves, and the cerci of the American cockroach are also adapted for this purpose. A hearing organ similar to that of the cricket is present on the front legs of the katydid (Fig. 60F), on the first abdominal segment of the short-horned grasshopper (Fig. 60G), and on the thorax of the water boatman, water scorpion, creeping water bug, and many butterflies and moths. Essentially, this hearing organ consists of a vibrating membrane similar to our own eardrum; but a vibrating membrane in itself is of no value unless the vibrations of the membrane are transferred in some way to the nervous system. Structures called chordotonal organs function to this end. They consist of rods, nerves, and nerve endings and are rather complicated, but the morphological unit is a more or less peglike rod contained in a tubular nerve ending.

We have all heard a fly hum or a bee buzz, but at the time we were probably not conscious of nuances in the hum or buzz. An experienced beekeeper, however, knows that bees often buzz at different pitches; he is familiar with the swarming sound, the hum of the queenless colony, and the angry note of a belligerent bee. It is likely that such notes can also be distinguished by the bees, so that they can govern themselves accordingly. Some insects doubtless produce a sound in an effort to ward off an enemy. When handled, many insects produce a sound that is hardly perceptible to the human ear but may cause birds and other animals to drop them. I have known people to become frightened by the loud buzz of the June beetle, and presumably many animals are similarly affected. When attacked, the little sand cricket faces its enemy and boldly defies it with a distinct rasping sound.

Generally, however, the sounds produced by insects are sexual calls. In the tree cricket the singing of the male evokes a sort of indirect response on the part of the female, for she does not have an ear and cannot hear him. But when the male raises his wings, a gland is exposed which secretes a liquid with a characteristic odor that the female can detect. And this is all she needs; she responds by climbing on the male to drink the liquid, and as she does so, mating takes place. Thus in this particular instance, the mating process depends not on the sound signals but on a chemotropic response.

The hot sun of August takes its toll of the vegetation, particularly if rainfall is at a minimum, and the leaves of the trees, once so green and tender, are now discolored and ragged. The sun cannot be blamed entirely; some of the wear and tear on the leaves can be laid at the door of the insects that feed on the tissues, both externally and internally. Since leaves are virtually paper thin, it almost passes belief that there are creatures so small that they can live and grow between the upper and lower surfaces of a leaf and yet be visible to the naked eye. But hold a leaf which is more or less discolored with whitish or grayish blotches (Fig. 61A), or which has long twisting lines (Fig. 61B), up to the light and you will see the blotch or twisted line inhabited by a tiny, wormlike animal.

These twisting lines and blotches are passageways and tunnels which the larvae of beetles, flies, sawflies, and moths excavate in the tissues of the leaves. Known as mines, they form a variety of pat-

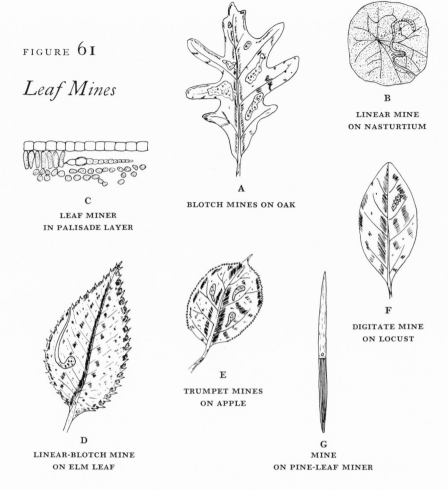

FIGURE 61

Leaf Mines

C
LEAF MINER
IN PALISADE LAYER

A
BLOTCH MINES ON OAK

B
LINEAR MINE
ON NASTURTIUM

F
DIGITATE MINE
ON LOCUST

D
LINEAR-BLOTCH MINE
ON ELM LEAF

E
TRUMPET MINES
ON APPLE

G
MINE
ON PINE-LEAF MINER

terns, each so characteristic of the insect which made it that people familiar with them can usually identify the insect by the form of the mine. As one writer put it, "they write their signatures on the leaves."

I have often wondered if James Russell Lowell had these little animals in mind when he wrote:

> *And there's never a leaf nor blade too mean*
> *To be some happy creature's palace.*

Whether they are happy is a moot question, but assuming they are (if shelter and plenty of food alone are conducive to happiness), they do not live an altogether carefree existence. For sometimes they find the veins of the leaves a barrier to further progress and must cut through them or confine their operations to a limited area. They must be careful, too, not to cut the latex cells, lest the secretions pour out and drown them. But their gravest problem is one of waste disposal. Some miners, as the insects are called, merely distribute the wastes over the floor of the mine, when they appear as black masses. Others go to the trouble of excavating side chambers in which to dispose of their refuse. Many merely leave their wastes behind as they eat their way forward. And many others have developed the habit of cutting holes in the surface of the leaf through which they push out their fecula; in some instances the frass emerges in the form of small pellets that adhere to one another like minute sausages linked together.

As may be seen with a lens, the miners as a rule are flat, legless, and lack setae or bristles; if these structures are present, they are much reduced. The head is often wedge-shaped (Fig. 61c), thus being an efficient device for separating the two epidermal layers of the leaf as the insect moves forward; sometimes the head is rotated and telescoped into the thorax. In some species the antennae and eyes are reduced, and in others the eyes are arranged alongside the head. The jaws are usually sharp, and operated by powerful muscles. In the sap-feeding species, the mandibles are platelike, with many sharp teeth for cutting through the plant cells to make the sap flow. Hard plates or tubercles are sometimes present to help the insects maintain a firm hold as they feed within the mine or tunnel.

Leaf miners are essentially immature insects, remaining within the leaf only until they have become full grown, when they transform into winged adults. If you follow one of the twisting passageways from the beginning, you will note that it starts almost as a pinpoint, where the egg hatched, and grows progressively wider as the insect increased in size.

Entomologists recognize two general types of mines: the twisting or linear mine, made by the insect as it moves forward, and the round or blotch mine, made by the insect as it moves around within the leaf. Modifications of both occur as the linear-blotch (Fig. 61D), trumpet (Fig. 61E), digitate (Fig. 61F), and tentiform.

A little study will show that miners vary in feeding habits, which to a large extent determine the location of the mine. Some species feed only on the palisade layer—the layer of cells directly beneath the upper epidermal layer; others feed only on the parenchyma cells, which lie below the palisade layer; still others feed on both. Hence the first are visible only from the upper side of the leaf and the second only from the lower side, but the third are visible from both sides.

Not all leaf miners feed exclusively on the cells; some, in fact, live entirely on the sap throughout their larval existence; and still others may start out as sap feeders and then become tissue feeders. One of the most unusual leaf miners, the basswood leaf miner, is also known as the polygon leaf miner from the shape of its mine. During the first two instars, it feeds on the liquid contents of the cells, voiding little or no fecula. Then, after the third molt, it goes back over the area previously covered and strips all the cells between the upper and lower epidermal layers. Since the insect feeds from the outside of the mine toward the middle, it must, before it starts feeding, mark the final area to be mined; otherwise its food supply would become exhausted before it matured. Moreover, as it deposits all waste material in minute pellets at the outer edge of the mine, the insect, always feeding toward the center, has a continual supply of food uncontaminated by fecal matter.

Most linear miners feed on the cell sap, blotch miners on the tissues. Frequently an entire family feeds together, making a large and ugly blotch mine. If you can find a mined leaf of dock or beet while it is still green, you should see several insects working, each making a bag in the tissues and all joining together to make a great blister. Finally, some species have developed the habit of entering new leaves when their food supply has become exhausted or when the leaves wilt or otherwise become undesirable.

There are many species of leaf miners, and they attack nearly all families of plants. Even a pine needle (Fig. 61G) has its little inhabitant. If you can find a pine needle occupied by the pine-leaf miner and hold it up to the light, you will see the little creature running up and down in its tunnel. A small hole near the lower end of the needle marks the place where the larva entered the leaf and through which, after it has enlarged the hole slightly, it will emerge as an adult moth.

The leaves of trees and shrubs are not only disfigured by leaf miners; they are also deformed or otherwise made unattractive by the many insects that curl, roll, fold, or tie them together with silk to serve as shelters. In hydrangea shrubs we often find a little green caterpillar, the hydrangea-leaf tier, that, having sewn two terminal leaves together and thus having enclosed the flower bud, lives

FIGURE 62

Two Leaf Rolling Insects

A

HYDRANGEA LEAVES
SEWN BY
HYDRANGEA LEAF TIER

B

BASSWOOD LEAF ROLLER

within the two leaves while it feeds on the developing flower and on the inner surface of the leaves (Fig. 62A). A leaf-green caterpillar with a brown head, the silver-spotted skipper, ties the leaflets of the locust together and lives concealed within them, emerging only to feed. And the leaves of the linden or basswood may be found rolled together (Fig. 62B) to serve as a temporary home for a bright-green larva with a shiny black head; I say temporary home, for when the larva becomes full-grown it departs and makes a smaller one in which to spend the winter. Other insects also use such shelters to hibernate in; the promethea moth (see Fig. 7F), the viceroy butterfly (Fig. 7C), and the grape-berry moth are a few examples. Leaves rolled and tied by the brown-tail moth serve as a communal nest in which a number of caterpillars secure a refuge from the low temperatures and storms of winter. You will find curled, rolled, or folded leaves on a great variety of plants, and many are well worth examining with the lens or reading glass. If you want to see how an insect sews the leaves together, remove an occupant and place it on an untied leaf.

In his *Natural History* Pliny tells us about galls, but much earlier Theophrastus knew of their medicinal and curative properties. So galls were familiar to the ancients, though they did not know what caused them; no one did, in fact, until 1686, when Malpighi came up with the answer. And he was only partially right, for not only are galls caused by insects, as he said, but also by fungi, nematode worms, mites, and probably in some instances, mechanical irritation.

Even today we have much to learn about the physiology of gall formation; nor do we know why any particular gall should have a distinctive shape. We know that the egg of a gall-making insect is laid on the host plant or inserted in the tissues, and that when the larva emerges, it makes its way to the meristematic tissues—that is, to the cells capable of dividing and multiplying—because galls are formed only in such tissues. Once the larva has reached these tissues, it presumably secretes some substance that stimulates the cells to greater activity, causing them to multiply more than they would normally. The result is an abnormal growth, or gall.

Meanwhile, as the gall develops the insect feeds on the tissues that compose it, at the same time secreting an enzyme which changes the starch of the cells to sugar in the same way that the plant enzyme normally converts starch to sugar. This sugar in turn is used by the plant to form new cells, that not only provide food for the insect but also promote the growth of the gall, since more cells are produced than the insect can possibly eat. So both the plant and insect profit from the association.

The study of galls and gall insects is a fascinating one, though in some respects it is complicated because we cannot always be sure that the insect which emerges from the gall is the one that made it: many insects do not make galls but lay their eggs in those made by others. Such insects are called "guests" or inquilines. Furthermore, both the makers and the inquilines are attacked by parasitic hymenoptera, which only adds to our confusion, for it is not always

C

OAK APPLE IN SUMMER
CUT OPEN
SHOWING GRUB

FIGURE 63

Some Representative
Insect Galls

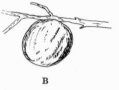

B

OAK APPLE

E

OAK APPLE IN WINTER
CUT OPEN

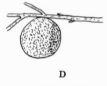

A

PINECONE GALL

D

OAK APPLE

easy to determine the interrelations of these insects. Many galls are complicated communities; for example, the pinecone gall (Fig. 63A) is a veritable insect apartment house. This particular gall has been studied in considerable detail: in addition to its maker, as many as 31 different species, represented by 10 inquilines, 16 parasites, and 5 transients, have been found living in it.

There is no evidence that the form of a gall is of any adaptive importance; it may be that the formation of any specific shape is purely mechanical. But the remarkable thing about galls is that those made by the same species of insects always have the same form, always occur on the same species of plant, and always appear on the same part of the plant, so that those versed in gall lore may know the identity of the gall maker by merely looking at the gall.

Galls occur on a wide variety of plants, but they are especially numerous on willows, oaks, roses, legumes, and the composites. They may be found on any part of the plant—root, branch, stem, leaf, blossom, fruit, and even the seed. Some are rather inconspicu-

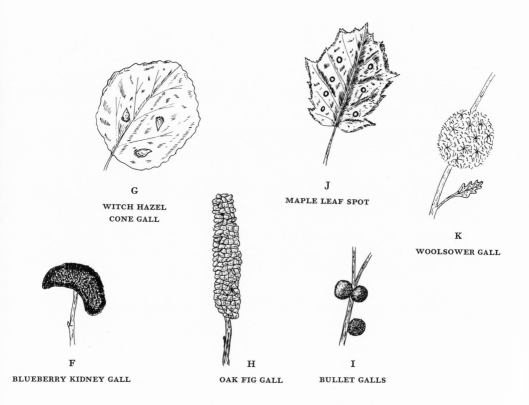

G
WITCH HAZEL
CONE GALL

J
MAPLE LEAF SPOT

K
WOOLSOWER GALL

F
BLUEBERRY KIDNEY GALL

H
OAK FIG GALL

I
BULLET GALLS

ous, merely blisterlike swellings; others are hairy or pilelike; still others are suggestive of bullets; while those that occur on flowers or in flower clusters often make them look as if ravaged by fire. A few are grotesque and eye-catching.

Few galls attract our eye as quickly as the oak apples, which may be so numerous on an oak tree as to suggest a fair crop of fruit on an apple tree. They are globular in shape (Fig. 63B), green while in the process of being formed, tough and firm in texture, and an inch or two in diameter. To all appearances they look like small unripe apples. At first glance they may seem to grow directly from the buds, but a closer examination shows that each one is a deformed leaf, hanging by the leaf stem or petiole. If one is cut open with a sharp knife or razor blade and viewed with the lens, it is seen to contain a juicy, spongy white substance and a large central larval cell holding a small grub or young gall wasp (Fig. 63c). The larval cell in an oak apple spotted with red (Fig. 63D) is supported by radiating fibers instead of a spongy substance (Fig. 63E).

Some galls are very shallow, such as the blister or spot galls found on asters and goldenrods. The spangle oak galls are thicker and look like shallow saucers. A few galls, such as the blueberry kidney gall (Fig. 63F) and the gouty oak gall, are quite woody. Galls often found on the leaves of the witch hazel are conical, green, or green and reddish-tipped, and rather pretty (Fig. 63G). My rose-bushes invariably have a few spiny galls that are green or reddish and somewhat globular. The cockscomb gall is an irregular but comb-like greenish, red-tipped elevation formed on the veins of an elm leaf, and resembling the comb of a cock.

The oak fig gall, a cluster of many individual galls, resembles a cluster of figs (Fig. 63H), and as seen with the lens, is covered with fine short hairs. Bullet galls (Fig. 63I) may appear singly or in clusters; they are yellow, sometimes tinged with red, and occur invariably on oaks. Sometimes maple leaves are thickly spotted with circular yellow, eyelike spots margined with cherry red (Fig. 63J). They are rather attractive. Few galls, however, have the eye appeal of the woolsower (Fig. 63K). Many naturalists consider it one of the most beautiful objects in nature. Found on the twigs of various oaks, it is woolly, creamy white, and admirably set off with pinkish-red blotches, the woolly growth containing seedlike grains. This is one object that must be looked at with the lens.

Did you ever touch a petunia leaf, find it sticky, and wonder why it is sticky? Or have you ever examined one closely to learn the reason? Look at the lower surface with the lens; it is covered with numerous fine hairs. Hold the leaf up to the light and the hairs seem to glisten. The tips are glandular and secrete a sticky substance that reflects the light, making them sparkle or glitter. A geranium leaf is sticky too, for the same reason. So are the leaves of the tarweeds, tobacco, Chinese primrose, and various members of the gourd family.

Hairs are outgrowths of the epidermal or surface-layer cells. They may consist of only one cell (Fig. 64A), when the epidermal cell and hair are one, or they may consist of many cells (Fig. 64B), in which case the hair is a filament that decreases in size from the base to the apex. A glandular hair (Fig. 64C) bears a rounded head at the upper end. Hairs entrap air, which reflects light and makes them appear white, or nearly so.

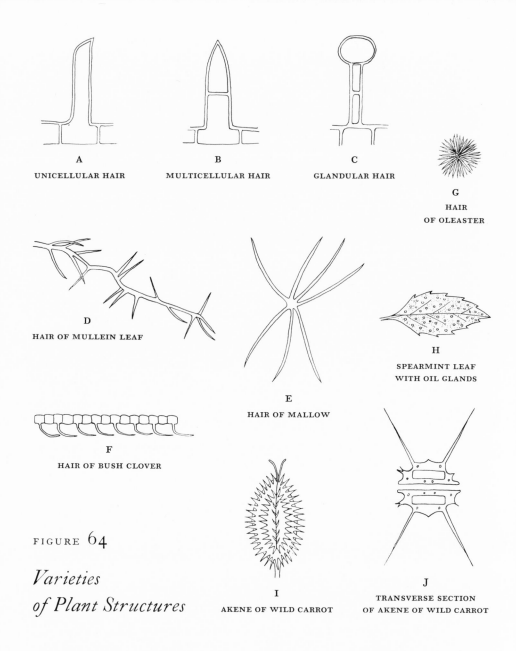

A
UNICELLULAR HAIR

B
MULTICELLULAR HAIR

C
GLANDULAR HAIR

G
HAIR
OF OLEASTER

D
HAIR OF MULLEIN LEAF

E
HAIR OF MALLOW

H
SPEARMINT LEAF
WITH OIL GLANDS

F
HAIR OF BUSH CLOVER

FIGURE 64

*Varieties
of Plant Structures*

I
AKENE OF WILD CARROT

J
TRANSVERSE SECTION
OF AKENE OF WILD CARROT

Of course, not all plant hairs are sticky; in some species they may form only a slight down covering, in others a woolly or feltlike mass. The leaves of the mullein are so woolly and make such a dense network that the thrips find a good winter home among them. With the lens, a single mullein hair (Fig. 64D) is seen to be branched. The mullein is a very common and picturesque plant of rocky pastures, roadsides, and waste places. It has a long stem which may extend as high as 6 or 7 feet, yellow flowers, and velvety-appearing leaves that, at first glance, look like good eating; but if you have the courage to bite into one, you will quickly find out why sheep and other grazing animals leave them alone. Hairs are something of a protective device, though they are not always effective. It used to be said that they retard transpiration or reduce water loss through evaporation, but research has shown that, although dead hairs may accomplish something in this respect, living hairs are useless.

The pearly everlasting, which is now just beginning to bloom, is another densely woolly plant. We find it in old fields and along roadsides. The stems as well as the leaves are profusely covered with hairs, and the leaves particularly have such a dense layer that all the veins except the midrib are hidden from view. Lamb's ears, also known as bunnies' ears or woolly woundwort, of our gardens is also so densely covered with hairs that it appears entirely white. When examined with the lens, the hairs are seen to be smooth and silken, and when you rub them with your finger they feel like fur. Another garden plant of interest is the vervain mallow or European mallow, because the hairs are star-shaped. Each hair consists of a short stalk, from the upper end of which a number of branches radiate at right angles like the points of a star (Fig. 64E).

The leaves of the bush clover, a common plant of pastures, thickets, and open woods, are covered with finely appressed hairs on the lower surface (Fig. 64F). Note the bristles at the tips of the leaflets. In some plants, particularly the thistle, the hairs have become stiff and bristlelike and doubtless are a most effective deterrent to animals. The hairs that occur on the leaves of the buffalo berry and oleaster are curious structures. They are shield-shaped and consist of a single scalelike expansion at the end of a short stalk (Fig. 64G).

Other specialized groups of plant cells are the glands that secrete the aromatic oils used by flowers to attract insects and those that secrete the bitter acrid substances designed to discourage animals from eating the leaves and stems. At one time the aromatic oils were used extensively to prepare perfumes and essences and to flavor various foods, drinks, and confections, but today they have been largely replaced by synthetics. Still, these substitutes can never replace the living flowers and aromatic herbs that add charm to our gardens. Examine the leaves of such herbs as spearmint, peppermint, thyme, majorum, and sage and you will find them dotted with glands (Fig. 64H).

The aromatic oils that give such plants as dill and fennel, caraway and coriander, their characteristic odors and tastes are not produced by glands in the leaves but by oil tubes in the seeds, or more strictly speaking, in the fruits or akenes. All these plants are members of the parsley family, but if you do not have access to any of them, the wild carrot will serve as a substitute. An akene of the wild carrot appears under the lens as a miniature green barrel beset with spines (Fig. 64I), reminding us in a way of the porcupine. It is sculptured with longitudinal ribs, between which are located the oil tubes or vittae, as they are called. If one is cut transversely with a razor blade, the ends of the tubes are quite visible (Fig. 64J). Unlike other members of the family, the oil of the wild carrot is not particularly pleasing to our taste.

It seems that caterpillars are more abundant in August than during any other month, for we now find them everywhere. To many people caterpillars are repulsive creatures, but some of them are quite beautiful, such as the white-marked tussock caterpillar with its four white tussocks, three long pencils of black hair, coral-red head, and two small red swellings. I strongly suggest you look at it through a reading glass or the smaller lens. Equally worth examining is the snow-white black-dotted, black-tufted caterpillar of the hickory tiger moth. It may be found on the leaves of hickory and butternut.

In August the second brood of the mourning cloak appears. The spiny caterpillars are gregarious and have the unusual habit, for a butterfly, of ranging themselves side by side on the leaves of willows, elms, and poplars, with their heads toward the leaf margin as they feed on the surface tissues. When one leaf has been completely skeletonized, they march in procession to another leaf. Each caterpillar spins a silken thread as they go, and all the threads collectively form a sort of carpet on which they walk. This is also the time of the year when the little velvety-black caterpillars of the painted beauty may be observed feeding within their silken nests on the leaves of the everlasting. Yellow bears too are common in almost any garden.

Almost every day I see the beautiful little eight-spotted forester flying about my house. The spotted pelidnota is a frequent visitor to the grape arbor, and at night stag beetles and prionids are attracted to the lighted windows. Treehoppers, too, are abundant on the trees and shrubs in the garden.

Treehoppers, which belong to a family of insects called the Membracidae, are appropriately named, since most of them live on trees and hop vigorously when disturbed. Some live on bushes, however, and some on grasses and other herbaceous plants. We find them in fields and meadows, along the roadside, in the woodlands —almost everywhere. They are popularly known as insect brownies; you have only to look one full in the face if you want to know why (Fig. 65A).

The brownies, you will recall, are believed to be good-natured goblins who do various household chores by night. They have been pictured as whimsical little people with quaint, if not grotesque, faces. Comstock once remarked that "Nature must have been in a joking mood when treehoppers were developed." Doubtless nature had a hand in forming them into the many strange and bizarre shapes for which they are famous, but nature plans nothing idly, and I am not sure she was in a facetious mood when she designed them. On the contrary, I think she had another purpose in mind, for most of them look so much like spines and other plant structures that they are not readily seen and thus escape detection. There is one species in particular—the two-marked treehopper—that occurs on the bittersweet vine and looks so much like a thorn that it is very hard to find.

Opinion differs greatly on just how much insects benefit by protective resemblance. The popular belief is that they benefit considerably, but this view is largely based on the assumption that the senses of animals are about like ours. It fails to take into account the birds' superior ability to detect insects. The truth is that in spite of such "protection," insects do not entirely escape detection. More

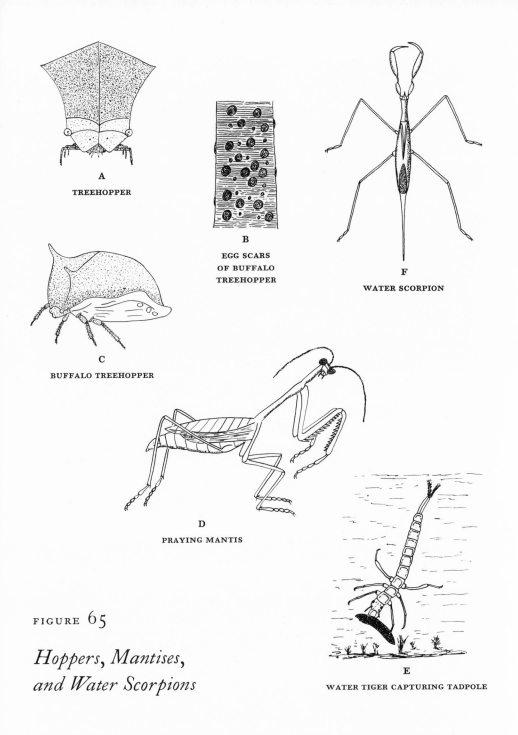

A

TREEHOPPER

B

EGG SCARS
OF BUFFALO
TREEHOPPER

F

WATER SCORPION

C

BUFFALO TREEHOPPER

D

PRAYING MANTIS

E

WATER TIGER CAPTURING TADPOLE

FIGURE 65

*Hoppers, Mantises,
and Water Scorpions*

than twenty different birds prey on the geometer caterpillars that try
to pass themselves off as twigs. However, if protective resemblance is

not 100 per cent effective, it doubtless can be regarded as advantageous, since any adaptation is better than none.

Although the two-marked treehopper commonly lives on the bittersweet, it may also be found on trees, shrubs, and other vines. The species is gregarious, and both adults and the immature forms are usually found clustered together. No matter how the vine twists and turns, you will find that they always rest with their heads toward the top so that the sap, which they suck, can flow more easily down their throats.

Treehoppers do not occur in sufficient numbers to be of any economic importance, except possibly for the buffalo treehopper and one or two other species. The female buffalo treehopper often injures young orchard trees and nursery stock by making scars in the stems in which to lay her eggs (Fig. 65B). This treehopper is easily recognized even with the naked eye, for it is green, somewhat triangular, with a characteristic two-horned enlargement at the front that reminds us of the buffalo (Fig. 65C).[2]

I also find leafhoppers in my garden, but like treehoppers, they occur wherever there is vegetation, being especially abundant on grasses. They are small insects; though visible to the naked eye, their anatomical details can be seen only with the lens. They are longer than they are wide and are slender, often spindle-shaped; though a few are rather plump. The antennae are inserted in front of and between the eyes and the tibiae of the hind legs are curved and armed with a row of spines. Leafhoppers are capable of leaping quickly covering tremendous distances for such small insects. You have only to walk through some fairly tall grass to scatter hundreds of them in all directions. There are more than seven hundred species in the United States. Unlike treehoppers, many of them are destructive pests to grasses and grains.

Among the many insect visitors to my garden—and my house as well, for I often find it climbing up the screening of my porch—is the praying mantis (Fig. 65D), a kind of uninvited guest that I tolerate with amused indifference. To be sure, it does feed on injurious insects (though in this respect it is not as valuable as the

[2] —Or strictly speaking, the bison.

ladybird beetles that specialize in keeping plant lice and scale insects under control); but it feeds on useful ones as well. However, it doesn't eat enough of these to do any harm, and so I let it remain, for it is an interesting insect and a fascinating one to watch, particularly when catching its prey.

That the praying mantis is well adapted for a predatory existence is obvious from just a casual glance at its front legs. They are exceptionally large, and capable of being extended in various directions in a most amazing fashion. Moreover, they are armed with sharp, toothlike spines that clamp a victim with traplike finality.

The mantis is not, of course, the only insect eminently endowed for a predatory existence. We have already mentioned several others; two more that come to mind are the water tiger (Fig. 65E) and the water scorpion (Fig. 65F), both of which may be found in almost any pond. The water tiger, which is the larva of the diving beetle, is most bloodthirsty in its habits. It is an elongate, spindle-shaped insect with a large oval, rounded, or flattened head and large sickle-shaped, hollow jaws admirably fitted for grasping prey and sucking the body juices of its victims.

There are two kinds of water scorpions: one, Nepa is oval, flat, and thin, the other, Ranatra, is linear and cylindrical. They are highly carnivorous, with front legs modified for predation. Both have a long respiratory tube at the end of the abdomen. By projecting the tube above the surface of the water, they can rest on the pond bottom or among rubbish or plants. They are sluggish creatures, often remaining motionless for hours on the muddy, leaf-covered bottom, and when they move, they crawl rather than swim. The lens shows details, such as the structure of the respiratory tube, not visible to the naked eye.

Any unpolluted pond, at this time of the year, is a treasure-house of animal life. Every phylum of the animal kingdom is represented except the echinoderms; even the coelenterates that are essentially marine are represented by several species (hydra and jelly-fishes). Every plant phylum, too, is present, though there are only a few aquatic ferns. Since all the plants and animals of a pond society show some nice adjustments to an aquatic existence and a nicely balanced series of interrelations, a pond furnishes an interesting excursion into the nature world—as indeed does the study of any biotic environment, even a weed patch. We are not so much concerned with these relationships and adjustments, except incidentally, as we are with the plants and animals themselves.

Most of the sponges we use today are man-made synthetics; if we think of a natural sponge at all, the bath sponge comes to mind. Although sponges live principally in the sea, where they occur in many shapes and colors, some forty or more species are found in freshwater. Generally dull-colored, greenish, and mosslike, they are not often seen; even when found, they are usually not recognized as sponges.

A sponge, whether a saltwater sponge or a freshwater form, consists of colonies of animals living together. When seen with the lens, a freshwater sponge has a rough surface peppered with numerous holes or pores (Fig. 66a); hence the name of the group Porifera (pore bearing) to which the sponges belong. The holes, which are of two sizes, are the openings of a network of canals and chambers through which water enters and leaves the sponge. It enters through the small openings (ostioles), passes through the incurrent canals (Fig. 66b) and then into chambers, and from these chambers moves through the excurrent canals into a central cavity called the gastral cavity, where it mixes with water from other canals and finally passes out through the larger openings (oscula). The water is made to flow through the canals by briskly waving flagella in the chambers (flagellated chambers). It is in these chambers that food is removed from the water for the cells (animals) that constitute the sponge, and it is in these chambers that waste materials pass into the water and eventually are carried out by it. Sponges can live only in clean water well supplied with food, and they will quickly smother should the water suddenly become polluted.

All sponges, whether marine or freshwater, have a latticelike framework or skeleton that functions as a support for the soft tissues. In the freshwater species this is composed of small transparent needles of silica. The needles, known as spicules, the largest of which measure about one hundredth of an inch, vary considerably in form and may be straight or curved, smooth or covered with

FIGURE 66

Sponges
and Sponge Structure

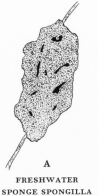

A

FRESHWATER
SPONGE SPONGILLA

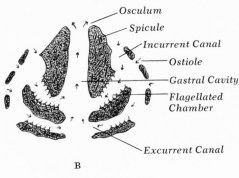

Osculum
Spicule
Incurrent Canal
Ostiole
Gastral Cavity
Flagellated
Chamber

Excurrent Canal

B

DIAGRAM OF SECTION
OF FRESHWATER SPONGE

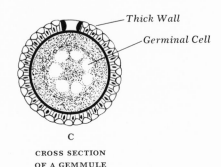

Thick Wall
Germinal Cell

C

CROSS SECTION
OF A GEMMULE

brierlike points, or dumbbell-shaped and sculptured; but they are always hooked or bound together in interlacing chains. Although a sponge may have spicules of several shapes, they are fairly constant for any particular species and thus serve as a means of identification.

Sponges are abundant in lakes, ponds, pools, and clear, slow-flowing streams even though they are not readily noticed. They are not free-swimming and are always sessile on water-soaked logs, the leaves of submerged water plants, and the undersides of stones. Some species branch out in slender, fingerlike processes that suggest plants in form as well as in color; others may cloak submerged twigs in spindle-shaped masses or spread over stones in mats 10 to 15 inches across and an inch or more thick at the center. When they live where they receive a considerable amount of sunlight, they are frequently colored green by small one-celled plants that live within them. Those which are found in the shade are a pale color. One of our more common species occurs on the surface of submerged logs; hence your best chance of finding a sponge is to look for a sunken log.

In the spring, sponge colonies start from small asexual units called gemmules (Fig. 66c). Gemmules survive drying and freezing and carry the sponges over the winter season or periods of drought. Once a colony has made its appearance as a tiny fleck of white on a submerged leaf, stone, or log, it gradually increases in size, reaching its maximum growth in July and August. In early fall it begins to shrivel and by October or November is dead. Meanwhile gemmules or winter buds have developed and, held among the interlocking spicules, are the only parts of the sponge to live through the winter. They are little masses of living cells contained within a tough, hard, and highly resistant shell, which is provided with a pore through which the living cells can emerge in the spring. The gemmules may be found in autumn and throughout the winter, and we can see their general structure with the hand lens, although we need a microscope to examine the spicules.

A freshwater sponge may play host to the parasitic larva of the Spongilla fly. The little parasite either enters the sponge through the ostioles or penetrates the soft sponge cells with its sharp mouthparts. Each of its slender mandibles is grooved on the inner side, so that when they are held closely together, they form a hollow tube through which the insect can suck the watery contents of the sponge cells. Because the larva is colored much like the sponge it is difficult to find. We have to search diligently with the hand lens, and even then we are not apt to discover it unless it moves.

Sponges often live in association with bryozoans. These animals occur in plantlike colonies of hundreds or thousands of individuals on lily pads, submerged twigs, the undersides of flat stones, which they cover with delicate traceries, and logs and boards, where they form a vinelike growth. As the colonies resemble moss, the bryozoans are popularly known as moss animals. In some species the individual animals, called zooids, live in brown tubes of lime; in others in a soft, transparent matrix of jelly.

The tiny animals may be seen with the lens or a microscope, but they are sensitive to any kind of disturbance and retreat within their tubes or jelly when molested. It is best to transfer a colony, or at

FIGURE 67

Zooids and Bryozoans

B

STATOBLAST
OF PLUMATELLA

C

PALUDICELLA

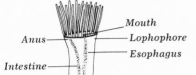

Mouth

Anus

Lophophore

Esophagus

Intestine

Statoblast

Stomach

A

INDIVIDUAL ZOOID

D

PLUMATELLA

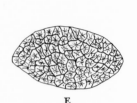

E

PECTINATELLA

least part of a colony, to a dish of water. Then, if left undisturbed for a few moments, the animals will slowly emerge and unfold a beautiful crown of tentacles. It's the next thing to watching a flower open with time-lapse photography.

The circle of tentacles surround the mouth (Fig. 67A), and each tentacle bears cilia that direct currents of water to it and such food particles as the water may carry. The so-called "head" region is called a lophophore, and when undisturbed the animals constantly rotate the lophophores or swing them back and forth; but at the slightest disturbance they all disappear.

The food canal of the bryozoans is a U-shaped tube that leads downward from the mouth and then turns sharply upward, with the anus just inside or just outside the circle of tentacles. The very flexible body is furnished with muscles that permit the animals to withdraw swiftly into their limy tubes or protecting caps of jelly. Once you have seen how quickly they vanish from view, you will realize how effectively they can protect themselves from their enemies.

In autumn the bryozoan colonies disappear, but the animals are carried through the winter as winter buds or statoblasts (Fig. 67B) that are similar to the gemmules of sponges. Resistant to drought and cold, they consist of groups of cells that form within the body of an individual bryozoan and are set free after the animal has died. Some statoblasts are enclosed in tough cushions that buoy them up like life preservers; others are provided with hooks that anchor them to various objects. Many are washed up on the shores of ponds and lakes, where they may be found in long dark ribbons on the sand; others float in brown films on the water surface. Those that have hooks are often carried about by animals, especially water birds, and are widely distributed.

I have illustrated a few of the more common bryozoans in Figures 67c, d, e. Colonies of Paludicella (Fig. 67c) are composed of delicate jointed, branching, club-shaped, recumbent or partly erect tubes placed end to end and separated from one another by partitions. They commonly occur on sticks and stones. Plumatella (Fig. 67d) colonies, which are also composed of tubes, form either a vinelike growth or massive clumps in a variety of places. Sometimes they are so abundant on sluice pipes and weirs of reservoirs that they can be removed with a shovel. Fredericella, which grows on the undersides of sticks and stones, usually in dark places, forms branching colonies that are partly creeping and partly erect, with antlerlike branches. Pectinatella (Fig. 67e) consists of many associated colonies in rosette groups clustered on the surface of large masses of jelly that cling to sticks and stones or hang from twigs and water plants. Colonies of Cristatella, which are elongate or oval, ruglike, gelatinous masses, measure 1 to 2 inches long. They occur most frequently on the undersides of leaves, especially lily pads, and are remarkable because they are capable of a slow, creeping movement, the entire colony acting as a unit. Looking at a colony of bryozoans through the lens as they wave their tentacles about is like looking at a miniature garden of delicate flowers responding to a fleeting zephyr. A more beautiful sight is difficult to find with the lens.

Because leeches suck blood, they are generally regarded as repulsive creatures. They have, however, certain features to commend them. Their habits, for one thing, represent a high degree of efficiency for their particular mode of living. Also, many are beautifully colored, with soft green, brown, and yellow tints that are actually concealing colors, because they serve to make the animals inconspicuous among the shadows and water-soaked leaves of their environment. Despite such concealing coloration, they often fall victim to birds, reptiles, flatworms, and fishes. Perhaps their sensitivity to any change in their environment and their curious nature are partly to blame for the numbers that are devoured. For even a shadow passing over the water excites them, and when they go forth to investigate, they become easy victims for an enemy predator.

This extreme sensitivity enables us to collect them with relative ease; all we have to do is wade out in the water wearing a pair of rubber boots, and within a matter of minutes one or more leeches should be attached to them. Leeches are fairly abundant in most ponds and sluggish streams, particularly the latter, where they live on stones, logs, and water plants or on the bottom mud. Sometimes they may be found attached to frogs, turtles, fishes, and wading animals. Leeches are segmented worms, hence relatives of the earthworms; but unlike the annelids, they lack setae except in one genus. The body is pressed toward the ends, terminating abruptly toward the posterior and tapering toward the anterior, which is necklike and extremely supple. A distinct head is absent.

The presence of a muscular sucker at each end is doubtless the leech's outstanding characteristic. These suckers function in the same way that a plumber's helper does, or the small suction cups used to display merchandise on a store window. The mouth is situated in the anterior sucker, and through the lens you can see that it is provided with three jaws armed with chitinous teeth for biting (Fig. 68a). When feeding, the leech first attaches its posterior sucker to an animal, then swings its anterior end back and forth until it locates a spot where the skin is broken or is well supplied with blood vessels. On finding such a spot, it attaches its anterior sucker, and with its triradiate jaws makes a wound in the flesh from which it sucks the blood by the dilatation of its muscular pharynx.

At the same time that the jaws open a wound, a small amount of the animal's saliva seeps into it. The saliva contains an anti-coagulant, hirudin—the source of the intense itching caused by a

A

ANTERIOR SUCKER
AND TRIRADIATE JAWS
OF LEECH

FIGURE 68

*Leeches, Bloodsuckers,
and Flatworms*

C

COMMON BLOODSUCKER

B

TURTLE LEECH—
VENTRAL SURFACE SHOWING SUCKERS

leech bite. If a leech is allowed to obtain its full quota of blood, however, no discomfort should be experienced, because by that time the hirudin should be completely removed with the blood.

The blood that the leech sucks from an animal is not immediately digested but is stored in the lateral pouches of an enormously large crop. The stored matter is the solid content of the blood; the fluid part was drawn off through the kidneys as the leech fed. Since a leech can ingest three times its own weight of blood, and since digestion is an unusually slow process, it may take as much as nine months for a meal to be completely digested; hence a leech doesn't have to eat very often.

Not all leeches live on blood; many of them feed on worms and insect larvae; a few are scavengers. Even the bloodsuckers are curiously varied in their eating habits, for a leech may feed on snail meat one time and suck turtle or frog blood the next. As a group, leeches exhibit both parasitic and predatory habits, and it may be that as a group they are wavering on the brink of parasitism.

There are some 250 species, but those we are most apt to see are the turtle leech (Fig. 68B), a broad, fat species with a smooth surface and greenish color; the brook leech (Fig. 13A), flat and mottled, greenish spotted with yellow; the horse leech, smooth and soft, generally some shade of olive-green blotched with irregular spots of light grays and dark browns and black; the worm leech, a small species varying in color from a light chocolate brown to jet black; and the

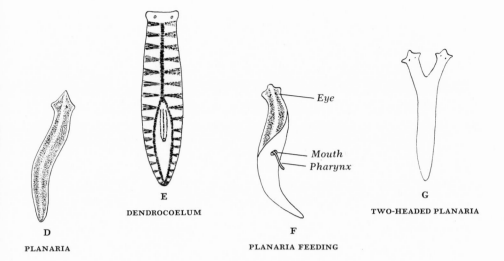

D
PLANARIA

E
DENDROCOELUM

F
PLANARIA FEEDING

Eye

Mouth
Pharynx

G
TWO-HEADED PLANARIA

common bloodsucker (Fig. 68c), dull or olive-green, with a row of some twenty red spots through the middle of the back and a row of black spots near each side.

If it were not for high school biology courses, in which flatworms are mentioned briefly, most people would not know that such animals exist. Of course, they may have heard of the tapeworm, but even then the name does not mean much to them. Most flatworms are parasitic but certain free-living species abound in our freshwaters. They belong to a group called the Turbellaria, in allusion to the turbulent lashings of the thousands of cilia that cover the lower surface of their bodies and that help in locomotion. Many turbellarians are so minute that they are not often discovered. Even the larger species, the planarians, are not often seen unless one is acquainted with them and knows where to look.

These worms are not called flatworms without reason. They are distinctly flattened, and all are rather uniform in shape, though some have earlike flaps called auricles. According to the species, they may be brown, gray, black, brick red, creamy white, or mottled. Our most common species is *Planaria maculata* (Fig. 68D). It measures about a quarter of an inch long and tapers into a pointed tail. At a casual glance it appears to be entirely black, but examine it closely and you will see its back irregularly spotted with black with smaller black spots scattered among the large ones, so that the entire surface looks dark. In appearance it can hardly be compared with Dendrocoleum (Fig. 68E), which is actually a beautiful animal, its creamy white color set off by contrast with the brown tracery of the food canals.

Planarians live on stems, leaves, algae-covered stones, and other submerged objects, and we can usually count on their being members of almost any aquatic society. They move with a characteristic gliding motion effected by almost imperceptible contractions of three sets of muscles: longitudinal, circular, and dorsoventral. These muscles make possible all sorts of agile bending and twisting movements that enable the animals to escape from almost any unfavorable environmental condition. They cannot swim and can move only when in contact with a solid object or the underside of the water's surface film.

The mouth, visible with a lens, is located near the middle of the ventral surface and opens into a cavity (pharynx sheath) that contains a muscular tube called the pharynx. When a planarian feeds, it extends the pharynx (Fig. 68F) some distance out of the mouth, uses it for exploratory purposes, and finally sucks food through it. The food, which consists of small animals, living or dead, passes into a large branched intestine sometimes visible as a vinelike tracery on the back. Undigested food materials are ejected through the mouth, but the waste products of metabolism are eliminated by a complex network of small tubes. If you want to observe a planarian feed, place it on its back in a small amount of water in a dish, and hold a small piece of snail meat or beef liver within its reach. It has to be hungry to perform for you, however. Needless to say, you can observe it much better with a reading glass than with the naked eye.

Planarians are extremely sensitive to various stimuli. Touch or prod one and it shows a negatively thigmotropic response—that is, moves quickly away from the direction of the stimulus. Or place a drop of ammonia or acetic acid in a dish of water containing several of these animals and see what happens. They dislike bright light and appear to be able to distinguish between small differences in light intensity. The two eyespots on the dorsal surface near the anterior end are specialized sense organs for light reception. Each one consists of black pigment supplied with light-sensitive nerve cells that connect with the brain. The pigment shades the nerve cells from light in all directions except from the source. You can observe the reaction of a planarian to light by shining a flashlight on it. Planarians also respond to water currents, and in flowing streams orient themselves upstream. Certain species have remarkable powers of regeneration: if an individual is cut in two transversely, the anterior end will regenerate a new tail and the posterior end will develop a new head. Pieces of one animal may also be grafted onto another animal, and by successive graftings many odd-shaped planarians can be created (Fig. 68G). A five-headed planarian is not unusual.

Those of you who peer into every pool of water you meet in your rambles have doubtless seen little animated specks of red, orange, green, yellow, brown, or blue. The dictionary defines a mite as anything small, and few words fit these animated specks any better. All we need do is add the word "water," and we have the name of these tiny animals (Fig. 69A).

Water mites are rotund or oval in form and not much larger than the head of a pin, though the bright red Hydrachna is about the size of a small pea—a real giant. Needless to say, we can see little of them without a magnifying glass. The skin of these little animals is soft and easily broken. The upper surface may be entirely smooth, may have a few scattered fine bristles, or may be densely clothed with short hairs. The eyes—two or four in number —are generally on the upper surface near the front border. Somehow, mites can remain under water for long periods, though they have no means of removing the dissolved oxygen and must breathe atmospheric air. On a mite's ventral surface one or more very small dark spots indicate the external openings of the tracheae, that ramify throughout the body, as in insects.

Water mites are active and restless and always seem on the move, either running about on the surface film or submerging and swimming in the water. Their eight legs, fringed with hairs, are effective propelling organs. The strokes are made in rapid succession and have the effect of moving the animals smoothly forward in a sort of gliding motion. They often descend to the bottom, where they creep about on the mud and sand and over the submerged plants. All are carnivorous and suck the body juices from whatever animals they can capture. The mouth is a complicated organ with short jointed palpi. Eylais, which has a red, oval body, is one of our more common species. Sometimes water striders may be seen carrying around a half dozen or more red specks, the immature young of the mite Limnochares, attached to their bodies.

We think of spiders as essentially terrestrial animals, and yet there are several species more at home in a pond than on land. They run over the surface of the water with all the grace of the water striders and even dive with facility when necessary. Indeed, one of them is so adroit at it that it is called the diving dolomedes. Doubtless the tarsal hairs give spiders a certain amount of buoyancy and help them to skate, but a great deal of their ability to run over the water is due to their lightness.

The diving dolomedes is a wandering spider and as a hunter is probably on a par with the wolf spiders. The various species of dolomedes are commonly referred to as the fisher spiders. It is not unusual to see one capturing a small fish or tadpole, though their usual food is insects. When diving, all the dolomedes carry down with them a bubble of air that permits them to remain submerged for fairly long periods. But like backswimmers, they must cling to some underwater support, for the moment they release their hold, their light weight sends them bobbing up to the surface like a cork.

I suppose there is some psychological reason why women dislike spiders and are afraid of them. Perhaps it is because they are poisonous and thus are believed dangerous. True, they are poisonous—that is how they kill their prey—but with the exception of the tarantulas of the Southwest and the black widow, none of our spiders can inject enough venom into a wound to cause any trouble, unless the person is allergic to its venom. They are considerably less dangerous to handle than many of our common insects; we have more reason to fear the sting of a bee or wasp or the bite of a mosquito—which, incidentally, are more disposed to attack us.

Presumably it is for the same reason that spiders are regarded as ugly and repulsive. Personally I find many of them beautiful, and some exceedingly quaint. Spiders are of value because they destroy many harmful insects, though they are not as important in this

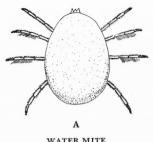

A

WATER MITE

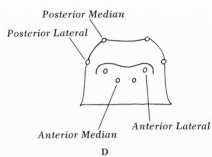

Posterior Median
Posterior Lateral
Anterior Median
Anterior Lateral

D

EYES OF A SPIDER

B

WEB OF ORANGE GARDEN SPIDER

E

EYES OF DOMESTIC SPIDER

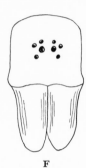

F

EYES OF GRASS SPIDER

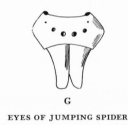

G

EYES OF JUMPING SPIDER

CHELICERAE OF SPIDER

H

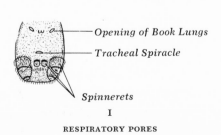

Opening of Book Lungs
Tracheal Spiracle
Spinnerets

I

RESPIRATORY PORES

C

JUMPING SPIDER

FIGURE 69

*Webs, Spiders,
and Spider Structures*

respect as some of the parasitic and predatory insects or the insectivorous birds and mammals, because they do not destroy
enough of them.

Spiders, moreover, are infinitely interesting when you know
them. For one thing, they are skilled civil engineers. Watch a spider
at work and you will marvel at the precision with which it con-
structs its web and the uncanny skill with which it measures angles
and calculates stresses and strains. Many spiders spin several kinds
of silk. One, which they spin as they move about (dragline), they
use to lower themselves from an elevated spot and also to form the
permanent frame of their webs. Another is a viscid sort of silk with
which they entrap their victims and then there is a third kind in
which they swathe them. Finally, they have a fourth kind—a thick,
often brightly colored silk that they use for wrapping their eggs. All
spiders, however, do not spin these different types of silk.

Originally, spiders used the silk only as a covering for their
eggs. Then they found a purpose for it as a lining for their retreats,
and after awhile they learned to build platforms from it outside of
these retreats. Snares were only a step away. At first they were quite
primitive, but as time went on, the intricate, beautiful webs that
now excite our attention and admiration were gradually evolved.

Snares differ greatly in construction. Those of the domestic or
house spider are merely a maze of threads extending in all directions without any method—the well-known cobwebs. Others consist of a more or less closely woven sheet on a single plane, with
threads extending in all directions without regard to any kind of
pattern. Such a snare is built by the hammock spider. Still others,
such as the funnel-shaped web of the grass spider and the orb web
of the garden spider, are built with a definite design.

August seems to be the heyday of spider life. To be sure, there
are probably as many spiders in spring as in summer, but those of
spring are mainly immature forms that are not easily seen, nor are
their webs. It is not until midsummer that both the spiders and their
webs have become full-sized and conspicuous; in fact, they seem to
be everywhere. I have only to go out into my garden to find the
garden spider (Fig. 69B) and its web, and directly outside my
window I can see the web of the shamrock spider stretched between

the branches of a honeysuckle shrub. On almost any morning I can look out and see my lawn covered with innumerable webs of the grass spider. Few of us realize what an immense number of webs are spun by this species until the dew condenses on them and makes them visible:

> And dew-bright webs festoon the grass
> In roadside fields at morning.

At times our lawns and the fields appear to be covered with a continuous carpet of silk.

Should you be curious about these webs and examine one, you will find it shaped somewhat like a broad funnel with a tube extending downward at one side. The tube is used by the spider as a hideout in which it lies in wait for its prey, or as a place of refuge in time of danger. It is open at both ends so that the spider can escape by the back door, if necessary.

Many lines of silk, which cross one another irregularly, compose the grass spider's web. Collectively they form a firm sheet held in place by many guy lines attached to grass stems. Above the sheet is usually an irregular open network of silken threads to catch flying insects or to impede their progress so that they will tumble onto the sheet. Touch the web lightly and you will see the waiting spider spring out from its hiding place to seize an unfortunate victim, but jar the web roughly and the spider will become frightened and speed out the rear door. The uses to which the grass spider puts its web show that spiders employ their webs not only as snares, which is doubtless their primary function, but also as shelters, and in some instances as supports for the egg sacs and nests for the young.

Unlike other web-spinning spiders, the orb weavers build their nests vertically, since they are thus more apt to be in the path of flying insects. The radii are usually fastened to a silken framework, which is supported by guy lines to surrounding objects such as plant stems. These radii are connected by a continuous spiral thread, spaced regularly except at the hub, which is of more solid silk and is usually surrounded by an open space.

The radii, the guy lines, the framework, and the center of the web are all made of dry, inelastic silk. The spiral line, however, is very viscid and elastic and adheres to any object with which it comes in contact. This is the trapping part of the web. Any unfortunate insect that touches one of the spirals and tries to escape becomes inexorably entangled in the neighboring lines and is held fast until the spider can reach it. When examined with the microscope, these threads of the spiral line are found to consist of two strands bearing a series of globules and looking much like a string of pearls. The strands are the elastic part, the globules the sticky portion.

When they have completed their webs, some orb weavers rest on the hub and there await their prey. Others have a retreat or den above or to one side of the orb. These spiders spin what is called a trapline from the hub to the den; it serves as a means of passage to and from the web and vibrates as a signal when an insect has been trapped.

The moment an insect is caught in the web, the spider rushes toward it, wraps it in a band of silk called the swathing band, and rolls it over and over. If hungry, the spider feasts at once; otherwise it leaves the trapped insect for a future meal. You can observe the swathing process by tossing a grasshopper into the web. As the spider rushes toward its victim, you will observe that it travels on the dry, inelastic radii and not on the viscid spiral, in which it would become entrapped. As strong as they are, the webs are frequently damaged; a spider may make a new web every twenty-four hours.

Not all spiders spin snares. Crab spiders lie in wait for their prey; wolf spiders chase their victims over the ground. Jumping spiders (Fig. 69c) are also hunters. They are small or medium-sized, rather stout, with conspicuous eyes and a body usually thickly covered with hairs or scales. They have thick front legs, which in the males are covered with peculiar bunches of hairs that serve as ornaments. Many jumping spiders are brightly colored, even iridescent.

These spiders may be seen jumping about on plants, logs, and fences. They can move sideways or backward, and can leap surprisingly long distances. All of them, of course, spin a dragline,[3] which

---

[3] All spiders of all ages spin draglines except a small group of primitive forms of the family Lipistiidae.

they fasten to the jumping-off place and by which they can regain it if they wish. Jumping spiders are very ardent in their lovemaking. The males put on a display that compares favorably with the courtship of many of the more advanced animals. At mating time the males dance before the females and strike the most singular postures, holding their legs extended forward, sideways, or above the head to show their ornaments, or moving them about to attract attention.

Although spiders normally have eight eyes, some have less, and so there are eight-eyed, six-eyed, four-eyed, and two-eyed spiders. The eyes vary in size and in their position on the head. Normally they occur in two transverse rows of four each. The two intermediate eyes of the first row are called the anterior median; the two intermediate eyes of the second row the post median; the eye at each end of the first row anterior lateral; and the eye at each end of the second row posterior lateral (Fig. 69D). Spiders that live in the dark or frequent shady places have what are called "nocturnal eyes"; these are pearly white in color and are presumed to function in faint light. The more conventional or "diurnal" eyes typical of most spiders lack this pearly luster and are variously colored.

If you have a decided aversion to spiders, you may not care to look at their eyes through your lens. But if you can overcome your dislike for them and do so, you will find that in the domestic or house spider they are arranged in the normal two rows, the anterior row being straight or nearly straight. You will also observe that the two pairs of intermediate eyes are the same size and are separated from one another, while the lateral eyes are contiguous—in other words, adjoining or in contact with one another (Fig. 69E). In the orange garden spider, the eyes are all alike, although the second row is strongly curved. In the grass spider, the two rows of eyes are so strongly curved backward that the anterior median and the posterior lateral eyes form a nearly straight line (Fig. 69F). The eyes of the crab spider are all small, dark-colored, and arranged in two curved rows.

Three rows of eyes are the rule in wolf and jumping spiders (Fig. 69G). In wolf spiders the first row consists of four small eyes and the second and third rows of two large eyes each, the eyes in the third row being situated far behind those of the second row. The first row of eyes in jumping spiders is somewhat curved and consists of four eyes: the anterior median, which are very large, and the two anterior lateral, which are smaller. The two eyes in the second row are so small that you have to look closely to find them; the eyes in the third row are much larger.

The eyes of spiders are but feebly developed and resemble the ocelli of insects. Except for hunting spiders, which have large eyes and relatively keen vision, most spiders are short-sighted. They depend almost entirely on their sense of touch, since their receptors for chemical stimuli, such as smell and taste, are poorly developed if not actually absent. But the receptors for touch are numerous and varied, and it is through these that spiders get to know their environment.

All spiders are predaceous and subsist on the body juices of living animals; they can rarely be made to accept dead food. The weapons of a spider are the anterior pair of pincerlike appendages called the chelicerae or jaws. Examined with the lens, they are seen to consist of two segments: a basal one, which is stout and usually margined by a toothed groove at the distal end; and a short, movable fang, which lies in the groove when at rest and is the part that is thrust into the prey (Fig. 69H). The venom, which is secreted by the poison glands located in the cephalothorax, is ejected through a tiny opening in the fang.

Spiders breathe by means of tracheae and certain structures called book lungs that are peculiar to arachnids. The book lungs are merely sacs of air that contain a series of leaflike folds of the body wall which are bound together like the leaves of a book. Their external openings may be seen as a pair of tiny slits on the lower surface of the abdomen (Fig. 69I).

To most people there isn't much difference between a lizard and a salamander, and the two are frequently confused. But they are dissimilar animals: lizards, like snakes, have a covering of scales; salamanders have a smooth and slimy skin. There are other differences too, of course. Although there is really no reason for mistaking a lizard for a salamander or a salamander for a lizard, sometimes one may be excused for mistaking a lizard for a snake. In fact, certain limbless lizards look very much like snakes.[4] The

---

[4] It is believed that snakes descended from lizards.

scales of lizards vary greatly in form, as the lens will show. Like snakes, lizards also shed their skins, but they do so in sections while snakes molt in one entire piece. There are anatomical differences between the lizards and snakes which we need not go into, as well as differences in habits. Lizards usually eat their food with some slight attempt at chewing; snakes swallow it whole. Lizards drink by lapping water, snakes by sucking it into the mouth.

As a group, lizards are probably less known to the average person than most other animal groups. This is probably owing, in part, to the misconception that lizards are venomous; actually there are only two poisonous species in the entire world. These are the Gila monsters, one of which is found in Mexico, the other in the southwestern part of the United States.

We have 127 native lizard species. Some of them are widely distributed, others occur only locally. Lizards are not common in New England; as far as I know, only one species, the common five-lined skink, is found in Connecticut and Massachusetts. This species, however, occurs throughout the Eastern and Central states south to Florida and the Gulf of Mexico. Also widely distributed over the same area are the six-lined racerunner, glass-snake lizard, greater five-lined skink, brown skink, and northern fence lizard. Doubtless the best known of all our American lizards is the Carolina anole, commonly called the chameleon, though it is not a true chameleon. It occurs throughout the Southern states.

Lizards are most interesting animals; why they have so little appeal is something I have never been able to discover. They live in practically all habitats except the strictly aquatic, the aerial, and the arctic, but even in these places they are found peripherally. Essentially diurnal in habits, they remain hidden in holes, beneath objects on the ground, and in similar places during late evening, night, and early morning, coming out only after the sun is fairly high. They even stay hidden on cool and cloudy days. In desert regions where the early part of the night is warm, some species are nocturnal to a degree. A few species are even night wanderers, but these do not remain active all night, becoming sluggish three or four hours after sunset. Since lizards cannot withstand low temperatures, they hibernate during the colder parts of the year. Only in the extreme southern part of Florida can we expect to find them during the winter.

Most lizards are insectivorous; a few are completely herbivorous, and a few others are omnivorous. When attacked or when threatened with danger, most lizards retreat into a hole or some other safe refuge; the arboreal species climb into a tree. In common with other animals, they fight when cornered; when all means of escape are cut off, many open the mouth and hiss, others bite and scratch. A few species "freeze" and, protectively colored, are then difficult to see. Some species even "play possum," becoming apparently lifeless. Horned toads have the curious habit of ejecting blood from their eyes. They also inflate their bodies to make swallowing by small snakes more difficult; they are not, however, the only lizards that can do this. Gila monsters protect themselves by the poison they secrete.

Red bats mate in August, and in a newly mown meadow, jumping mice search for new homes. And as night falls, raccoons steal from the woods to feed on the juicy kernels in the corn patch. Meanwhile flying squirrels engage in their nightly revels.

During the dry, hot days of August the red efts remain hidden beneath logs and stones or in the moist humus of the woodland floor, but on a foggy or rainy day they emerge and wander about in search of food—insects or other small animals. After a rainy night I usually find them crawling along the road or woodland path, occasionally by the hundreds. They are very pretty animals, especially when seen through a magnifying glass. They are very timid, however and will often peer out from among the leaves with an expression of alert shyness. Sometimes they remain motionless for so long that they seem carved out of stone; then without warning they dart away with lightning speed.

During my August rambles I sometimes come upon a garter snake and her newly born young. This is the snake which is supposed to take the young into her mouth and throat for protection against threatened danger, but I have never seen her do it. The brown snake also gives birth to young during the month, and the eggs of the black snake hatch. The green snakes, however, are just beginning to lay their eggs under flat stones and logs or in sawdust piles. A beautiful green, these snakes blend so well with their surroundings that they are usually found only by chance. And

a more docile and gentle animal you will never find. I have yet to
see one bite, and even when newly captured it will submit to the
most vigorous handling without showing the slightest sign of
fear.

The floral configuration of the August landscape is an odd mixture of the old and the new. Many of the July flowers are still in
bloom; some spring flowers even continue to blossom, and will do
so until autumn. Then there are others which are new on the scene.
We cannot categorically say that any one species appears in July or
August, for in some localities it may bloom in one month, in other
localities in the other; indeed, it may even have appeared in June.
At some time in mid or late summer, we see the false foxgloves
with their yellow flowers. These plants are parasitic on the roots of
the white oak and the witch hazel. The cancer root, an odd little
plant with a stem of scales instead of leaves, is also a parasite,
usually on the roots of the beech. A better-known pirate, the common dodder, winds its way among the shrubbery of a moist thicket,
its bright threads showing like tangled yellow yarn. With the lens
look at the numerous sharp suckers (haustoria); they penetrate the
stems of other plants and hold on so tenaciously that it is difficult to
dislodge them.

Follow almost any woodland trail and you should find the tick
trefoils, with their magenta butterfly-shaped blossoms. And if you
are not averse to walking where the ground is a little damp, you
may see the evergreen leaves of the goldthread, its small, solitary
white flower perched at the tip of a slender scape some 6 inches
high. Dig it up if you want to know why it is called the goldthread.

The wild peanut is not designed to attract one's attention, for it
is much like any other vine trailing its way over the shrubbery of
the roadside. Unless you stopped and looked at it closely, you
would not know that it has two different kinds of flowers which
develop into two different kinds of fruit. The small lilac blossoms,
in drooping clusters, are succeeded by many small pods about an
inch long which contain three mottled beans; the solitary fertile,
petal-less flowers, borne on a threadlike creeping branch at the base
of the plant or from the root, develop into a pear-shaped pod with
one large seed—the peanut.

In shady places, in the woodlands, in thickets, on river banks— in almost any kind of habitat one can think of except the frozen north—we find various skullcaps. We need only to glance at the flowers to know they belong to the mint family and are relatives of the horehound, catnip, bugleweed, and pennyroyal, all also now in bloom. Among the sedges and cattails of the marsh, the blossoms of the swamp rose mallow flutter like banners in the breeze. The pale-pink, or sometimes white, flowers are as much as 4 to 7 inches across. There are many mallows; the round-leaved mallow or

FIGURE 70

Seed Vessel of Mallow

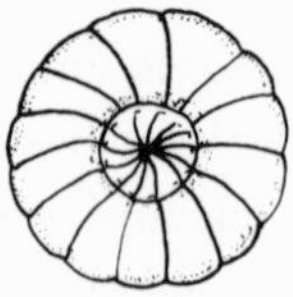

cheeses, sometimes also called cheeseflower, always appears in my garden, for it is an exceedingly common weed and quickly takes possession of any piece of ground, however small. I think the flowers will remind you of miniature hollyhocks when you look at them with your lens. The name cheeses refers to the round cheeselike form of the seed vessels (Fig. 70).

Whoever named the purple-flowering raspberry probably did not know that no member of the rose family can put out a true purple flower. Seeing it for the first time you might take it for a wild rose, but one look at its fruit and you will know that it is not a rose. You will find it in stony woodlands and along the shaded roadside, often in company with the bellflower, long since an escape from the garden. The bellflower is a plant of a rather rank and rigid habit, with hairy leaves and a one-sided raceme of purple bell-shaped flowers. It is altogether lacking the grace of its more exquisite cousin the harebell, whose dainty blossoms on tremulous threadlike stems are the playthings of every breeze. But like the dandelion, the harebell stems, though frail-appearing, can withstand the strongest air currents. The flowers of both should be examined closely, for bellflowers have welded their once-separate petals together to ensure fertilization by the insects, which must enter the blossoms only where the stigmas come in contact with their pollen-laden bodies. The odd-shaped flowers of the turtlehead are also designed for insect pollination, specifically for the bumblebee. If you study the flowers carefully, you will see why.

Despite its reputation, August is not altogether a month of heat and sultriness for we may have a night of showers, and the next morning will be cool and clear. The day may warm up as the sun climbs higher in the sky, but the next few hours should be pleasant for a walk abroad. Perhaps we shall take the woodland trail or simply stroll the country road; perhaps we will pause by the brookside and by the meadow to watch the redwings gather in flocks or the goldfinches break up the silvery cushions of the thistles for the down to line their nests. But wherever we wander, we see the boneset and the joe-pye weed, the vervains, the tansy and burdock, the butter-and-eggs—and suddenly we realize that the sunflowers are lifting their yellow heads and that the goldenrods are waving in the breeze. Autumn is near at hand.

There are other signs that fall is approaching. We need only look at the cranberries and blackberries to see how they are beginning to deepen in color. The curved stems of the Solomon's seal are heavy with fruit. In the thicket the chokeberries are beginning to ripen, and in the shady tangles of the roadside the berries of the nightshade gleam like little red lanterns. The first transient birds are putting in an appearance: the olive-sided and yellow-billed flycatchers and the bay-breasted, Cape May, and magnolia warblers. We see, too, the Northern water thrushes as they pause by slow-moving streams on their southward flight, while the tree swallows, gathering in flocks, head for the seashore. There, drab-colored sparrows search for seeds, and nimble sandpipers scurry over the sands, chasing retreating ripples and skipping back out of reach of each advancing wave. And above the open sea, cormorants pass, and herring gulls and terns, having left their breeding grounds, wheel, dart, and tip from side to side with outstretched wings.

August is definitely on the wane, and September is fast approaching.

SEPTEMBER

"When Summer gathers up her robes of glory."

SARA HELEN WHITMAN

"A Still Day in Autumn"

SEPTEMBER

T HE FIRST PART OF SEPTEMBER IS MUCH like August. The days are hot and on the sultry side, and the nights, though a little cooler, are still too warm for comfort. But after a week or so, the days become less sultry and the nights perceptibly cooler:

The sultry summer past, September comes;
Soft twilight of the slow declining year.

Summer, of course, is quickly drawing to a close. The insects will gradually disappear with the passing days, though they are still abundant. Crickets and grasshoppers abound in the fields and meadows, and together with katydids and tree crickets, continue their chirping both day and night. It has been said that short-horned grasshoppers shuffle, rustle, or crackle; crickets shrill and creak; the long-horned grasshoppers scratch and scrape. If you have an ear for music, you can soon learn to differentiate between the various insect "songs" and to identify the singers much as an ornithologist can identify birds by their songs. Perhaps you have noticed that the chirping of the snowy tree cricket is somewhat consonant with the weather. On warm days the chirps are rapid and high-pitched; on cool days they are slower in tempo and somewhat like a rattle.[1]

Second broods of butterflies and moths, which have not been seen since early summer, appear about this time, and certain late

---

[1] There is even a formula to determine the temperature by the number of chirps per minute: $T = 50 + \dfrac{(N - 92)}{4.7}$. T equals temperature Fahrenheit, and N equals the number of chirps per minute. For the katydid the formula is $T = 60 + \dfrac{(N - 19)}{?}$.

3

larvae feed on the foliage of various plants. The caterpillars of the silver-spotted skipper may be found on the locust, where they tie the leaflets together with silk and live within the shelters. The silver-spotted is one of the very few skippers that winters in the pupal stage; most winter as larvae.

Caterpillars of the fall webworm are quite common on various trees, especially apple and ash, and their webs are a conspicuous feature of the September landscape.[2] The webs are often mistaken for those of the tent caterpillar but are much lighter in texture and cover all the leaves on which the larvae feed. Both the caterpillars and adults vary considerably in markings. Some years I have found hundreds of the red-humped apple worm (Fig. 71A) on the blackberry, though they are more commonly seen on the apple. These caterpillars have a coral head and a hump of the same color on the first abdominal segment. They are more or less gregarious, especially when resting; they crowd so closely together that they completely cover the branch.

There are some one hundred species of dagger moths in North America. The name refers to a daggerlike mark near the hind outer angle of the front wings, though this does not occur on all species. The caterpillar of the American dagger is densely clothed with yellow hairs. I have often seen it crawling along a city sidewalk in search of a place to pupate. And who has not seen woolly bears (Fig. 71B) hurrying over the ground at this time of year, in haste to find a cosy retreat for the winter.

Sometime when you are in the woods, break open a decaying stump or log or a rotten fence post, and you may bring to light a community of termites, or white ants, as they are often called (though they are not ants nor, in my opinion, do they look like ants; furthermore, the winged forms are dark). Much has been written about the social insects—bees, wasps, and ants—and their caste system. By definition a caste is essentially a division or class of society; but in entomology it is one of the polymorphic forms of the social insects, each form or caste having its particular share in the duties and work of maintaining the colony or community (division of labor).

[2] In the South they may be seen during the summer.

FIGURE 71

Apple Worms, Termites, and Beetles

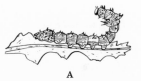

A
RED-HUMPED APPLE WORM

B
WOOLLY BEAR

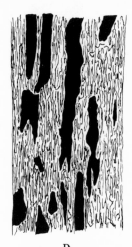

C
SOLDIER TERMITE

D
GALLERIES OF TERMITES

E
GALLERIES OF ASH-BARK BEETLE

Somewhat surprisingly, since they are not as high on the scale of insect evolution as bees, wasps, and ants, the termites have a more highly developed caste system. In most species there are four castes, each of which, unlike the castes of the other social insects, includes both male and female individuals. The four castes consist of workers; fertile males and females with fully developed wings; fertile males and females with undeveloped wings; and soldiers (Fig. 71c). The soldiers have undeveloped sexual organs and monstrous mandibles and heads. Their chief duty is to protect the colony, although they sometimes fail to do so. If you have never seen these insects or have never looked at them closely, examine them with the lens and compare them structurally with other insects. You will probably find many kinds of individuals, which might prove somewhat confusing; but the reason for such multiplicity of forms is largely that there is no pupal stage, hence immature forms of various sizes and degrees of development will be seen among the adults.

Although termites do not live exclusively on wood, those that live in wood use it as their staple diet. In some species, the wood is digested for them by one-celled animals (protozoa) which live in their intestines. The tunnels or galleries that termites excavate run parallel to one another and usually with the grain of the wood, but they do not form such an intricate series of tunnels and chambers as those found in the nest of the carpenter ant. The galleries (Fig. 71D) of termites also differ from those of ants or other tunnel-making insects by the grayish, mortarlike material (composed of excrement) with which they are plastered.

The next time you see large black carpenter ants entering or leaving a dead tree or log in an almost steady procession, break it open and you will find the galleries excavated by them. They form a rather complicated series of parallel concentric chambers which in an old nest become a veritable labyrinth of galleries, halls, and rooms. The corridors are usually excavated in parallel series of two, three, or more, separated by columns and arches or by thin partitions, and the rooms are somewhat crudely arranged in stories and half stories. Entrance to the nest may be by a circular or oblong door that opens into tubular circuitous galleries which communicate with the interior; or the entrance may be a spacious vestibule. Looking at these galleries through a reading glass that magnifies them several-fold, you wonder if you have suddenly blundered into some ancient ruin recently unearthed by an archeological expedition. On my desk is a section of a large nest I removed from a dead tree many years ago. As I look at it, I ponder the eternal mystery of that vast design in which an insignificant insect can fashion an intricate series of passageways and communicating chambers worthy of man's most elaborate architectural efforts.

Sometime when you are outdoors you might also pull the bark from a dead branch or trunk of a dead tree and examine the inner layer or sapwood. It may be ornamented with smoothly cut burrows similar to those shown in Figure 71E. The burrows are the work of small or medium-sized beetles, usually brown but sometimes black, with a body which is blunt at the hind end as if cut off transversely. They are known as engraver beetles or bark beetles. Like leaf miners, each species has its own characteristic pattern.

The initial burrow excavated by the beetles, which may be either simple or branched, is known as the egg burrow. Most species make niches in the sides of the tunnel that, since eggs are deposited in them, are called egg niches. When the larvae emerge, they feed on the bark, sapwood, or both, and thus fashion lateral tunnels which often extend parallel in a more or less regular manner. Most engraver beetles infest forest trees, but a few species, like the fruit-tree bark beetle and the peach-tree bark beetle, attack fruit trees.

Unlike the engraver or bark beetles, ambrosia or timber beetles excavate burrows in solid wood. Their burrows may be recognized by their uniform size, by the absence of wood dust and other refuse, and by the stained walls. The staining is due to fungi that grow on the walls and that the beetles eat; hence the name ambrosia. The galleries of these beetles form different patterns according to the species, but they are all essentially alike in having a main gallery, often branched. Usually there are lateral chambers termed cradles, in each of which an egg is laid and a larva reared, that extend deeply into the solid wood. All the galleries of both the engraver and the ambrosia beetles are visible to the naked eye, but they are better seen with a reading glass or lens.

On any September day clouds of winged ants may be seen, sometimes in such numbers as to shade the earth when they pass overhead. Some species fly during the summer, but the common little yellow ant whose rings of excavated sand grains decorate our garden walks, and the large black carpenter ant whose galleries we have described, both wait until now. Winged ants are mostly males in pursuit of young queens, and when the nuptial flights are over the males soon perish, falling victim to birds and other insectivorous animals or else dying from starvation. The females proceed to found new colonies, if they are not captured by workers of previously established colonies, or if they do not enter such formicaries of their own will. But whatever they do, they break off their wings or have it done for them by workers, and in the course of time lay their eggs. The workers hatching from these eggs become the basis for new colonies or for the replenishing of the old.

With September our mood seems to change. Perhaps the reopen-
ing of schools has something to do with it, or perhaps it is the sound
of the football, or perhaps an unconscious buckling down to the business of making a living after the letdown of summer and vacation
time. It could be, too, that the red maples are beginning to show
touches of color, reminding us that autumn is at hand. Or is it the
apples beginning to redden in the orchard, the peaches wearing the
blush of mellow ripeness, or the grapes hanging in dense blue clusters
in the grape arbor or along stone walls that mark the boundaries of
a farmland?

Autumn is harvest time, when store counters are laden with all
sorts of garden produce, and vegetable stands along the roadside
advertise their wares with temptimg displays. Nature puts on an
even more lavish display for the birds and mammals that now have
the opportunity to gourmandize on countless luscious fruits and
berries whose brilliant colors rival the flowers in the September
landscape.

Along the roadside the buttons of the pokeweed are so heavy
with juice that they drop and stain the ground, and in almost any
wayside thicket the stems of the wild spikenard bend from the
weight of dense berry clusters. In such places one may see, too, the
purple berries of the Indian cucumber perched above gaily painted
leaves.

Follow the winding woodland trail and you will come upon the
speckled berries of the Canada mayflower, the curious doll's eyes of
the baneberry, and the bright-red lanterns of the painted trillium.
Where the brook winds its way through the swampy woods you will
find the scarlet berries of the Indian turnip.

In the marshy meadow, the orange-scarlet beads of the black
alder glisten in the sunshine, and in an upland field the brilliant
fruit of the haw flashes against the sky. Everywhere wild cherries
and elderberries, the red, gray, and blue berries of dogwoods, and
the red and purple sprays of viburnums provide a banquet table for
all the wild creatures that want to feast.

How effective color is in attracting birds and mammals, or if it
has anything to do with it at all, is uncertain. Many brightly colored
fruits and berries seem to have little appeal, whereas many dull-
colored ones are eagerly sought. Apart from their value as food and

as vehicles for seed dispersal, fruits and berries add charm and beauty to the landscape. And when viewed with the lens, many of them have a richness of color and texture not apparent to the naked eye.

We speak of fruits and berries indiscriminately as if they were all the same, and in a sense they are, for a berry is a fruit per se. But so is an apple, peach, and bean pod, and therefore we come to the question of just what is a fruit. For obviously a berry, apple, peach, and bean pod are not quite the same.

A botanist will tell you that a fruit is a matured ovary of a flower, and that it includes one or more seeds and any part of the flower which may be closely associated with the ovary. Look at an apple, for example, and you will see the remnants of the sepals and the stamens. But this tells only part of the story, for there are simple fruits, aggregate fruits, and multiple fruits. Moreover, there are many kinds of simple fruits: berry, pepo, hesperidium, drupe, pome, legume, follicle, capsule, silique, akene, caryopsis, samara, schizocarp, and nut. A simple fruit is one developed from a single ovary; an aggregate fruit is one derived from a single flower having many pistils; and a multiple fruit is one formed from the ovaries of many separate yet closely clustered flowers, such as the pineapple. Both the strawberry and raspberry are aggregate fruits, the former, as we have already seen, consisting of a number of one-seeded fruits called akenes, and the latter consisting of a number of drupelets—small stony fruits. Compare both of these with the lens and you will see the difference.

The simple fruits differ from one another in certain structural features and in the number of ovaries. It is not necessary for us to examine their botanical distinctions except briefly. A berry is a fleshy or pulpy fruit containing a number of seeds; actually, the fleshy portion represents the ovary wall. Such seemingly diverse fruits as the grape, banana, tomato, and blueberry are all berries. Both the pepo and hesperidium are a type of berry having a hard and leathery rind respectively. They are represented by the squash, cucumber and watermelon, and the orange, lemon, and grapefruit. A drupe is a one-seeded stone fruit, such as the cherry, peach, and plum. A pome, such as the apple, is a fleshy fruit containing seeds; it differs from a berry in that the fleshy portion represents the receptacle.

The legume (bean, pea, locust), follicle (milkweed, columbine, peony), capsule (lily, iris, poppy, tulip, violet, evening primrose) and silique (mustard, cabbage, and other crucifers) are dry dehiscent fruits—that is, fruits that split open when mature. They differ essentially in the manner in which they open. The akene (sunflowers, buttercup, dandelion), caryopsis (corn, wheat, rice, oat, barley, and other grasses), samara (ash, elm, maple), schizocarp (carrot, parsnip, celery) and nut (acorn, hazelnut, chestnut) are also dry fruits, but they do not open when ripe (indehiscent). Most of them are one-seeded fruits and differ in certain botanical details.

There are a number of different kinds of apples—Northern Spy, Delicious, Baldwin, MacIntosh—yet one apple is much like another apple, just as one peach is much like another peach; that is to say, a pome or drupe shows little variation. Botanically all akenes are structurally much alike, but externally they vary in shape and other characteristics, so that we can often tell by looking at an akene the name of the plant which produced it. The akene of the dandelion, for instance, is somewhat spindle-shaped, with four or five ribs and one end prolonged into a very slender beak from which extends the parachute tuft of silken hairs that catch the air currents (Fig. 72A). The akene of the buttercup (Fig. 72B) not only lacks the silken hairs, but is compressed and has a very short beak. The akene of the early buttercup has an awl-like beak (Fig. 72C).

Some akenes, such as those of the field or sheep sorrel (Fig. 72D) are entirely beakless. So, too, are the akenes of the related buckwheat (Fig. 72E), which are similar in form but granular and marked with lines. Another relative, the smartweed or water pepper, has a three-angled akene (Fig. 72F) that is broadly oblong or ovoid. Figure 72G shows an akene of the common cinquefoil. In some instances akenes are furnished with plumose appendages. Such akenes are found in the clematis (Fig. 72H) or virgin's bower, a beautiful trailing vine commonly found draped over bushes in copses and by moist roadsides. When viewed through the lens these akenes appear like tiny twisted tails. All akenes are more or less visible to the naked eye, but details are hard to see unless they are magnified.

Akenes with appendages are frequent among the composites. An appendage, called a pappus, may take various forms. In the

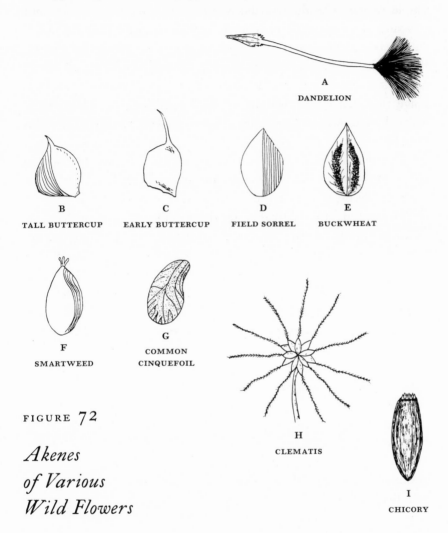

FIGURE 72

Akenes
of Various
Wild Flowers

chicory it is a shallow cup (72I); in the sunflower it consists of two deciduous scales (73A) and in the sneezeweed of five scales (Fig. 73B); in the sow thistle it is composed of delicate, downy hairs (Fig. 73C) while in the dandelion it is made up of a number of silken hairs (Fig. 72A).

As I wander about the countryside at this time of year, I never cease to marvel at the bewildering variations on what is, after all, a simple theme. The fleshy fruits show some difference in form and structure—compare a berry, a hesperidium, a drupe, and a pome—but all berries are much alike, and so, too, are drupes and pomes. But the dry fruits, such as the akenes, vary considerably and are as different as the plants that produced them. In size and shape they range from the small oval silicles[3] of the pepper grass (Fig. 73D), which I occasionally place in my mouth for their peppery taste, and the slightly larger triangular silicles of the shepherd's purse (Fig. 73E), to the long-curved, cylindrical, cigarlike capsules of the catalpa (Fig. 73F) or the shiny, leathery-looking, maroon-brown pods of the honey locust (Fig. 73G), which may be 16 or 18 inches long.

Most of us are familiar with the capsules of the seedbox, in which the seeds become loose and rattle when the plant is shaken. We also know the slender, curved, violet-tipped capsules of the fireweed. But not many of us, I daresay, are acquainted with the inflated, prolate-spherical pod of Indian tobacco (Fig. 73H), the three-sided capsule of bellwort (Fig. 73I), or the awl-tipped, egg-shaped pod of wild indigo (Fig. 74A). These are only a few of the countless dry fruits that may be found with a little looking and studied with the aid of the lens. And since many of these fruits remain after the flowers have faded and fallen, they serve as reminders of your summer flower companions.

Apart from providing food for both man and animals, fruits have a twofold purpose: they are a protective covering for the maturing seeds and a vehicle for seed dispersal. Many pulpy or fleshy fruits adhere to the beaks and feet of birds. The pulp dries and the seeds drop off as the birds fly about or rub their beaks against twigs and branches. Many seeds, too, are voided with excrement, especially seeds that are indigestible, such as the hard-shelled pits of cherries and haws.

Fruits which do not appeal to birds and mammals as food sometimes obtain free transportation by having structures that catch in the fur of the mammals as they come in contact with them. They also catch in our clothing. Probably the most familiar of these hitchhikers

[3] A silique broader than it is long.

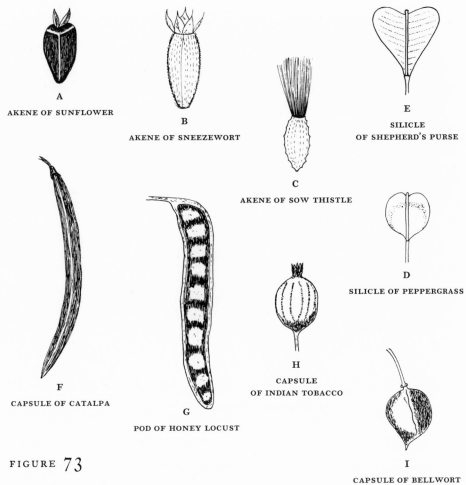

A
AKENE OF SUNFLOWER

B
AKENE OF SNEEZEWORT

C
AKENE OF SOW THISTLE

E
SILICLE
OF SHEPHERD'S PURSE

D
SILICLE OF PEPPERGRASS

F
CAPSULE OF CATALPA

G
POD OF HONEY LOCUST

H
CAPSULE
OF INDIAN TOBACCO

I
CAPSULE OF BELLWORT

FIGURE 73

Fruits
of Some Common Plants

are the burs of the burdock (Fig. 74B), a common plant of roadsides and waste places. Examine one of the burs and it becomes apparent why they cling so tenaciously to our clothing. Each bur consists of a number of akenes that are oblong, three-angled, and ribbed, with one end truncate and the other in the shape of a hook.

These burs have one thing in their favor: they remain intact and can be removed in one piece from your clothing. Brush against the beggar ticks, and your skirt or trousers will be covered with hundreds of barbed seed vessels (Fig. 74C); it takes an inter-

FIGURE 74

More Fruits

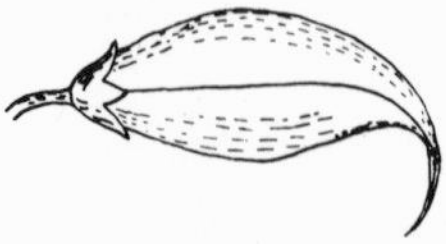

A

POD OF WILD INDIGO

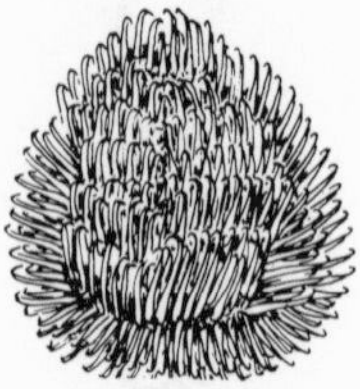

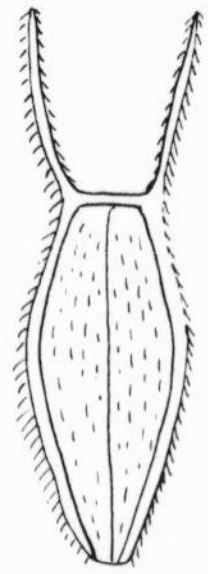

C

SEED VESSEL
OF BEGGAR TICK

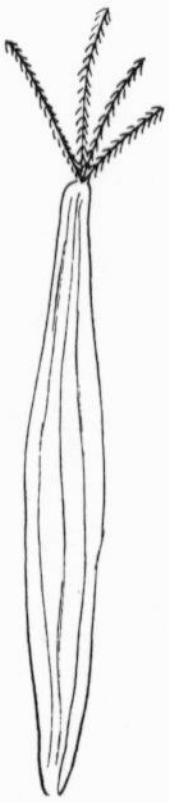

B

BUR OF BURDOCK

D

SEED VESSEL
OF SPANISH NEEDLE

minable time to get them all off. The seed vessels of such related plants as the bur marigold, tickweed, and Spanish needles are also barbed (Fig. 74D). They are all akenes and vary in form, being wedge-shaped, linear, or oblong. Some of them have two prongs, others have four.

Equally as effective in hitching a ride, and just as exasperating to remove, are the pods of the tick trefoils (Fig. 74E). Looking at them with the naked eye one might wonder how they can cling to the clothing, but examine them with the lens and you will see that they are covered with minute hooked hairs. I had known the cocklebur some time before I actually saw the burlike fruits, and when I looked at all the hooked prickles with the glass, I thought the name quite descriptive (Fig. 74F). Even more descriptive are such names as catchweed, cling-rascal, wild hedgebur, stick-a-back, and grip-grass. All these names refer to a single plant commonly called cleavers (Fig. 74G). I don't know whether the names apply to the prickley stems or to the small burlike fruits that occur in pairs and are covered with short, hooked bristles; but it doesn't really matter —both get caught in your clothing.

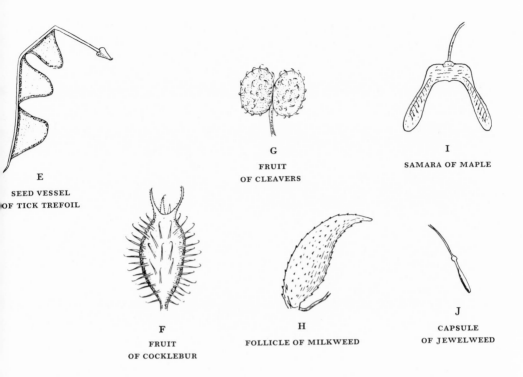

E

SEED VESSEL
OF TICK TREFOIL

G

FRUIT
OF CLEAVERS

I

SAMARA OF MAPLE

F

FRUIT
OF COCKLEBUR

H

FOLLICLE OF MILKWEED

J

CAPSULE
OF JEWELWEED

One of the most delightful ways to spend a few hours outdoors at this time of the year is to wander about the countryside studying the countless variations in fruiting structures and the many ways the plants have contrived to scatter their seeds. Observe how the September breezes break up the silvery cushions of the thistles and how the shimmering, gossamer-winged seeds are borne into the air. Open a milkweed follicle (Fig. 74H) and you will find symmetrical packs of golden-brown seeds, each with a tuft of silken sails. Better still, watch a pod open and see how the silken sails spring out, catch the faintest breath of air, and sail off to unknown destinations. Or watch how the samaras of ash and maple (Fig. 74I) are caught up and whirled about by the wind, and how the wild cucumber shoots its seeds out from its miniature cannon. Perhaps you might prefer to trigger the jewelweed into firing its tiny projectiles (Fig. 74J) at an imaginary foe. The ingeniously contrived mechanism that propels the seeds is there for you to study. You can also watch fruits that float on water and are carried downstream by rivers on their way to the sea, which many of them doubtless reach, though many others find a place to rest somewhere along the way.

Fruits should be only a part of such an outdoor exercise; the seeds, too, should not escape your close scrutiny. But seeds are such common objects that we accept them for what they are and give them little thought except when planting time comes. Then we read the directions on the packets we bought from the neighborhood store or from some seedsman. We give the seeds themselves a cursory glance as we drop them into a furrow or scatter them on the ground, trusting that a benevolent nature will do the rest; somehow she usually does better than we deserve.

Unless you are an apartment dweller and live in the heart of a city, you have only to walk beyond your doorstep to find seeds in abundance. If you have a garden, you have a happy hunting ground. Collect them at random and look at them—not with your naked eye, for your eye alone cannot do them justice—but with your lens, and you will be amazed at their infinite diversity. If perfection is beauty, then the form of a seed is a thing of beauty, for one of the most striking features of seeds is the perfection of their simple forms. Silhouettes of seeds reveal this most distinctly and evoke an intellectual and emotional response.

Seeds (Figs. 75A to H) are typically more or less globular or oval in shape, yet there are seeds that are extremely thin and flat or greatly elongated. They may be smooth or wrinkled, pitted or angled; they may be furrowed with ridges like those of Doric columns and with geometric patterns in miniature relief. Only your lens will reveal such refinements of sculpturing and unexpected beauty.

There are seeds that are twisted or coiled or otherwise distorted. There are seeds that are covered with hairs or supplied with silken strands or delicate membranes to make them wind-borne. Some are as fine as dust; others are several inches in diameter; and there are all sizes in between. They may be a shining jet black; or they may be blue, red, yellow and other bright colors, or their color may be the less striking and more somber brown and gray. They may be a single solid color, may have two or more colors scattered about at random, or may show a definite pattern or design produced by a blending of various colors. In short, seeds are a source of wonder, not alone for how they look, but also for what they are—the very mainspring of life.

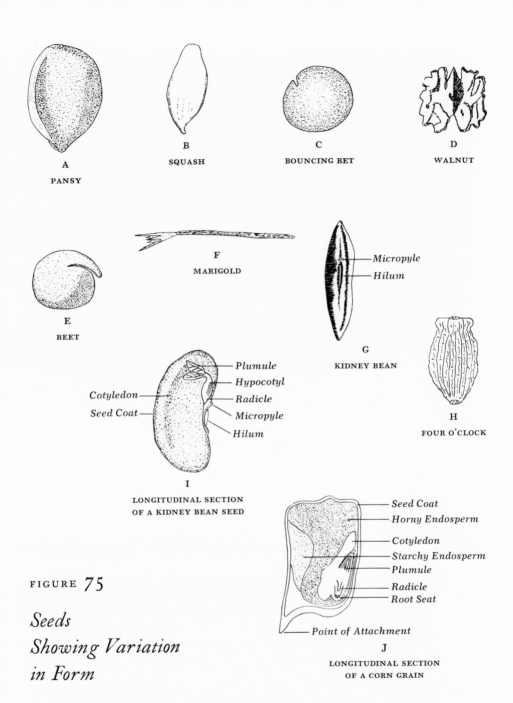

A
PANSY

B
SQUASH

C
BOUNCING BET

D
WALNUT

E
BEET

F
MARIGOLD

Micropyle
Hilum

G
KIDNEY BEAN

Plumule
Hypocotyl
Cotyledon
Radicle
Seed Coat
Micropyle
Hilum

I
LONGITUDINAL SECTION
OF A KIDNEY BEAN SEED

H
FOUR O'CLOCK

Seed Coat
Horny Endosperm
Cotyledon
Starchy Endosperm
Plumule
Radicle
Root Seat
Point of Attachment

J
LONGITUDINAL SECTION
OF A CORN GRAIN

FIGURE 75

*Seeds
Showing Variation
in Form*

Botanically a seed is a ripened ovule and consists of an embryo plant, some stored food, and a protective covering or coat. The unripe ovule is a small structure or body in the ovary of a flower. It may readily be seen if you cut the ovary open with a razor blade or pocket knife. In most cases you will find many ovules. Each ovule contains an egg cell, and when this egg cell has been fertilized by a sperm cell of a pollen grain, the ovule develops into a seed.

Split any seed lengthwise and you will find a tiny replica (embryo) of what would eventually have become a mature plant if the seed had been allowed to germinate. For practical purposes a kidney bean is best, because it is readily obtainable, it is large enough to handle easily, and the embryo can be clearly seen. First soak it in some water to make it easier to open, and then divide it lengthwise into two equal halves. The embryo will be seen attached to one of them, as shown in Figure 75I.

Note the two tiny leaves folded over one another. They are immature foliage leaves and are attached to a short rudimentary stem called the epicotyl. A small. bud lies between the leaves; known as the plumule, it normally develops into the stem that bears the leaves. Until such time as the plant can make its own food and become self-supporting, the growing plant draws on food material stored in the seed. This material occupies the greater part of the seed and is called a seed leaf or cotyledon. (In some seeds, such as the corn, it is known as the endosperm (Fig. 75J). Though some seeds have only one cotyledon, there are usually two of them, one in each half. Each is attached to the main axis of the embryo at a place called the hypocotyl. Extending downward from the hypocotyl is the rudimentary root or radicle, which normally develops into the root.

Although seeds show minor differences, they are all basically alike. And though they may show a great diversity in surface markings and configurations, they all agree in having a minute pore or pit and a scar (Fig. 75G). The pore or pit marks the position of the micropyle, an opening in the ovule through which the sperm cell entered on its way to the egg cell, and the scar or hilum marks the point of attachment of the seed to the ovary. Both of these structures may be seen with the hand lens; if you are unable to locate the micropyle, soaking the seed in water should bring it into view.

The wild flowers of September are essentially those of August. For the most part, those that are predominant on the September scene are mostly composites—tansy, boneset, joe-pye weed, sunflowers, goldenrods, and asters. Although some of them appeared in July, it is at this time of the year that they seem to take possession of the fields and woods and to give the characteristic tenor to the landscape that we associate with late summer and early autumn. Together with the verbenas and mints, they add, in royal purple and gold, a final touch of imperial splendor to the declining year and provide a fitting climax to the succession of brilliant blooms that began so modestly in March.

In describing a dinner, Pepys speaks of a second course as consisting of "two neat's tongues, cheese, and tansy." The tansy was a sort of cake or fritter made from the leaves of the tansy and apparently was well thought of at the time. Until recently tansy tea was a favorite beverage for colds and similar ailments in this country; perhaps it still is.

The tansy is one of the more common plants of summer, its rather flat, clustered, dull orange-yellow flower heads a familiar sight along roadsides and in fields and waste places. They resemble the flower heads of the daisy except for the white rays. In early times the boneset was supposed to have medicinal properties, and a nauseous tea was made from the leaves. Beetles seem to like the dull white, odorous flowers, but the butterflies not at all; they prefer the dull magenta-crimson flowers of the joe-pye weed, said to have been named after an Indian medicine man who was supposed to have been able to cure typhus fever and other diseases with a decoction made from the plant.

From a distance asters appear as central yellow disks surrounded with an outer ring of petals which are variously colored in white or tints of blue, violet, or purple. I believe there are some two hundred and fifty species of asters, most of which are found in North America; few of us know them all. But look at almost any one of them with the lens, and you will see that the central disk is composed of many erect perfect tubular blossoms (Fig. 76A) which are yellow at first and change with age to purple or brown. The pistillate rayflowers are elongated and strap-shaped (Fig. 76B).

We associate goldfinches with thistles, but they visit the flowers only after the seeds have been formed. It is the butterflies and bumblebees that visit the flowers when they are newly opened, for the densely clustered florets are rich in nectar. The dense, matted, wool-like hairs that cover the stems and leaves discourage the climbing pilfering ants, and the spiny leaves are equally as effective against grazing animals.

There are several species of thistles, all more or less common in pastures, in waste places, and along the roadside and fencerows. In such places we also find the burdock. Its globular flower heads, hook-bristling green burs with magenta or nearly white perfect tubular flowers, are a standing invitation to butterflies, which delight in magenta, and to bees of various kinds. To appreciate the depth of coloring of both the thistle and burdock, it is necessary to examine the flowers with the lens.

At first glance the pearly everlasting may not seem a composite, for the flowers look like miniature pond lilies, the tiny petal-like scales surrounding the central staminate flowers being arranged not unlike the petals of the water plants, but a composite it is, as may be seen by a study of the flowers. And there is little question that the ironweed is a composite too, though composed only of tubular flowers, madder-purple, and clustered in a thistlelike head. The flowers remotely resemble bachelor's buttons without petals, or from a distance they could even be asters. A common species, ironweed prefers moist ground, where butterflies of many kinds visit it, chief among them the tiger swallowtail. Bees and flies are also frequent visitors.

I doubt that many have ever looked closely at goldenrods simply because, like so many other things that are common, we cannot be bothered with them. But detach a flowering stem and examine it with your lens. You will be surprised at what you see—a row of tiny yellow goblets (Fig. 76c). It is said that no flower attracts so many insects as the goldenrod. I don't know how true this is, but countless insects do visit the blossoms. Among the more frequent are the locust borer, a beautiful black beetle with numerous wavy yellow bands, and the blister beetle, a black beetle which is frequently found in such numbers that the golden plumes appear as if sprinkled with soot.

The deep red blossoms of the cardinal flower, a deeper red than the bird of the same name, are the same color as the hat worn by a prince of the Roman church. They almost seem to kindle into flame the moist thickets in which the flower grows. Not many in-

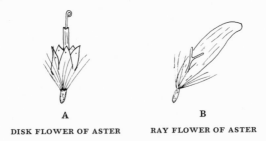

A
DISK FLOWER OF ASTER

B
RAY FLOWER OF ASTER

FIGURE 76

*Flowers of Aster
and Goldenrod*

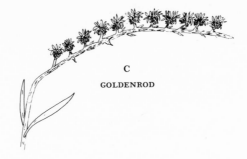

C
GOLDENROD

sects visit the blossoms, for their tongues are not long enough; but the hummingbird can reach the nectar. Why doesn't the humming-bird visit the cardinal's twin sister, the blue lobelia? The flowers are built on much the same plan, though the lobelia is slightly adapted for the bumblebee, which is a frequent visitor. Is it the color of the blossoms? Study the habits of the rubythroat and you will find that its visits are confined chiefly to such flowers as the painted cup, Oswego tea, coral honeysuckle, columbine, and garden salvia, fuschsia, and phlox.

The celandine appears in April but is still in bloom. It will continue to flower among its seedpods until the cool air of autumn kills or sends to cover the gnats and other small insects that serve it well. The four yellow petals suggest the mustards, but pick the blossoms and they droop poppy-fashion. On second thought, don't pick the blossoms or break the stems, unless you don't mind being covered with a yellow juice that stains whatever it touches.

When the snow lies on the ground and the sun shines brightly, the dried stalks of the blue vervain etch delicate shadows on the snow. To many people this is the plants' greatest appeal. Certainly the purple blossoms are too small to be attractive, nor do the slender stalks appeal to us, though they branch upward like the arms of a candelabra. The reason is that they have buds at the top, flowers in the middle, and brown nutlets at the bottom, an arrangement that does not please our sense of the aesthetic. The name of the plant is misleading, for the flowers are not blue, nor do they approach any semblance of blue. Flowers, however, were not designed for our enjoyment but to attract the insects, and so many bumblebees, honeybees, and other bees are usually seen about the blossoms that I sometimes wonder if they even sleep on the blossoming spikes. There is also a white species of vervain.

Practically dwarfed by the taller composites and verbenas and virtually hidden by them, the low-growing rabbit-foot clover and the milkwort actually have to be searched for. Both are delightful little plants, the former with fuzzy flower heads, the latter with globular cloverlike heads of tiny magenta flowers. One has to look at them with the lens, however, to appreciate their beauty.

It is a lamentable fact that of all our common plants, the most common are the ones that most of us know the least. Yet we come in almost daily contact with these plants that are economically the most important members of the plant kingdom. From the moment the March sun begins to warm the earth, the grasses, in green tenderness, give us the first intimation of spring. They tinge the brown hillsides even before the snows have ceased, and from then until the frosts of autumn take their toll, there is never a day when they are not in bloom. We find them along the wayside and on the woodland trail, in gardens and orchards, along the banks of winding streams, and in waste places, fields, and meadows. But so intent are we on growing our garden flowers or collecting those of the fields and woods that we overlook flowers which are not as brilliantly colored as the more familiar blooms, yet which are just as beautiful in their rose and lavender, purple and green tints.

When we examine the tiny blossoms of grasses through our lens, we are amazed at their seemingly delicate and fragile quality, and we wonder how they can survive the buffetings of wind and falling raindrops, or withstand the merciless rays of the hot summer sun. We need not have the soul of the artist to be enchanted by the grace of swaying stem and drooping leaf. There are grasses so tall that they rise above our heads, and others that barely extend above the earth; there are grasses whose flowering spikes are hardly noticeable, and others whose panicles are half a yard in length; there are grasses which are stout and robust, and others so slender that their stems are like golden threads.

The grasses no less than the flowers contribute to the September countryside. Where the farmer has tilled the ground, old witchgrass lifts its blossoming heads like shower fountains, and in meadows and open woodlands purple spikelets of the purpletop glitter in the sunshine. Foxtails (Fig. 77A), both yellow and green, decorate the wayside, and in fields and similar places where the September sun has turned the smaller grasses a golden brown, the purple lovegrass spreads a reddish-purple mist above the ground. Even about the dooryard the crabgrass, bane of every gardener, extends its spikes like the fingers of a hand.

When a grass blossom is examined closely, it may seem at first to bear little resemblance to a conventional flower. Compared with a lily, the two apparently do not seem to have much in common; yet if you were to select a lily that blooms in a spike, and were to imagine that the lily suddenly crowded the flowers and reduced the petals to mere scales, you would have a lily with a reasonably grasslike appearance. The flowers of grasses occur in clusters called spikelets (Fig. 77B). Spikelets vary in size and may be composed of one, several, or many flowers. The short stem on which the flowers of a spikelet are placed is called the rachilla. Sometimes the rachilla is prolonged as a tiny thread lying outside the uppermost flower.

A grass flower stalk usually has many spikelets, and collectively they form an inflorescence which may be either a spike (Fig. 77C) or a panicle (Fig. 77D). If the flower stalk is unbranched and the spikelets are attached directly to it, the inflorescence is called a spike; if, however, the flower stalk is branched and the spikelets are attached to the branches, then the inflorescence is known as a panicle. In a spike the lower spikelets bloom first, or from below up-

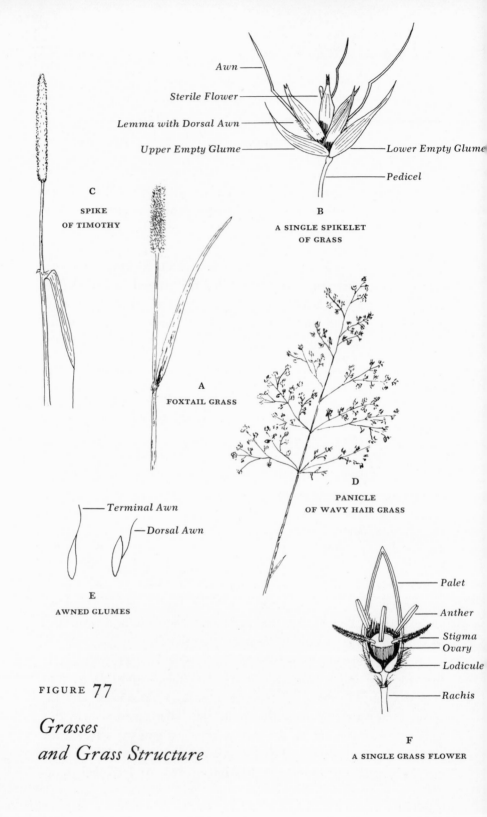

Awn

Sterile Flower

Lemma with Dorsal Awn

Upper Empty Glume

Lower Empty Glume

Pedicel

B

**A SINGLE SPIKELET
OF GRASS**

C

**SPIKE
OF TIMOTHY**

A

FOXTAIL GRASS

D

**PANICLE
OF WAVY HAIR GRASS**

Terminal Awn

Dorsal Awn

E

AWNED GLUMES

Palet

Anther

Stigma
Ovary

Lodicule

Rachis

FIGURE 77

*Grasses
and Grass Structure*

F

A SINGLE GRASS FLOWER

ward; but in a panicle the uppermost spikelets are the first to blossom, followed successively by those below.

As we study a grass flower carefully with the lens, we find that the sepals and petals of a conventional flower appear in the form of modified leaves called bracts (bear in mind that sepals and petals are also modified leaves), and that the rachilla bears a number of these chaffy overlapping bracts. The two at the base of the spikelets (glumes) are larger than the others (Fig. 77B) and enclose the rest of the spikelets. Each flower of the spikelet is enclosed between two bracts, which are usually similar to the glumes but smaller. The lower of these bracts is known as the lemma, the upper one as the palea or palet.

Lemmas often have a bristlelike appendage called an awn. Awns are not always present, but when they do occur they may be straight, bent, or twisted. In some lemmas they are attached at one end and are called terminal awns; in others they are attached to the back and are called dorsal awns (Fig. 77E). Many lemmas are rather flattened and folded so that the two edges are brought closely together; then the midvein becomes prominent as a ridge on the back. Such lemmas are said to be keeled. When the veins are conspicuous, the lemma is said to be three-nerved, five-nerved, or nine-nerved, according to the number.

Unlike the lemma, the palea is awnless, and is usually two-nerved with two keels. At the base of the ovary and within the lemma and palea are commonly two (in rare cases, three) minute, thin, and translucent scalelike structures, or lodicules (Fig. 77F). They probably represent two reduced perianth segments. They are rarely seen except at the time of flowering when, for a short time, they become swollen with sap and by forcing the lemma apart, allow the flower to open.

Most grass flowers are perfect, with one to six (usually three) stamens. The anthers are lightly attached to the slender filaments, and when they tremble in the wind they release the pollen grains. As grasses depend on the wind for pollination, many pollen grains fall to the ground and are wasted; hence to ensure sufficient seed, they are produced in vast numbers. A single anther of rye, for instance, contains no less then twenty thousand pollen grains.

We must not confuse the grasses with the sedges and rushes, which they resemble superficially, especially in color. Sedges and rushes belong to different families. With few exceptions, the stems of sedges are solid and in many species sharply triangular. The flowers are small and arranged in spikelets, as the grasses are, but unlike the flowers of the grasses, each one is protected by only a single scale, though a perianth is sometimes present.

Sedges, for the most part, prefer wetter places than grasses—marshes, swamps, the edges of ponds and streams—and are protean in form. Some leafless are like green bayonets tipped with blossoming cylindrical heads; others, broad-leaved and branching, seem like transplants from the tropics. Some rise only a few inches above the ground; others extend as high as one's shoulders, if not higher, and have many flowering heads. Few are of any economic value; though the papyrus from which the ancients made their paper was at one time highly valued.

There are many sedges: yellow nutgrass, bristle-spiked cyperus, pond sedge, spike rush, chairmaker's rush, meadow bulrush, fox sedge, fringed sedge, and woolgrass are among the more common species, and you can find them throughout the summer. At this time of the year the woolgrass is plentiful and conspicuous with its spikelets in terminal umbels clothed in dull gray wool.

Who has not read how the floors of houses in early England were covered with rushes, and was not the stage of Shakespeare strewn with these plants? Erasmus did not think much of this rather "barbaric" custom and considered these floor coverings pestilential —with good reason, too, since the lowest layer was often left unchanged for years. The rushes which were used for this purpose do not appear to be true rushes but sedges. True rushes resemble both grasses and sedges in their general appearance, but one should have no difficulty in distinguishing them from either, since their flowers are like miniature lilies and show a perfectly six-parted perianth with three to six stamens and three stigmas. The bog rush is the largest of our common species. It grows in moist places and has the distinction of being one of the few meadow plants that remain green until late autumn; even in winter we often find low tufts of its dark-green stems along winding brooks.

FIGURE 78

Mosses
and their Life Histories

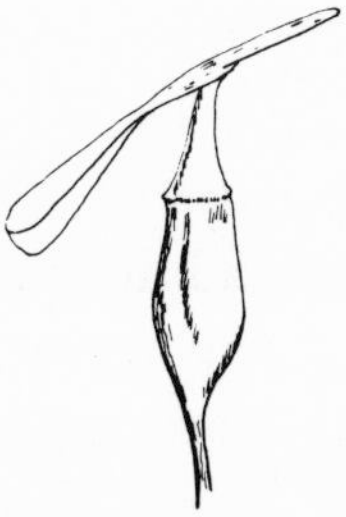

D
CAPSULE
OF COMMON DICRANELLA

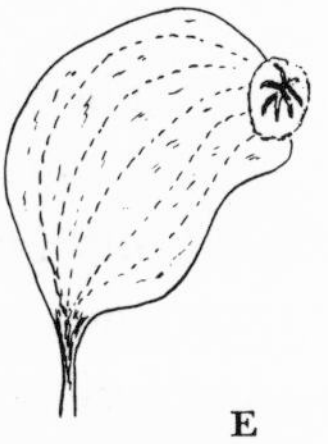

E
CAPSULE
OF PHILONOTIS

Capsule with Calyptra

Capsule without Calyptra

Operculum on Lid

Seta

F
CAPSULE
OF URN MOSS

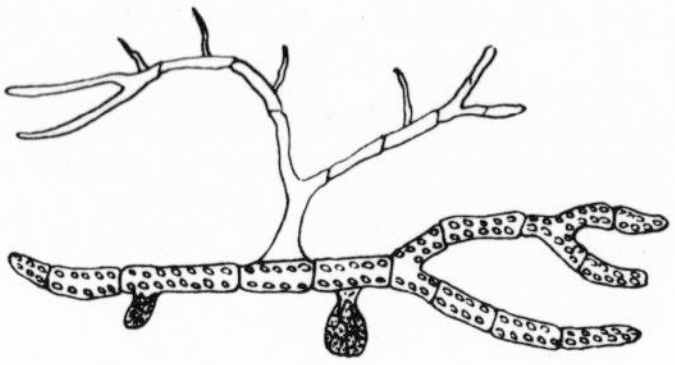

C
PROTONEMA

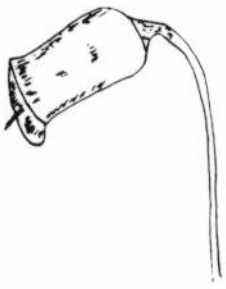

B
CAPSULE
OF HAIR CAP MOSS

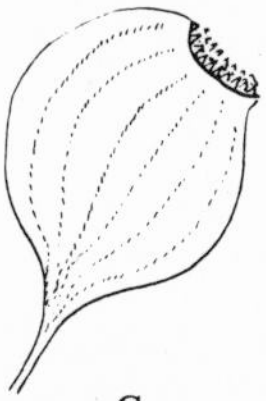

A
MOSS PLANT

G
CAPSULE
OF APPLE MOSS

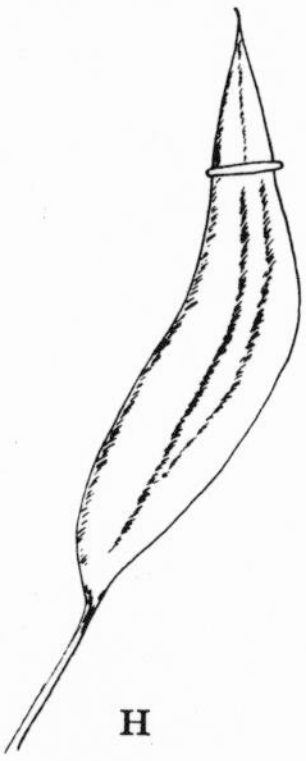

H
CAPSULE
OF YELLOW DITRICHIUM

Like the grasses, the mosses are largely ignored except by a few who find the study of these plants a delightful pastime, and for much the same reason which is largely the misconceived idea that they are difficult to understand. And, of course, they are not as glamorous as the larger flowering plants, though this may be a matter of opinion.

An individual moss plant is a rather inconspicuous form of vegetation, and yet when a number of them grow together they collectively provide a bit of greenery to places that would otherwise be uninviting and barren. They cover the woodland floor with a soft green carpet and are the first plants to appear on the naked sides of ditches, clay banks, and other unsightly spots. As little cushions they fill in the crevices of pavements and relieve the harshness of rocks and boulders. A decaying log is never so attractive as when covered with mosses and lichens.

A moss plant consists of a vertical slender stem which bears spirally arranged, very thin, very small green leaves and which is anchored to the ground by rather stout rootlike hairs, or rhizoids (Fig. 78A). Compared with the higher plants, both the stem and the leaves are simple in structure, since the various cell layers are not greatly differentiated from one another.

The most striking part of the plant is undoubtedly the fruiting portion (Fig. 78A) or sporophyte. The spore case, in which myriads of green, dustlike spores are formed, is a thin-walled capsule (Fig. 78B) set on the tip of a flexible stalk (seta), the two resembling a tiny Turkish pipe. In certain stages the spore case is covered with a conical light-brown hairy cap or veil (calyptra). When this veil falls off, the case is seen to be tightly closed by a round lid, the edges of which fit closely about the rim. In some species it looks like a miniature tam-o-shanter and in others a miniature duncecap.

When the spores become mature, the lid is pushed up by the swelling of a ring of beadlike cells (annulus),[4] and when it falls off a number of tiny teeth (peristome) are revealed bordering the rim case and bending inward.[5] They are attached to the rim of a membranous disk (epiphragm) that covers the mouth of the spore case, and they are hygroscopic in nature. When the humidity is high, they hold the epiphragm in such a way that the spores cannot escape, but when the humidity is low, they are so modified as to form a ring of holes beneath the rim of the epiphragm through which the spores may pass. The capsule or spore case is, in fact, a miniature pepper box with a grating around its upper edge instead of holes in the cover.

The spores are disseminated by air currents, and when they find a suitable place to germinate, they develop into a branching creep-

[4] Not present in all species.
[5] There may be one or two rows of these teeth, according to the species; and in some species they are entirely lacking.

ing filament called the protonema (Fig. 78c), which spreads over the ground to form the tangled green felt so often observed where mosses grow. If some of it is examined with the microscope, little knots or enlargements may be seen near the base of the branches. These are buds that normally develop into the leafy moss plants.

At the tips of the leafy stems, sex organs develop. In some species they occur on the same plant; in other species on separate plants. The male organs (antheridia) are spherical, oval, or club-shaped bodies with long or short stalks; the female organs (archegonia) are flask-shaped with long necks or canals. When the two-tailed spiral, free-swimming sperms are mature, they leave the antheridia and swim through the dew which often covers the moss plants, or through the water in which the plants may be submerged, or in rain drops splashing from one plant to another, until they are attracted to the archegonia by a chemotactic substance. Then they swim down the canal until one of them unites with the egg. Upon the union of sperm and egg, the egg begins to divide, and following a succession of further cell divisions, an embryo is formed. It gradually grows vertically into the elongated stalk bearing, at its distal end, the spore case or capsule, the latter covered by a cap or veil that was torn from the neck of the archegonium and carried up by the apical growth of the embryo.

Some species of mosses may be recognized by the characteristic fashion in which the leaves grow, but most of them can be better identified by the shape or form of the spore case, which varies according to the species. Unfortunately, the spore cases only occur during certain times of the year. Mosses can be separated into two groups on the basis of their habit and method of producing these cases. In one group the plants grow more or less erect and develop capsules on stalks which extend from the end of the stem. In the other group the plants are creeping, and the stalks grow out laterally from the main stem.

The study of mosses may easily become a rewarding pastime. Most of us, however, will probably be satisfied to examine the capsules, which come in all shapes: balls, eggs, urns, horns, bells, and crook-necked gourds (Fig. 78B,D,E,F,G). But since they are small (about the size of a match head), the lens is necessary to give them definition.

While you are searching for mosses, you might look along the bank of a brook or stream or among the mosses in damp woods, and you will probably find flat, ribbonlike plants growing close to the ground. They are papery thin and may be either long and slender or repeatedly lobed and forked. Called liverworts, they are very simple plants without stems or leaves. Of little economic importance, they are only interesting because they represent the transition stage from a water-living habit to a land-living habit; in other words, they bridge the gap between the fundamentally aquatic algae and the higher flowering terrestrial plants.

Some of our better known species are Anthoceros, Riccia, Pellia, Conocephalum, and Marchantia, which is the one you are most likely to find. It has a peculiar dull-green color with a broad ribbon-shaped thallus (a thallus being a plant body without true roots, stems, or leaves) which is generally forked once or twice. The upper surface is divided into angular areas, in the center of which we can distinguish an air pore, if we use the lens. Numerous fine hairs (rhizoids) extend from the lower surface into the ground and anchor the plant.

Sometimes we may see umbrella-like structures extending upward from the upper surface of the thallus. The umbrella part may be either flat, shield-shaped, and radially lobed (Fig. 79A) or divided into deep fingerlike lobes that usually curve downward (Fig. 79B). These umbrella-like structures contain the sex organs and are known respectively as the antheridial and archegonial disks. Because they do not occur on the same thallus, it becomes necessary for the sperms to swim to the eggs; this explains why liverworts always grow in wet or damp places. As with ferns and mosses, the sperms are oriented in the right directions by chemotactic substances released by the archegonia, a phenomenon called positive chemotaxis. Occasionally little cup-shaped or saucer-shaped structures with toothed margins occur on the upper surface of the thallus. Called gemmae cups (Fig. 79A), they produce budlike bodies (gemmae) that become detached and are washed out or blown away and develop into new plants.

Without stopping to look more closely, you have doubtless seen stumps and logs in swampy woods covered with a continuous green carpet. Such a carpet is likely to be a scale moss. Scale mosses, which are not true mosses, are prostrate plants, like the liverworts, but the thalli are differentiated into stems and leaves of a very primitive nature; actually the leaves are scalelike. The scales are in pairs, and the plants have a distinct upper and lower surface in

FIGURE 79

Thalli of a Liverwort

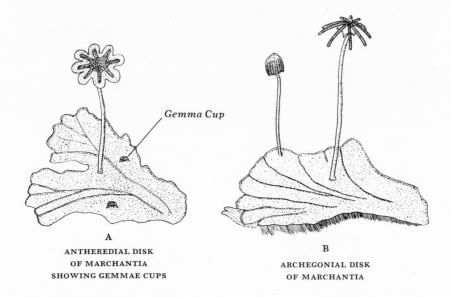

Gemma Cup

A
ANTHEREDIAL DISK
OF MARCHANTIA
SHOWING GEMMAE CUPS

B
ARCHEGONIAL DISK
OF MARCHANTIA

contrast to the radial symmetry of the true mosses. On the bark of birch and beech trees we frequently find delicate brownish or reddish traceries of the tiny Frullania, one of the more common scale mosses. Its scales can be seen only with the hand lens. Porella is larger, and is found on rocks and soil as well as on the bark of trees. The scales are deeply lobed, and the tips of the stems bend upward. With the lens the scales of the pale green Trichocolea appear like clusters of algal filaments, for each is deeply lobed and branched into many hairlike divisions. These scale mosses are very pretty primitive plants, though few of us look at them or even know they exist.

A few more species of mushrooms have appeared, and now these plants are abundant everywhere. When in the woods I now look for the *Russula hygrophorus* beneath a mat of leaves, or the sharp-scaled Pholiota that grows in dense tufts on prostrate trunks or old stumps. Occasionally I find the oyster mushroom or other Pleuroti jutting out from the stumps and trunks of trees, and I still become enchanted with the delicate, violet-tinged beauty of the masked Tricholoma. Less conspicuous than many of the mushrooms that grow in the open, the conelike boletus (Fig. 80A) is difficult to find among the bewildering lights and shadows that play upon the dead leaves covering the woodland floor. Puffballs (Fig. 80B), small or large, are now common in fields, woodlands, and along the roadside, ready to send their puffs of spore clouds into the air like smoke from miniature volcanoes. They perform for you if you touch them ever so lightly.

The mushrooms we see throughout the year—at least one or two species may be found during the winter—are the fruiting part of the plant. They produce spores that vary in shape, size, and color, though all are microscopic. When the spores become ripe, they are released and carried on the wind until, by the force of gravity, they fall to the ground. If they land in a spot favorable to growth, they germinate and through cell division send out a tiny chain of cells that penetrate the substratum suitable to the species. This tiny chain, absorbing nourishment from the organic matter in the substratum, grows longer through further cell divisions until long chains of cells are formed. These appear to the unaided eye as tiny threads (hyphae) but when seen with the microscope they look like the drawing in Figure 80c. These threads in turn branch, and the branches eventually interlace with each other, forming webby mats that ramify to a considerable distance through the substratum. These mats are often mistaken for root fibers, but they constitute the vegetative body (mycelium) of the mushroom. Mushroom growers call these mats spawn.

Ultimately small outgrowths, called buttons, appear on the mycelial threads. No larger than pinheads, they increase in size, and as they grow, minute stems develop that gradually lift the buttons above the surface of the substratum. There they develop into the characteristic mushrooms we know so well.

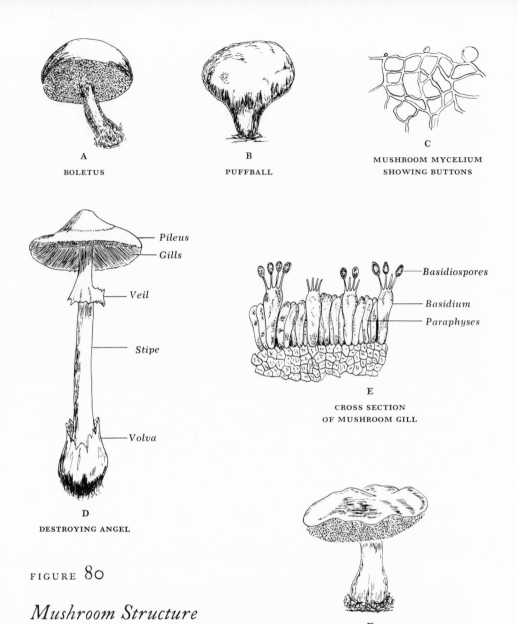

A

BOLETUS

B

PUFFBALL

C

MUSHROOM MYCELIUM
SHOWING BUTTONS

Pileus
Gills

Veil

Stipe

Volva

D

DESTROYING ANGEL

Basidiospores

Basidium

Paraphyses

E

CROSS SECTION
OF MUSHROOM GILL

FIGURE 80

*Mushroom Structure
and Varieties of Fungi*

F

HYDNUM

A typical mushroom is shown in Figure 80D. It consists of the
following structures: the stem or stipe; the cap or pileus; the gills,
which are thin plates that radiate beneath the cap from the stem to
the edge of the cap, like the spokes of a wheel; the annulus or ring;
and the volva. The ring and volva are remnants of two thin membranes, one of which extended from the margin of the cap to the

stem and covered the gills, and the other of which covered the rest of the cap while it was in the button stage. As the cap develops, expands, and becomes extended upward, both membranes are torn, and in some species they remain attached to or surround the stem.

A mushroom consists essentially of mycelium, except on the surface of the gills. If a very thin section is cut across the gill and examined with the microscope, it will be found that the central portion is made up of loosely tangled mycelium threads and, just outside this loose mycelium layer, a layer of shorter cells (hymenium) from which extend club-shaped bodies. Some of these bear two to four and, in some species, as many as eight little prongs or stalks (sterigmata). Some sterigmata develop a spore, while others remain sterile. The spore-bearing bodies are called basidia, the sterile ones paraphyses (Fig. 80E).

Unlike the conventional type of mushroom, some species such as the Boleti produce spores on the walls of the tubes that extend downward from the lower surface of the cap; others produce them on clubs or teeth. When we look at the lower surface of a Boletus both with the naked eye and the lens, it reminds us of a much-used pincushion. Each hole is the mouth of the tube through which the spores finally are liberated. There are a number of species, and many are quite beautiful.

Mushrooms that produce their spores on teeth are the club or coral mushrooms. They are fleshy fungi of upright growth which have their spore-bearing surface exposed on the apices of branching or simple clublike forms. Many resemble corals of exquisite shades of pink, violet, and yellow. Species that bear their spores on awl-shaped teeth which project downward are known as the Hydnums (Fig. 80F); two common forms are the bear's head hydnum and the hedgehog mushroom.

Puffballs, earthstars, and the like are known as the pouch fungi. They produce their spores in a stemless, spherical sac covered with a rind which, in different species, has various characteristic ways of opening to permit the spores to escape. Earthstars, which are actually puffballs, are the most picturesque forms of these odd plants. One species is a veritable barometer, for if wet weather is in the offing, the points of the star open up; otherwise they remain closed.

If we look in a damp, shaded place, we may find what appears to be a slimy, moldlike mass on a piece of rotting wood or on decaying leaves. It is anything but attractive and yet, when examined with the microscope, may reveal an unexpected beauty. The slimy mass is a plant, a slime mold; more accurately, it is the plant body or plasmodium and is simply a mass of naked protoplasm, in consistency much like that of an egg white.

The slime molds are a unique group of plants and have few characteristics in common with other plants except that they produce spores. Some species may be compact in form and only a few millimeters in diameter; others may form an open loose network or an irregular film or sheet several inches across. Lacking chlorophyll, they obtain their nourishment from the damp organic substratum on which they live, ingesting solid pieces from which they later digest the nutritive materials.

It may not seem apparent at first, but these plants move, somewhat in the manner of the amoeba. We need only keep one under observation for a few hours to find that it has traveled a considerable distance. Ordinarily the plasmodium is sensitive to light; while in the vegetative state it stays close to moist and shady places, but when it is about to produce spores it emerges from the shade into drier, more elevated, and more brightly lit places.

The spores are produced in structures, sporangia, such as is shown in Figure 81A. A sporangium consists of a stalk[6] and an enlarged upper portion, containing, within a hardened envelope, a lacy, delicate framework called the capillitium (Fig. 81B) in which the spores are enmeshed. In some species the sporangial walls are beautifully colored: white, violet, purple, orange, and brown. When the spores have matured, they are released, and on being deposited in a moist place, at once begin to germinate and form one-celled, free-swimming organisms called zoospores. They swim about for awhile and then form what are called swarmcells. These swarmcells divide several times, creating more swarmcells which unite with others to form eventually a new plasmodium.

Very often we notice on plants such as the lilac and the rambler

[6] In some species the sporangia lack stalks and develop directly from the plasmodium as spherical or ovoid bodies.

rose, as well as a host of others, a delicate, grayish, cobweblike coating on the leaves. It looks like a film of dried soapsuds, but when examined with the microscope is seen as a branching mycelium from which short hyphae (haustoria) extend into the leaf tissues. Other hyphae may also be seen extending vertically upward. These aerial hyphae bear spores—conidia—in the form of chains, which are formed one at a time by cell division. The spores readily break apart from one another and are dispersed by the wind. If they fall on a plant slightly covered with a film of water, they germinate and develop into mycelia, which in turn produce haustoria and conidia, thus repeating the life cycle on a new host.

This fungus plant is a powdery mildew. Powdery mildews are usually mild parasites, inflicting slight damage on the host plants, but a few species are sufficiently virulent to cause significant damage. In autumn, minute black spots, each smaller than the point of a lead pencil, may be seen amidst the grayish mycelium. Under the microscope these spots appear as hard black spheres (cleistothecia), from which hyphae radiate like the spokes of a wheel (Fig. 81c). Each hypha ends in a hook of characteristic form. Cleistothecia contain one or several thin-walled sacs called asci in which two to eight spores, ascospores, are formed.

The formation of the cleistothecia is the plant's way of surviving through the winter. The following spring they crack open, freeing the asci, which in turn rupture and release the ascospores. On finding the proper host, the ascospores germinate into new mycelia, and a new cycle is begun.

I doubt if there is anyone who has not had bread become moldy. The delicate fuzzy growth is a vast network of mycelium threads, analogous to those we found in the mushroom. The small white spheres that form on the aerial hyphae are the sporangia; as they mature, they turn black (Fig. 81d), the walls rupture, and the spores are liberated. When they encounter favorable conditions for growth, they germinate, each producing a hypha which, by extensive branching, forms a new mycelium.

There are other kinds of molds—the blue and green molds that occur on cheese, jellies, jams, leather, and wallpaper. Superficially they resemble the black bread mold, but there are some differences,

FIGURE 8I

Concerning Molds

A

SPORANGIA
OF SLIME MOLD

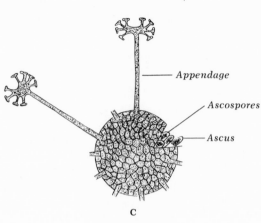

—— *Appendage*

—— *Ascospores*

—— *Ascus*

C

CLEISTOTHECIUM
OF POWDERY MILDEW

—— *Capillitium*

—— *Stalk*

B

SINGLE SPORANGIUM
OF SLIME MOLD

Sporangium

Sporangiophore ——

Stolon

Mycelium

D

BREAD MOLD

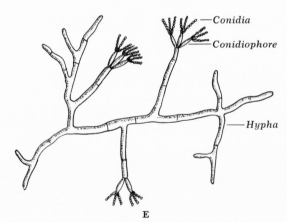

—— *Conidia*
—— *Conidiophore*

—— *Hypha*

E

COMMON BLUE MOLD
(PENICILLIUM)

notably in the color of the spores. There are some two hundred and fifty species; many are harmful, some are useful as Penicillium (Fig. 8IE).

We have relatively little rain during the month of September, and the days are bright and sunny. The sun is getting lower in the sky, however, and the hours of sunshine are decreasing. Perhaps we may not notice it, but the animals and plants, more sensitive than we are, become aware of it and respond in various ways. As the nights become progressively cooler, the toads stop feeding and hop restlessly around or sit in protected corners. They are among the first of the amphibians to go into hibernation, and before many days pass they will dig into the soft soil of the garden or in the banks of ponds in which they breed.

By now the woodchuck has become so fat and sluggish from stuffing himself with grasses and tender green succulents that he seems hardly able to move. No longer do I find him waddling along in the field or by the woodland border; instead I see him sitting by the entrance to his burrow, a picture of listlessness and immobility. Not many more days of wakefulness remain to him; before the month is out he will have gone into his burrow, not to reappear until next spring.

The bear, too, has become fat from stuffing himself, because, like the woodchuck, he spends the winter in his den. The chipmunk, of course, is busy filling his underground granary, and the white-foot is following his example. The mouse is partial to beechnuts—as many as a peck have been discovered in his nest—but he is also fond of basswood seeds and the berries of the black alder; I have seen him on occasion climb up the twigs on a moonlit night to gather the berries.

Meanwhile the muskrats have begun to plan their winter homes, and deer, which during the summer come out of the thick woods at night to feed on water lilies and other succulent water plants, have forsaken the water courses and taken to the deep woods where there is now plenty of food, especially beech mast, on which they get amazingly fat. Before long the bucks, having rubbed the velvet from their antlers, will begin their wooing and will battle one another for a mate. Both sexes have started to shed their summer coats and to take on the warmer "blue" pelage of winter; and the spots of fawns are disappearing.

Other mammals are getting ready to shed their summer "furs," and the birds, too, have begun to exchange their bright plumage for the duller hues they use for traveling and for the winter season. The birds are now more in evidence than they have been for some time; they are gathering in flocks and gorging themselves on ripening seeds and fruits in preparation for their southward journey.

In the early morning I can see ducks flying toward the coast, where they will spend the winter in the salt marshes. Later in the day I can watch the neatly uniformed redwings wheel and advance in military platoons over the marsh. Woodland birds flit from thicket to thicket, quickly scudding to shelter as the ominous form of a hawk appears overhead, for now that the mating season is over, the hawks have more leisure to spend in the air.

Many birds have already left for the South, and those whose breeding grounds are farther north are passing through. We have only to go out on a moonlight night to see hundreds of them flying across the bright face of the moon at heights of a quarter mile or so. If the night is dark and cloudy, we can hear them calling to one another in an effort to keep together. Not all birds migrate by night, of course.

Insects migrate, too. Most of them travel only a short way, such as from a tree or shrub into the ground, or from a field to a nearby house or barn. Some, however, journey considerable distances. The most notable example is the monarch butterfly: as cold weather approaches, monarchs gather in swarms of thousands of individuals and, like birds, fly south. How these butterflies find their way, or how they endure such long journeys, or why they migrate at all, we do not know. But they are not the only ones to do so. The painted lady, buckeye, purple wing, and cloudless sulphur, among others, also migrate, though not so regularly.

Rather surprisingly for such a late date, the copperhead snake brings forth its young alive, although sometimes it will do so toward the latter part of August. The young, which number from five to ten, are about 10 inches long and have bright sulphur-yellow tails. The rattlesnake also gives birth to a dozen or more young, which are born with one button on their tails. In both cases the baby snakes are born from egglike envelopes retained within the mother's body until the embryos have fully developed and are able to tear their way out and fend for themselves.

As the September days pass, there is a gradual but perceptible change in the colors of the leaves. Chlorophyll is no longer being made, and as it begins to fade, other pigments come to the fore to replace the chlorophyll green with reds, yellows, browns, and all the brilliant colors that conspire to make the autumnal foliage the spectacle that it is. We need only a few cool nights to hasten the process, and these we usually have toward the end of the month. And as we observe the deepening of these colors and the patterns beginning to take shape, we know October is upon us.

OCTOBER

"A pomp and pageant fill the splendid scene."

HENRY WADSWORTH LONGFELLOW

"Autumn"

OCTOBER

T HE RAYS OF THE SUN NOW REACH THE EARTH at more of a slant than they have, bringing somewhat cooler days and nights—invigorating days, and nights of refreshing sleep—to say nothing of putting an end to insect annoyances. The temperature is now more equable, with golden days and clear nights following one another. Rain is at a minimum, and yet we need not fear drought, for crops and gardens are no longer in need of quickening showers. Suppose the water table does drop a mite; November rain and the winter snows will take care of it, and replenish the springs in time for the plants to renew their growth and the farmer to plant his crops.

The glories of the sunrise are repeated in the sunset glow, and during the day the landscape is painted with colors that rival those of spring. Now the colors are not of flowers, however, but of leaves that nature paints with pigments from her palette. Much of the brilliance of the autumn coloring is due to the red maples. As in the spring, when the opening crimson and scarlet flowers spread a glow over the landscape, the scarlet and crimson leaves now stand out against the azure October sky and wrap the woodlands in a scarlet cloak. The sumacs, too, no less wonderful, contribute their share to the autumn pageant. They gleam in scarlet and gold, often deepening to crimson and orange. The staghorn sumac makes thickets of its own, and brightens waste places and neglected fields that otherwise would remain a barren void, even as its smaller brother the smooth sumac flings its magnificent beauty along fencerows, over deserted fields, and up rocky, gravelly mountainsides,

> *Like glowing lava streams, the sumac crawls*
> *Up the mountain's granite walls.*

Not to be outdone in magnificence, poison ivy adds its brilliant crimson to shady copses and hidden nooks, while Virginia creeper, with sprays of cardinal red and maroon, decorates the somber boles of dark evergreens and the shadowy outlines of stone walls. Along some bordered stream and on the distant hillside, the red oak flaunts its rich, dark, purplish red, while here and there the scarlet oak appears as if on fire.

But on the uplands the beech, its trunk a mixture a pearly white and bluish gray, and its leaves of the palest Naples yellow, serves notice that all is not crimson and scarlet on the October scene. The birches, too, their trunks spotted with shadows of violet-blue, take up the challenge with golden-yellow. And by the wayside and in the farmyard stands the sugar maple, with leaves of yellow or crimson, scarlet or orange, green with a spot of crimson, or crimson with a spot of pink; they may even have a patchwork of yellow, purple, and scarlet. One never knows what the sugar maple will do.

Of course, wild flowers add a bit of color to the October scene, and the berries contribute their share. We may not see all the complexities of color, nor catch the bluish shadows of the roadside, the browns and grays of weatherbeaten fences and stone walls, the play of light on the distant farmhouse.

Goldenrods and asters are, perhaps, the most conspicuous of the wild flowers still in blossom, and surprisingly there are quite a few of them at this date when you total them up. Daisy fleabanes, whose kinship to the asters and the daisy is apparent at a glance, are still in flower in fields and along the roadside, their heads nodding in the breeze like village gossips. By stone walls and fencerows the bright-yellow flower heads of the Jerusalem artichoke rise like miniature suns. Both the blazing star and the bur marigold, having appeared in August, still continue to flower, the former in fields and waste places, the latter in swamps and ditches. Both are composites, the blazing star with tubular florets only, the bur marigold with disk and ray flowers. Bees, flies, and butterflies are frequent visitors to both. The florets of the blazing star show many beautiful variable tints, which are easier to see with the lens than they are with the naked eye.

A pernicious weed of the grain fields and gardens, the pigweed or lamb's quarter, is succulent and tender if cooked when it is small. As a dish for the table it is the equal of spinach and beet greens, its cultivated relatives; some people prefer it to spinach when it is served with butter and a little lemon juice. Its small, insignificant flowers are not particularly interesting; but the small, lens-shaped black seeds have the merit of remaining viable even though they lie dormant in the soil for years.

FIGURE 82

Two Common Flowers of Summer

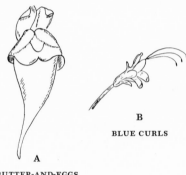

B

BLUE CURLS

A

BUTTER-AND-EGGS

The seeds of the ragweed also retain their vitality after long periods of dormancy. This is one of the more common of our weeds, intruding almost everywhere. It is a notorious cause of hay fever, though in this respect its pollen is no worse than the pollen from several other plants. A related species, the giant ragweed, normally attains a height of about 10 feet; under unusually favorable conditions it may grow to 18 feet.

Ladies' tresses and butter-and-eggs are of considerably more interest than either the pigweed or ragweed. A marsh orchid and the last of the orchids to flower, the plant called ladies' tresses has a peculiarly twisted or spiral flower spike with small, translucent, yellowish-white or variably cream-white flowers that are cleverly contrived for insect pollination. With your lens watch what happens when a bee alights on one of the blossoms. Watch, too, when a bumblebee visits the butter-and-eggs (Fig. 82A): notice how, when guided by the gold-orange palate to the place where the curious flower opens, its weight depresses the lower lip and makes an entrance through the gaping mouth large enough for it to pass into the throat of the corolla. You will be able to see how its back brushes off the pollen from the stamens overhead as its tongue seeks the nectar deep within the spur, and how the gaping mouth of the flower springs tightly shut after the bee has taken its departure. A weed it may be, but a more beautiful weed is hard to find, with its canary-yellow and orange cornucopias and gray-green linear leaves. Were this plebeian perennial less satisfied to grow everywhere without cultivation, it might well have become a garden ornamental instead of decorating a city lot or some equally unattractive place.

Despite its frail appearance, the chickweed is probably the hardiest plant on earth, and one of the most persistent. I know of no other that blossoms every month of the year in the North: I have often found it blooming in secluded spots during midwinter thaws. Everyone is acquainted with it, but you should look closely at the blossoms if you never have. You will find them white and starlike, and apparently set inside larger green blossoms, though this is not quite true, since the white stars are composed of the petals cleft down the center and the green ones are formed of the sepals. One might wonder what practical purpose could be served by this straggling little plant which is such a pestiferous weed in our gardens, but everything has its use; the seeds are given to caged songbirds.

Anyone who has ever had a garden need not be reminded of the purslane. Were it not such a troublesome plant, seemingly impossible to get rid of, we might regard it with more favor, for its thick, fleshy, dark-green leaves, its succulent, often terra-cotta pink stems, and solitary yellow blossoms are not unattractive. But we look upon it with dislike and esteem its relative the garden portulaca more highly.

Cats love the catnip so well that they will seek it wherever it grows, sometimes traveling considerable distances to become half-crazed with delight over its aromatic odor. Although it is a plant of waste places, I have had it growing about my dooryard. We find other mints, too, at this time of the year—the wild mint and blue curls (Fig. 82B), for example. The latter is remarkable for the extraordinary length of the violet stamens; spirally coiled in the bud, they extend in a curving line far beyond the five-lobed corolla when the flower is open. As for the Indian tobacco, the most stupid of animals know enough to leave it alone, but not man, who used it to make a quack medicine; how the Indians could have smoked and chewed the bitter leaves, I'll never know. Chew even one of the green bladder pods and you will experience a feeling of nausea.

When one sees the fall dandelion in bloom for the first time, he begins to wonder if the common dandelion of spring has not become aberrant. But a second and closer look shows that it is not the same species; indeed, the two do not even belong to the same genus. Yet in a way it can be looked upon as a smaller edition of the larger and more robust dandelion we know so well, for both have certain features in common; their leaves, for example, have the same backward-turned, sharp-pointed lobes, or "lion's teeth," that originally suggested the name for both plants.

Despite all these late-blooming flowers, October really belongs to the gentian:

> *Thou waitest late and com'st alone,*
> *When woods are bare and birds have flown,*
> *And frosts and shortening days portend*
> *The aged year is near his end.*

These words of Bryant's[1] are not strictly accurate; but no matter, it is still a delight to find the gentian in blossom, for it is becoming much too rare. I have never forgotten the day when I chanced, for the first time, upon this gay and lovely flower in a shady copse that fringed a woodland pool.

October is the nutting month. Acorns, beechnuts, hickory nuts, and butternuts are fast coming to maturity, and as I wander through the woods they fall from the branches and beat a tattoo upon the ground. When I was a boy I went nutting every year, but now I prefer to leave nut gathering to the chipmunks and the squirrels. The chipmunk has few more days above the ground if the weather holds to its normal pattern; before the month is gone he will have disappeared, although if the days remain unseasonably warm he may wait until December to retire for the winter. In any

---

[1] From his poem "To the Fringed Gentian."

event, he is busy storing seeds and nuts in his underground granary, and the gray squirrel is no less industrious.

The red squirrel does not follow his cousin's example of burying nuts; though he does eat nuts, he prefers seeds, which he stores away in caches here and there. Should you venture where these squirrels are abundant, they will bark and spit at you from the branches and observe your every move, as if resentful at your trespassing upon what they consider their own private domain.

FIGURE 83

Fruiting Structures of the Oak, Beech, and Butternut

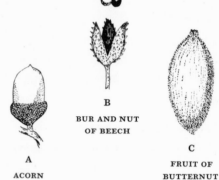

B
BUR AND NUT
OF BEECH

A
ACORN

C
FRUIT OF
BUTTERNUT

An acorn (Fig. 83A) is a nut, botanically speaking, and seemingly one acorn is much like another. But they are quite variable in shape, size, and form of the cup. With a little study one can easily learn to correlate the species of an oak with any given acorn. Aside from the cup, the lens will show little besides what can be seen with the naked eye; but if the cup is examined with the glass, it is seen to be woody in texture and made up of a number of tiny scales which have grown together, either entirely or with free tips.

I have read that beech trees have to be at least forty years old before they can bear fruit. I don't know how true this is, but since the trees grow to a venerable age—300 to 400 years—it might be so. The tree is not consistent in bearing fruit, but produces large crops at irregular intervals. The fruit is a tiny burr (Fig. 83B) and, as seen with the lens, has soft, spreading recurved prickles. It opens in four sections and remains until after the nuts have fallen.

The nuts, brown, shining and triangular, are sweet and edible; there are two, sometimes three, to a bur.

Unfortunate, indeed, is the person who has not gathered a few hickory nuts and eaten the sweet kernels. The nuts of the shagbark, pignut, and mockernut do not appear much different when seen with the naked eye and the lens, but examine the greenish-bronze husk of the butternut (Fig. 83c) through the glass and look at the small, clammy hairs that are sticky to the touch; then break it open and look at the nut: it is deeply sculptured. The husk of the bitternut, too, is covered with a sort of scurfy pubescence; note the four prominent winged sutures reaching halfway to the base, and observe that the apex shows the remnants of the stigma. You might examine both the husk and the nut of the black walnut, too.

About the second week of October, the brook trout begin to move upstream to their spawning grounds. The life of the pond or stream is much as it has been all summer, and it will continue so until lower temperatures cool the water. Then many members of the pond community will retire for the winter, while others, unaffected by temperature changes, carry on their normal activities. As one stands by the water's edge and looks into the water, one may see whirligigs and water striders, diving beetles and backswimmers, and many others of the insect clan; in addition, there is perhaps a fish or two or an amphibian, and maybe species of other phyla. But countless others cannot be seen—animals that swim in the water, lurk among the submerged vegetation, or crawl on logs and stones. Some are so small that we need a microscope to know they are there at all; others appear as tiny specks; still others look like the tiny threads which most of us assume they are.

Such minuscule threads are the hydras, said to be named after the fabled monster slain by Hercules. As you recall, the monster had nine heads, any one of which (except the middle one) would grow two in its place if cut off. The hydra has somewhat the same ability, for if the head end is split in two and the parts separated slightly, a "two-headed" hydra results. I don't know how many heads a hydra can grow; theoretically, I suppose there is no limit.

The hydra is a common inhabitant of ponds, slow streams, and still pools, and we usually find it hanging from such water plants

FIGURE 84

Some Common Inhabitants of Freshwaters

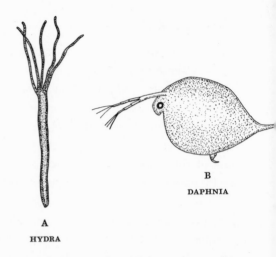

A
HYDRA

B
DAPHNIA

as nitella and elodea or from the undersides of water lilies. It looks like a small piece of thread, frayed at one end, and about three quarters of an inch long, when extended (Fig. 84A).

Since hydras are too small to be observed in their natural habitat, it is best to transfer a spray or two of nitella or elodea to a glass bowl or jar of pond water. If any hydras are present on the plant stems, they will soon hang out into the water. Looking at one of them through the lens or a reading glass, you can see that what appears to be the frayed ends of the thread are tentacles. They surround the mouth and seem to wave about continuously. The "thread" itself is the body of the animal, and as you watch, you will see that it can contract and expand at will. So, too, can the tentacles. Tap the bowl or jar, and the entire animal contracts to the size of a pinhead.

The tentacles, used to capture and convey food to the mouth, are furnished with stinging cells, which also occur on all parts of the hydra except the basal disk. There are four kinds of stinging cells. Though they differ somewhat in structure, basically all, when activated, discharge a threadlike structure which either penetrates or entangles a prospective prey. At one time it was believed that these stinging cells, nematocysts, were triggered to explode by touch or some other mechanical stimulus, but it has since been shown that they are discharged by a chemical agent. They are to some extent defensive, but are probably of more value in capturing some of the livelier organisms on which the hydra feeds.

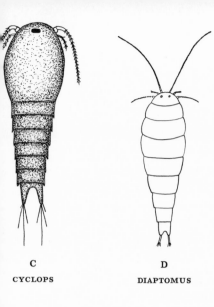

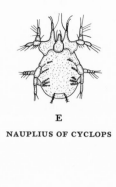

E

NAUPLIUS OF CYCLOPS

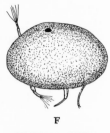

F

OSTRACOD, CYPRIS

C

CYCLOPS

D

DIAPTOMUS

The hydra's prey includes small crustaceans such as daphnia, cyclops, and cypris, minute worms, immature insects, and small clams. Hydras even capture and swallow small tadpoles and fish at times. After being swallowed, the prey passes into a capacious cavity or stomach, where it is digested (you can see this with your lens). The useless parts are ejected through the mouth, since there is no special organ for this purpose. The hydra has an insatiable appetite and will eat until its body becomes distended.

The daphnia, cyclops, and cypris that serve as food for the hydra are crustaceans belonging to various orders and referred to collectively as the entomostracans. The daphnia (Fig. 84B) is what is popularly called a water flea, of which there are several hundred species. Water fleas swarm through the waters of our ponds in amazing numbers. It was Swammerdam who first called them water fleas when, in 1669, he described the common water flea as *Pulex aquatiticus arborescens*—the water flea with the branching arms— and water fleas they have remained ever since.

Most water fleas—which incidentally, have long been favorite animals for study and are known scientifically as the cladocerans— are minute, even microscopic; although a few, such as the beautiful and fierce carnivore Leptodora, may measure as much as ¾ inch in length. Usually their bodies but not their heads are enclosed in a shell or carapace which is thin, finely reticulated, striated, or sculptured (in some species with conspicuous spines), and so transparent that we can see the internal organs.

All cladocerans have five pairs of leaf-shaped appendages used in creating currents of water that pass over the respiratory valves at the bases of the legs and that also carry particles of food to the mouth. In addition they are equipped with two pairs of antennae, the first pair often minute, the second pair larger and used in swimming. They forage on submerged plants and subsist on bacteria, the lesser green algae, diatoms, and minute particles of organic matter.

Daphnia, one of the more common species, is particularly abundant at this time of the year. With a microscope you can watch the ingestion of food and follow it on its way to the digestive tract. Also visible are the rapidly beating heart, circulation, respiration—in fact, there is little you can't see. Daphnia has less privacy than the goldfish in a bowl and can keep no secrets from anyone who wants to know all about its private life.[2]

Equally abundant in our ponds are water hoppers or copepods, small, pear-shaped animals with a body tapering to a forked tail, four pairs of swimming legs, and a pair of antennae that are primarily sense organs but are often used in locomotion. They look not unlike miniature lobsters. A jar of pond water may show small white specks moving jerkily about that, when magnified, will probably prove to be water hoppers, in all likelihood cyclops (Fig. 84c), which is the most common species. This little animal was named after the cyclops, the mythological race of one-eyed giants, and when you look at it with your lens, or preferably the microscope, you will see why.

Water hoppers feed on microscopic organisms and any kind of decayed matter, but they prefer decomposing plant tissues. Some of them live among the plankton of open water, and these forms are apt to be slender and transparent; others that live near the shore and among the submerged water plants are usually shorter and often dark-colored. Many of the water hoppers are variable in color; species of diaptomus (Fig. 84D), for instance, may be red, blue, or purple, and when they occur in large numbers, they color the water.

[2] Since water fleas and other small aquatic animals are fairly active, I suggest you add a drop or two of a solution made of two parts of 1 per cent chloretone and five parts of water to slow them down a bit.

Females may be recognized by the pair of egg pockets hanging from their sides. During the mating process the male clasps the female with his antennae, and the transfer of the sperms takes place as they swim about. Young copepods, nauplii, do not resemble their parents at all; you can get a better idea of what they look like from the illustration (Fig. 84E) than from a description. They molt several times before they acquire the typical adult form.

When you first look at an ostracod (Fig. 84F), you may have to do so with the lens or a microscope, for they average only a millimeter in length, which is about the size of a pencil dot. Then you may think you are looking at a miniature clam or a very young one, for like the clam, ostracods live within a bivalve shell, the two halves being connected by a hinge which is opened and closed by muscles. But here the similarity ends. Unlike clams, that move by means of a muscular foot, ostracods have a number of jointed appendages that they extend between the two valves and, by kicking movements, use to propel themselves forward.

Some ostracods are free-swimming, but most of them are given to creeping about in the soft ooze or climbing on submerged water plants. They are omnivorous scavengers, being especially partial to decayed vegetation, and where such food is in abundance, they frequently swarm and appear as countless moving specks when we look down into the water. By immersing an empty bottle and allowing the water to flow in, we can collect hundreds of them.

You won't find the shell of the ostracods as transparent as that of the water fleas, but you will often find it sculptured or otherwise marked in patterns of contrasting colors. Many ostracods are yellowish-white, others are brown, orange, or violet, and those that live among the algae are often green. A few species have no eyes, some have one, and still others have two. If you have never seen them, I think you will find them quite fascinating.

The entomostracans are an important source of food for other animals, especially small fish. But many no doubt die a natural death, and many others fall victim to the traps of the bladderwort, which are more marvelously contrived than those of the sundew, pitcher plant, and Venus's-flytrap. The bladderwort (Fig. 85A) is rather delicate and vinelike in appearance as it floats beneath the surface among the water lilies and pondweeds. It has no roots, but twiglike rhizoids serve the same purpose. The slender stem bears leaves finely divided into threadlike segments and bladders that become filled with air. When the plant is in flower, the bladders function as pontoons to keep the yellow blossoms out of the water where the insects can visit them, but at other times the bladders act as traps.

As seen with the lens, a bladder (Fig. 85B) is a slightly compressed sac with a slitlike opening. Armed with teeth and bristling hairs, this opening is guarded by a valve or trap door that only opens inwardly. How does the trap operate? When an unwary animal swims into the opening, the valve is stimulated and opens. As it does, water flows into the bladder so that its walls, which are ordinarily compressed inward, are now pushed outward under the pressure of the water. This outward movement creates a suction that draws more water into the bladder, and the water carries with it not only the animal that tripped the trap but others that may have been

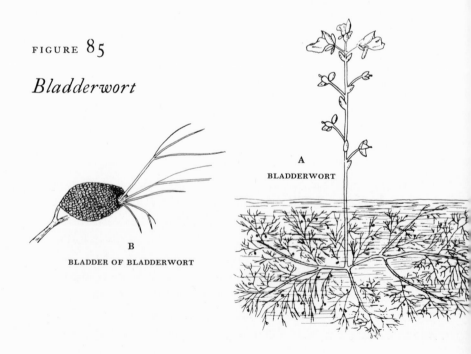

FIGURE 85

Bladderwort

A
BLADDERWORT

B
BLADDER OF BLADDERWORT

nearby. The action is much like that of a sinking ship which pulls down with it anything within the area affected by its suction. Once within the bladder there is no escape, because the valve is strongly elastic and shuts tightly once the animals are inside. What is the fate of the imprisoned animals? They are digested by fluids secreted by the bladder walls and are used by the plant as food.

Imagine, if you can, a group of animals equally at home on the open surface of lakes, where they live in vast hosts; in the cool waters of a perennial spring; in the shallows of ponds and bogs, where they live in the bottom ooze or on the stems and leaves of water plants; in the most transient of pools; or in a rain barrel, a rainspout, or stone urn, where they are subject to desiccation between rain showers. Imagine, if you can, a group of animals that are swift-moving, or slow, or sessile; or that live symbiotically within the tissues of submergents, or that build tubes to live in. In short, they are so diverse in form and habits, and appear in so many varied habitats, as to satisfy the most discriminating investigator.

These unusual animals are the rotifers or wheel animalcules. They are generally microscopic in size, though a few can be seen with the naked eye. Many are brilliantly colored, and sometimes these are so abundant that they color the water. Many are solitary and move freely through the water, the swifter feeding on microscopic animals, the slower living on vegetable matter. Others form colonies that are often conspicuous and that, when attached to leaf tips, give the semblance of flowers. Frequently these forms become detached and go bowling through the water like small white rolling spheres sufficiently large to be visible to the naked eye. They are among the hardiest of animals and can withstand extremes of drought and cold. In the fall they produce tough shells in which they lie dormant through the winter; they can even be frozen in the ice for a long time and still be able to resume their normal activities when thawed out.

Rotifers come in an infinite variety of forms, yet all have one feature in common: circlets of cilia. Look at a rotifer with the microscope and you will see the cilia beating incessantly. The circlets look like two rapidly rotating wheels; hence the name Rotifera, or wheel bearers. The two circlets rim the head region or corona, and the cilia are used for locomotion and collecting food. The mouth is at the lower edge of the corona and opens into the pharynx, the lower end of which forms a grinding organ called the mastax. You can see it and more, for the rotifers are transparent, and you can witness all that goes on within them.

The mastax, which is a sort of food pouch, is equipped with chitinous teeth or jaws that crush or grind the food materials, such as algae and protozoa, swept into the mouth by the cilia. A short esophagus leads into a glandular stomach where food is digested, though in some species it is digested in both the stomach and the intestine. Undigested food materials pass into the cloaca and out through the cloacal opening (anus). The excretory system consists of two long tubes and flame cells, the latter with flickering cilia that send currents of fluid down the tubes. These tubes connect with a bladder that contracts at intervals, forcing the contents into the cloaca and then out through the cloacal opening. The bladder also helps the animal to maintain a proper water balance. Figure 86A will help you locate all these organs.

At the lower extremity you will observe what is commonly called the foot, which in some species is forked and might better be called a tail. The animal uses it as a means of attaching itself to plants and other objects, though a cementlike substance, secreted by cement glands, does the actual adhering. Male rotifers are extremely minute, and we seldom see them. In some species the eggs develop without fertilization, and there is a question whether males exist at all. Where they do occur, they are usually smaller than the females and degenerate.

In Figures 86A,B,C, I have shown some of the more common species. Melicerta builds a tube by cementing small pellets together. It lives on the lower surface of lily pads, on hornwort, and on milfoil, where it can usually be found in fairly large numbers. It is a beautiful and fascinating creature; indeed, all the rotifers are, and I can guarantee you many enjoyable and entertaining moments watching them.

Certain species of crayfish, which are abundant in most ponds, mate during October, though they do not spawn until spring. In most ponds there are also abundant freshwater mussels and clams,

FIGURE 86

Some Typical Rotifers

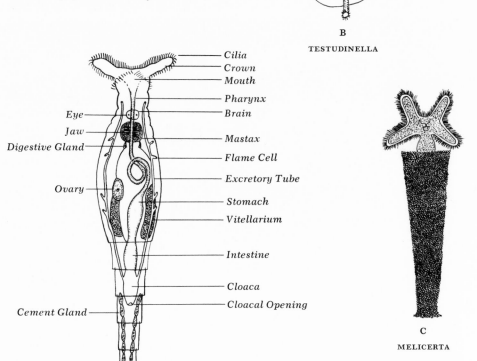

B

TESTUDINELLA

Cilia
Crown
Mouth
Pharynx
Eye
Brain
Jaw
Digestive Gland
Mastax
Flame Cell
Ovary
Excretory Tube
Stomach
Vitellarium
Intestine
Cloaca
Cement Gland
Cloacal Opening
Toe
Foot

A

PHILODINA

C

MELICERTA

papershells, fingernail clams, pill clams so small we need the lens to see them, and pond snails, both right-handed and left-handed ones.

Some of the pond snails breathe with gills; others have a lung sac and are air breathers. The land snails that often come down to the water's edge may be mistaken for the aquatic species, but the pond snails have two tentacles, the land snails four. The latter (Fig. 87A) are very hardy animals and survive cold far below the freezing point. They are still active in the woods and gardens, but as the month advances and the air gets cooler, they cease eating and begin to move beneath stones and into tree trunks, or bury themselves in the moss and leaves. They remain dormant during the winter and, as a rule, hibernate alone, though sometimes they gather in groups. When they retire for the winter, they withdraw within their shells and secrete a thin curtain of mucus over the aperture which prevents them from drying out.

Despite the way they move and the burden they carry, snails get about very well. Actually, the snail's foot is one of the most wonderful means of locomotion ever devised by nature. Place a snail in a glass tumbler and with a reading glass watch it climb up the side. Observe how the foot stretches out and holds onto the glass. Note, too, that a slime gland at the anterior end of the foot deposits a film of mucus on which the animal moves; it lays down its own sidewalk, as it were, and this sidewalk is always the same, smooth and even.

The most conspicuous feature of the snail is undoubtedly the shell, which is formed from a secretion that hardens on exposure to the air. Examine a shell with your lens and you will find it twisted on an axis called the columella, successive additions of new shell being shown by lines of growth. Also look at its skin with the lens—it is like that of an alligator, rough and divided into plates, with a surface like pebbled leather.

Each of the two pairs of tentacles is provided with round knob-like tips. On the longer pair they function as eyes, on the shorter probably as organs of smell. It is said that snails can detect food hidden as far away as 20 inches. They feed mainly on the soft tissues of plants and the algae on stones, which they scrape off with the ribbonlike rasp or radula. This is essentially a flexible file covered with rows of horny teeth and hooks. As the teeth and hooks vary considerably in shape and arrangement, they serve as a basis for classifying the animals. Place an apple before a snail and then, with a reading glass, watch it eat. With its radula and round tongue, which can be seen popping out, it soon makes a fair-sized hole. But the snail's table manners leave much to be desired; it is a hopeless slobberer.

A rain gutter would seem an unlikely place to find an animal, and yet as we have seen, rotifers are not uncommon there. The sediment of a rain gutter may contain another animal, one of the oddest of all, and one which has defeated all attempts to classify it. To be sure, it has been placed with the spiders and mites, but no one is sure that it belongs there. The animal may also be found

A

SNAIL

FIGURE 87

A Snail,
the Water Bear,
and the Harvestman

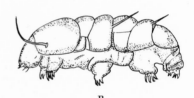

B

WATER BEAR

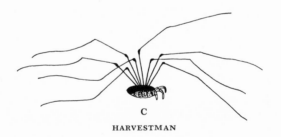

C

HARVESTMAN

elsewhere, as in the bottom mud of a pond or among water plants. It is so small, however, that we cannot see it with the naked eye; the only way to find it is to add a little water to the sediment of a rain gutter or the mud from a pond and examine it with a microscope.

This strange and most unusual animal is known as the water bear (Fig. 87B) or tartigrade because of its habit of moving slowly. It is soft, colorless, and transparent, with four pairs of short hooked legs, a small mouth at the part of the body that represents the head, and an eye on each side.

The water bear feeds on bits of decayed vegetation and is one of the toughest animals we may ever expect to find. Not only can it be dried and shriveled up into nothing more than a grain of powder, but after existing in such a condition for months it may be brought back to life when placed in a little water. It can also survive being heated to a temperature of 250 degrees Fahrenheit. As a curiosity, if for no other reason, the water bear is worth looking for.

Many spiders are still active, as yet unaffected by any change in the temperature. You have only to examine the goldenrods to find crab spiders lurking in the blossoms, or to look about in the fields to see wolf spiders running over the ground. This is the time of year when young spiders go ballooning, and on almost any warm and calm day we can see them sailing through the air. When about to take off, they climb to the top of a bush, a spike of grass, a fence post, or even the summit of a clod of earth. Facing the wind, they extend their eight legs, four on each side, and elevate their bodies at an angle of about 45 degrees, so that they are raised above the take-off perch. In this position they spin several silken threads or a brushlike cluster of threads, which catch in the air currents. That these silken threads exert a considerable uplifting force may be seen by the manner in which the spiders incline their legs and by the tension of their bodies. Somehow they seem to know when this uplifting force is sufficient to support them, for suddenly they let go of their perch and float off on their silken parachutes in whichever direction the air currents carry them.

The spiders are not entirely at the mercy of the air currents; they exercise a certain amount of control over their parachutes by climbing about on the silken threads, pulling in and winding up or streaming out more filaments. Perhaps you have seen gossamer threads shimmering in the sunlight or clinging to your clothing and wondered what they were.

Although in the South the harvestmen (Fig. 87c) hibernate during the winter, we have only one species in the North that does so. Other species die with the approach of cold weather. Eggs carry the species through the winter and hatch the following spring. The young are not only small but timid, and usually hide under stones and other objects until they become full grown in late summer. Then they become more venturesome and come out into the open. Because they are seen more often at this time of the year, they are called harvestmen, although when I was a boy we called them daddy longlegs. We see them most often in pastures running over the boulders, but they also frequent fields and woodland borders. I invariably have them in my garden. These rather odd animals are often confused with spiders, but differ from them in having the cephalothorax and abdomen broadly joined to form a single unit. They resemble mites in this respect, but unlike them have well-marked segments in the abdominal portion, as the lens will show. The lens will also show the two eyes, situated on a prominent tubercle, like a lookout tower.

It is astonishing how these animals can support their bodies on their long, thin, fragile legs, and how they can move over the ground and through the grass without getting them caught in the grass blades. Yet eight legs, I suppose, should be enough to carry a body not much larger than a grain of wheat; and doubtless there are times when they do get their legs entangled. Getting a leg caught, however, is a trifling matter, for if the animal cannot extricate itself it will merely "throw it off" and grow a new one. The legs are unquestionably much stronger than we suspect; strangely enough, they separate easily from the body. This ability to discard the legs at will, combined with their unusual length, has led some observers to believe that the legs serve as a sort of protective fence when the animals are attacked; that an enemy grasps a leg as the nearest thing to seize and is left with it as the harvestmen throws it off and escapes. But I have never seen them use their legs for this purpose, and I doubt if they would, as they have a scent gland that gives off an odor strong enough to discourage any would-be attacker. In fact, they have few enemies anyway, and probably for this reason.

On a bright sunny day I may hear the peeper, but before many more days have passed the diminutive frog will nestle down under the moss and leaves for its winter's sleep. Leopard frogs are not so common in their usual haunts as they were a few weeks ago, and as the days pass more and more of them head for the marshes, ponds, springs, and shallow pools, where they will bury down into the mud. The spotted salamanders, too, are beginning to crawl into holes beneath the leaf cover of the woodlands, and the reptiles are feeling the pinch of frosty nights. The garter snakes are congregating at their winter quarters, and wood turtles are digging into holes in the banks of streams.

During the first days of October, insects seem as abundant as ever. Gnats and flies dance in the golden sunshine; field crickets continue their nightly serenades; and the snowy tree crickets may still be heard from the treetops. But as the October days become shorter and cooler the insects gradually begin to disappear; the cicadas cease their incessant chorusing; ground beetles retire beneath logs and stones; and termites start their downward migration into the soil. Wasps and bumblebee colonies, too, die out, as the queens search for likely winter retreats and the worker honeybees drive the drones from the hives. Occasionally a red admiral butterfly may be seen visiting the still blossoming flowers, but most of the butterflies that last into the fall are the dull-hued species which are inconspicuous among the falling leaves.

Since the insects become less abundant, the October birds are chiefly those that subsist on seeds and berries: robins, bluebirds, purple finches, sparrows, and thrushes. Although a few, such as the phoebe, still remain, most of the insect eaters—the whippoorwill, nighthawk, chimney swift, kingbird, flycatchers, and swallows— have departed. The nuthatches, woodpeckers, and blue jays are winter residents and will remain.

The blue jays are particularly noisy, for their nesting duties are long since completed, and they are free to revel in the plenty that nature has provided. Whenever I walk in the woods they announce my presence with their raucous cries and fly among the treetops, a boisterous, rollicking crew, screaming as if in great pain or terror, but apparently for no other reason than to exercise their vocal cords. Yet their din is not without value, for as the hunting season opens it serves as a warning to the four-footed creatures to scurry for cover.

The catbirds are still about my house, and occasionally, on a warm day, I hear their whispering song. In the fields across the road, chipping sparrows have gathered in a flock, but before long both they and the catbirds will have gone. The southward flight of the hawks is now at its height, and flocks of grackles may be seen in a cornfield or apple orchard. Meanwhile our winter visitors are putting in an appearance, and transients are passing through: white-throated sparrows and ruby-crowned kinglets. Juncoes have already appeared in my garden.

We linger outdoors these days more than we were wont to do, for November is hurrying on the scene, and soon purple shadows will lie softly on snowy fields, winds will howl through leafless branches and about the eaves of the house, and we will have only memories of more pleasant days.

NOVEMBER

"When shrieked
The bleak November winds, and smote the woods,
And the brown fields were herbless, and the shades,
That met above the merry rivulet. . . ."

WILLIAM CULLEN BRYANT

"A Winter Piece"

ℐ OVEℳ B Eℛ

As I LOOK OUT MY WINDOW, I CAN SEE THE wind in swirling gusts blow the fallen leaves crazily over the ground. Only yesterday, or so it seems, they were resplendent in yellows and golds, russets and reds. Now, tattered and torn, they lie scattered about, the playthings of every wilful breeze, a poignant reminder that the winter season is just ahead. Only the rosy glow in the sky as the sun disappears below the horizon remains as the essence of October's brilliance.

Traditionally November is bleak and cheerless, but actually it is as capricious as March, rather a struggling mixture of summer and winter, its moods changing with the varying winds. For there are days when the sun shines brightly and the air is as warm and soft as that of May. Sometimes you can walk in the woods and see moths[1] flying about like so many tiny ghosts, or the caps of the Pholiota gleaming like unset jewels on a decrepit log. You can walk along a country road and see the golden dandelions and the purple asters, or hear the meadowlark singing one last song before he sets out on his southward journey.

But then tomorrow the skies may grow leaden and heavy with the threat that legions of cold may sweep down from Canada and bring a snowstorm with them. I remember as a boy we usually expected skating and coasting on Thanksgiving day, and only a few years ago we had an unusually severe and quite unexpected snowfall that seemed to set the pattern for the winter, since it developed

---

[1] These are the moths of the fall cankerworm. They emerge from their cocoons at this time of the year.

into one of the worst winters we have had in recent memory. Snow-storms followed one another with clockwork regularity, and snow-drifts formed on the fields and meadows like white tents, clogging country roads and woodland trails with seemingly impassable bar-riers.

Usually, however, the snows do not get beyond the border in November, and the most we can expect are flurries and frosts that whiten the ground temporarily. November rains, though, are often frequent and heavy. In the Middle Atlantic states and in New England, the month is generally mild and rather sunny; in the South it is on the summery side, though not with the high temperatures of July and August.

Although the trees and shrubs have become leafless, or are rapidly becoming so, and the grasses and other herbaceous plants have turned a sickly yellow and brown, the landscape is not entirely the picture of desolation that the poets would have us believe. Indeed, there is perhaps more color now than in early spring, though the colors are not the fresh, lush tints of spring but rather soft and mellow. As you walk in the woods you will find that they have a subdued aesthetic quality, particularly as the day grows old and purple and brown shadows begin to steal silently over them to blend into the twilight and the blackness of the night.

You might wonder, too, when you see the golden stars of the witch hazel, if time has

> *. . . grown sleepy at his post*
> *And let the exiled summer back?*
> *Or is it her regretful ghost*
> *Or witchcraft of the almanac?*

But beware of the large, bony, black seeds that may suddenly be discharged and hit you in the face, for they are propelled with a force that can carry them a distance of 45 feet (Fig. 88).

Houseflies still linger outdoors, at least during the early part of the month, and honeybees are still engaged in foraging expeditions. But the bees now make fewer and fewer trips afield, venturing forth only during the sunny midday hours, and the loads they carry become progressively lighter. The field crickets continue to chirp, and a red dragonfly occasionally wings its way across the fields; but most of the insects have snuggled into winter quarters. Insects that have spent the summer on trees and shrubs have sought refuge beneath the bark and in the stems and buds, or have descended to the ground cover of leaves and grass. Wasps have left their paper nests for secluded roof corners. And the ground-dwelling species such as the wireworms, the larvae of May beetles, and the mound-building ants, have burrowed below the frost line or have retired to the lower parts of their nests.

FIGURE 88

Witch Hazel

Other animals, too, have taken to their winter retreats. Except for the water spiders, which may be found in large numbers on bridges, spiders have largely disappeared; snails, slugs, and myriapods have crept beneath the leaf mold; and earthworms have huddled together in rounded chambers deep in the ground. Except for rattlesnakes, which may appear for one last sunning before vanishing for the winter, the snakes have slipped into deep cracks and crevices of piled-up rocks and ledges, where they have knotted themselves into torpid tangles. Moles have dug deeper into the soil; bats have retired to their caves or gone south; and other hibernating mammals have either sought their winter dens or will soon do so, depending on the weather.

Meanwhile deer are gathering in family parties and squirrels are completing their harvesting. Muskrats are putting the finishing touches on their winter homes, while the fur bearers are donning their heavy overcoats. And the varying hare and weasel are acquiring their winter colors.

Despite the lateness of the year, many of our avian summer visitors are still with us: the redwings and cedar waxwings, among others. Occasionally I see pipits in a marsh or pasture, but frosty nights quickly send them on their way. Sometimes I catch sight of a hermit thrush or myrtle warbler, but if I look for them the next day they are gone.

Sometime during the year I invariably visit the seabeach, not to lie on the sands in the hot sun, as so many seem to find pleasure in doing, but to stroll leisurely with an observant eye or to examine the tide pools for what I might find. It is doubtless more pleasant to visit the beach during the summer, but I prefer November, especially after we have had what we call in New England a "nor' easter," for one never knows what a storm may cast upon the sands. Needless to say, I always go there when the tide is low.

To one who visits the seashore for the first time, the stretches of sandy beach and the coastline of jutting rocks may seem desolate indeed, a barren and uninteresting waste that seems to serve no other purpose than to form a natural barrier between the sea and the land. To be sure, the beach is a vast sarcophagus, a final resting place for a host of creatures washed up by the waves. One sees innumerable empty shells mixed with rows of dead seaweed; dried sponges; the tests of sea urchins and sand dollars; the remains of crabs and fish; and the molted shells of horseshoe crabs (Fig. 89). Too, there are the long strings of saucerlike capsules which contain the eggs of the channeled whelk or the giant whelk; the collarlike sandy rings with the eggs of the moon shell; the yellowish capsules of the waved whelk the size and shape of a split pea, which overlap each other and are attached in a mass about the size of one's palm; and the so-called Devil's pocketbooks, the egg cases of the skate.

FIGURE 89

Horseshoe Crab

FIGURE 90

Some Marine Arthropods

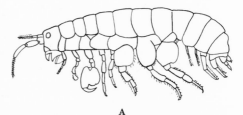

A

BEACH FLEA

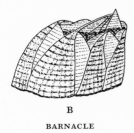

B

BARNACLE

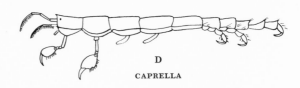

D

CAPRELLA

C

BARNACLE SHOWING APPENDAGES
EXTENDED FOR FEEDING

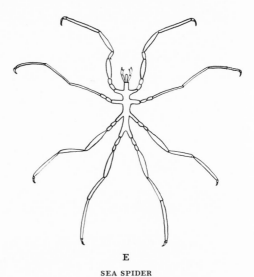

E

SEA SPIDER

But the beach is also a place teeming with living things. It was Thales, I believe, who first suggested that life may have begun in the water. Doubtless he was impressed, as who isn't, with the myriad forms that exist in the sea, only a small number of which we can find in the sands or the rock pools when the tide has ebbed.

Turn over some of the seawrack on the beach and you will see beach fleas (Fig. 90A) jumping in all directions. These little animals are relatives of the scuds and belong to the same group, the Amphipoda. They are usually brown or green in color to match the decaying seaweeds. The lens will disclose how the last three pairs of abdominal appendages are modified for jumping. The gills are attached to the first joints of the thoracic legs, and the antennae are long and hairy. There are other anatomical details that might interest you.

As you walk along the sands you will see jets of water spurt out of the sand between the tide marks, or as each wave recedes, little bubbles of air rise from the sand in its wake. If we dig where we see the spurts of water, we should unearth the soft-shell clam or sand clam (Fig. 91A). Or, if we are quick enough and dig deeply enough where we see the bubbles of air, we might get the razor-shell clam (Fig. 91B), although when disturbed ever so slightly, it vanishes into the sand as if by magic to a depth of 2 or 3 feet. Even if we are

FIGURE 91

Some Common Mollusks
Found on the Sea Beach

B
RAZOR-SHELL CLAM

A
SOFT-SHELL CLAM

C
MUD SNAIL

D
DOG WHELK

E
MOON SHELL

F
CHITON

able to dig it out, we are not sure that we have captured it, for it is quite likely to spring out of our grasp. Propelled by the strong steel-spring action of the foot, it can shoot through the air like an arrow for several yards and then, by a succession of jumps, reach the water and disappear. We should not leave it on the sand either, even momentarily, for it assumes a vertical position and vanishes into the sand in a matter of seconds.

One would not suspect the number of animals that live beneath the surface of the sand, for few of them give any evidence of their presence. We can, however, locate one or two by the little mounds of sand they push before them as they plow their way just below the surface. One of these, the mud snail (Fig. 91c), or basket shell, is so completely covered with fine revolving lines crossed by equally fine growth lines that it reminds us, especially when viewed with the lens, of a finely woven basket. A related species, the dog whelk

G
LIMPET

J
PERIWINKLE

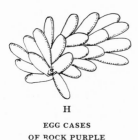

H
EGG CASES
OF ROCK PURPLE

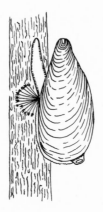

I
BLUE MUSSEL

K
OYSTER DRILL

(Fig. 91D), is somewhat smaller, with a shell covered by a series of spiral grooves and beaded vertical ridges that give it a woven or pimpled appearance. A third species, the common moon shell (Fig. 91E), has such a tremendous foot that we wonder how the animal can withdraw it completely within the shell.

Worms, too, inhabit the subsurface of the sand, but in their variety, beautiful colors, and interesting shapes you will not think of them as worms. Only the lens will show the beauty and delicacy of their structure and appendages.

The nereids, named presumably after the sea nymphs because of the graceful manner with which they swim, are among the most common of these worms. They are related to earthworms, bristleworms, and leeches, but they differ from them in a number of details, such as having a distinct head. The head, when examined with the glass, is seen to have four eyes, a pair of short, conical tentacles, a pair of palps that function as sense organs and are used for testing food, and four pairs of long tentacles (cerri) used as feelers. The mouth is on the lower side. When feeding, the worms seize their prey with their jaws, which are then withdrawn together with the pharynx and function as a gizzard to tear the food apart.

The body of the worm is rounded above and almost flat below. The segments are externally alike, except for those of the head and tail, and have on each side a projecting appendage (parapod). The parapods are flattened, fleshy lobes from which protrude bundles of horny bristles that are protective in character and that also anchor the animals in the burrows in which some of them live. The last segment is elongated and cylindrical, with a pair of cirri instead of parapods that extend outward and give the animal the appearance of having a divided tail.

There are several species of nereids which differ only in minor details. One of the largest and most common is *Nereis virens* (Fig. 92), commonly known as the clam worm. It attains a length of 18 inches and occurs along the New England coast. *Nereis limbata*, common from New England to South Carolina, is a smaller species reaching a length of only 5 to 6 inches.

The female of virens is a dull green tinged with orange and red. The male is steel-blue. This color blends into green at the base of the parapods, which are a bright green, producing a pleasing contrast with the network of blood vessels. In the sunlight the skin of both sexes reflects prismatic hues; hence the worms are brilliantly colored.

The clam worm usually spends the day in a burrow which it builds in the sand by the simple expedient of exuding a viscous fluid from glands along the body. The fluid hardens into a translucent sheath, and because it is sticky, grains of sand adhere to it, forming

FIGURE 92

Clam Worm

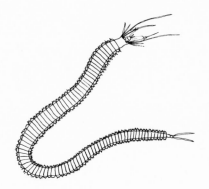

a flexible tube. The worm enters and leaves at will, and even moves around in it. At night it emerges and swims freely about, in the manner of an eel. Extremely voracious, it feeds on various marine animals, and in turn falls victim to fish such as the tautog and scup.

Along the edge of the water you may see, or you may have to dig for, the curious sand bug, Hippa (Fig. 93A). Hippa is a crab, but a most remarkable crab. When it folds its legs beneath the yellowish-white carapace, it resembles an egg. The eyes are minute, on long slender stalks, and the antennae are plumelike. These are some of the details seen with the lens; there are others.

Hippa is an active little animal, burrowing in the loose and shifting sands with a speed that is amazing. It likes to live along the water's edge, following it as the tide rises and falls. As the waves come in and cover its burrows, it emerges and scrambles to a higher level to bury itself anew, only to emerge again as larger waves come in and to scramble to a still higher level. In places where the crab is numerous, the edges of the advancing waves seem alive with them.

South of New England we have another amazing crab which is

astonishingly fleet of foot, something unusual for a crab. It runs sidewise on the tips of its toes, and if you think it isn't fast, try to catch one. Colored like the sand, it so nearly imitates its environment that when it suddenly stops running it seems to have disappeared; hence its name of ghost crab (Fig. 93B). It is an expert burrower and digs perpendicular burrows, often 3 feet deep, high up on the sandy beach. It is partial to sand dunes, and if you look about in such places you can sometimes see the sand perforated with hundreds of burrow openings on the sea side of the dunes.

Not all that burrow in the sand by the water's edge are worms or clams; in addition there is the mantis shrimp, an animal that suggests the lobster and yet is quite unlike it. It is of more than passing interest because of the terminal joint on the second pair of appendages. This joint has six sharp, curved spines which fit into sockets in a groove on the second joint. By this singular organ the animal can hold its prey securely in a way reminiscent of the praying mantis. Should you find the animal, be careful handling it, for the spines can inflict a severe wound in your hand.

Rocky shores are even more crammed with life than the sandy

FIGURE 93

Some Species of Crabs

A
HIPPA

B
GHOST CRAB

beaches that lie between them. For the rocks not only provide a place of attachment for the seaweeds, but their cracks and crevices give shelter to many animals that could not survive in exposed situations. Moreover, the rock pools furnish a habitat below low-water mark for animals that cannot withstand the ebb and flow of the tides.

Barnacles are the most conspicuous of these animal forms because of the countless numbers that are found attached to the rocks. At one time they were believed to be mollusks because of their shells, but they are now grouped with the anthropods, partly because of their jointed appendages. If we examine them (Fig. 90b) with the lens, we see that the shell is usually composed of six thick calcareous plates, rigidly attached to one another and to a fold of skin surrounding the body. At the top are two valves which open and close like double doors, and it is through these doors that the appendages, fringed like feathers, are extended when the animal is covered with water (Fig. 90c). They sweep through the water like a casting net in a regular rhythm, beating back and forth as many as a hundred times a minute, capturing small swimming creatures and organic matter that serve as food. As someone once remarked, "A barnacle is a little shrimplike animal standing on its head within a limestone house and kicking its food into its mouth with its feet."

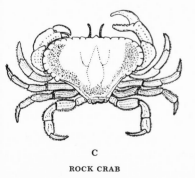

C

ROCK CRAB

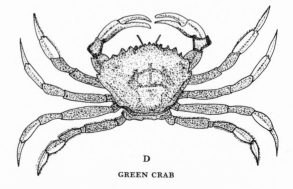

D

GREEN CRAB

If we look carefully we may find chitons (Fig. 91F), which are almost as firmly attached to the rocks as the barnacles. The most primitive of living mollusks, they differ from other members of the tribe in having a jointed armor of eight transverse plates that overlap like the shingles on a roof. In some species the lens shows a fine network of markings on the plates.

We may also find a limpet or two (Fig. 91G), which we can readily recognize by the conical or tent-shaped shell which completely lacks a twist or spiral. The tortoiseshell limpet is a common species on the New England coast and is so firmly attached to the rocks that it is difficult to remove without breaking the shell. However, if we can slip a thin-bladed knife under it, we can usually pry it loose.

The littoral zone is so crowded with organisms that there is a constant struggle for existence, and no one group can be said to have undisputed possession of a given place. Many of them are small forms and easily overlooked. Examine a frond of one of the rockweeds with the naked eye, for instance, and you will probably find it covered with plantlike growths that you might assume to be another species of seaweed. Many people have done so, and have even collected them as such. But look at these growths closely with the lens, and you will find that they are animals—hydroids, in fact.

Now, a hydroid (Fig. 94A) is a colony of associated animals that live a communal life. But this does not tell the entire story, for a colony, in a zoological sense, is a group of animals that are organically connected to one another. In a hydroid these animals (zooids) are invested with a transparent, cellophanelike, chitinous covering (perisarc) and are connected to one another by a fleshy tube (coenosarc) that is attached to the perisarc with small strands. In a colony there is a division of labor. Hence we find some of the zooids hydralike (hydranths) that is, with tentacles; these are the nutritive zooids, and they obtain food for the entire colony. Others, the reproductive zooids (blastostyles) are long, cylindrical, mouthless, and covered with the perisarc. There may also be small club-like dilatations; these are the immature zooids. The blastostyles give rise to buds that develop into small bewitching jellyfish or medusae. When mature, the medusae (Fig. 94B) become free-swimming and may produce either eggs or spermatozoa. When the eggs are fertil-

FIGURE 94

Sea Anemone, Pennaria, and Hydroid Obelia

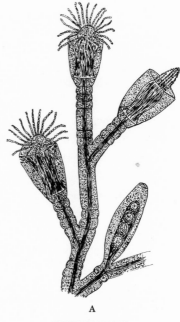

A

HYDROID OBELIA

B

MEDUSA OF OBELIA

C

PLANULA OF OBELIA

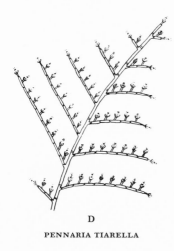

D

PENNARIA TIARELLA

E

SEA ANEMONE

ized, they develop into ciliated, free-swimming larvae called planulae (Fig. 94c). The planulae eventually attach themselves to an object, such as a stone, and by budding, grow into the colonial hydroids.

Typically a hydroid colony is attached to a support by a creeping stem with a vertical axis having short, lateral, alternate branches that terminate in zooids, though there is often more complex branching. In some species the zooids or hydranths are unprotected or naked; in others they are covered with goblet- or bell-shaped cups that either terminate the branches or are attached directly to the stem; that is, they are sessile. *Pennaria tiarella* (Fig. 94D) has the appearance of a plume, and the naked zooids, red in color, are arranged on the upper side of the branches like rows of diminutive bellflowers. The delicate and beautiful little Clava occurs in velvetlike carpets.

Obelia commissuralis, a delicate, bushy hydroid as much as 8 inches long with a zigzag stem, is quite handsome when seen with the lens. In this species the zooids are protected with a bell-shaped cup. So, too, are the zooids of *Sertularia pumila*, whose main stem creeps over the fronds of seaweeds, often crossing and recrossing in a tangled mass. These are only three of the many hydroids that occur along the East Coast but they are representative and will give you some idea of what these animals are like.

While examining a frond of the rockweed, you may also find it encrusted with a delicate calcareous lacework. You will see many similar patches on stones, shells, and other seaweeds, and they may be white, yellowish, or brick-red. When examined with the lens, these crusts appear to be shells of beautiful form and sculpture set closely together. As you watch, each shell opens a trap door, and tiny filaments emerge that expand into flowerlike circles of tentacles, often golden-yellow in color. They remind you of the moss animals of our ponds, and that is exactly what they are.

The individuals (polypides) of a bryozoan colony are small polyplike or hydralike animals living within a calcareous shell (zooecium) that is actually the cuticle or exterior of the organism. In some species the openings of the zooecia are surrounded with spines; in others they have lids that shut down when the polypides are retracted. In the genus Bugula the zooecium is singularly modified into a bird's head with a hooked beak.

Some species, such as *Bugula turrita*, instead of forming crusts like Membranipora and Crissia, grow up as fluffy treelike colonies composed of spiral branchlets with transparent zooecia, so that the polypides and their anatomy may be plainly seen. Still others, such as *Aetea anguina*, are composed of erect club- or vase-shaped zooecia mounted on creeping horizontal stems visible to the naked eye as networks of white threads. There are many species of bryozoa, and as you look at them with your lens you will be delighted with their exquisite, dainty, and beautiful appearance.

Perhaps you will also see on the rockweed tiny flat, shelly spirals that you will assume are minute gasteropod shells; but look closely with your lens and you will find them inhabited by the worm Spirobis. Maybe there will be a few of the pink urnlike or vaselike capsules of the rock purple (Fig. 91H) and the small jellylike masses

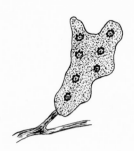

FIGURE 95

Compound Ascidian

of the compound ascidians (Fig. 95)—odd animals, but no stranger than many others we find along the seabeach.

Seaweeds, of course, are abundant on the rocks and in the tide pools, and though we may examine one or two closely, we are apt to glance at others cursorily and thus miss one of the oddest animals found in the rock pools, the little Caprella. Figure 90D will show you why. This peculiar animal, slender as a skeleton, so closely resembles the branching seaweeds in form and color that it is practically camouflaged. To heighten the deception, it sways in the water like seaweed. It reminds us of the walking stick in form or the measuring worm in behavior for, like them, it holds on to a support by its posterior feet and extends its body out rigidly. It even walks the way a measuring worm does, bringing its hind feet up to the front ones and doubling its body into a loop. The appendages at the front show considerable variation. The body segments may be smooth or ornamented with spines, tubercles, or a combination of these characters.

The person who explores a rock pool for the first time is naturally more attentive to the more conspicuous animal forms and probably overlooks the curious spiderlike animals known as sea spiders (Fig. 90E). They appear to be all legs as they crawl along slowly over the seaweeds and hydroids. Almost the entire body is composed of the cephalothorax, which bears a conical suctorial proboscis on the anterior end and on top a prominence with four eyes. There are four pairs of long, slender legs, with a smaller fifth pair that suggests the chelicera of the spiders. The jointed legs show the animal to be an arthropod, but other than this it is an aberrant form and its position in the animal kingdom is uncertain.

Few animals intrigue a visitor to the seashore as much as the starfish (Fig. 96A), which is not a fish but a spiny-skinned animal with a number of arms that radiate like the points of a star. Perhaps seastar would be a better name. At first the starfish may seem to be completely rigid, and to a certain extent it is; but it is capable of some bending and twisting. Its somewhat rigid structure is due to a meshwork of calcareous plates which are embedded in the softer parts of the body and from which project numerous spines, some of them moveable. The reason the animal is not completely rigid is that the plates are not united to form a complete shell but are joined by connective tissue and muscles.

Aside from their habit of preying on oysters, starfishes have little economic importance. Yet they and their relatives—serpent stars, brittle stars, sea urchins, sea cucumbers, sand dollars, and sea lilies—form a unique group of animals with many features of interest to the naturalist and zoologist. For example, take the water-vascular system, a sort of hydraulic-pressure mechanism that enables them to move about. If you look on the upper surface of a starfish, you will notice a small orange disk. Called the madreporite, it covers the opening to the vascular system and is perforated with a number of minute openings, as the lens will show; actually it is a sieve plate. Water enters through these openings and passes down an S-shaped canal, known as the stone canal because its wall is made rigid by a series of calcareous rings. This canal connects with a tube (circular canal) that encircles the mouth and that in turn is connected with the radial canals extending the length of the arms, one canal for each arm. Water passing down the stone canal flows through the circular canal and then along each of the radial canals.

On the lower surface of an arm there is a groove (ambulacral groove) extending its entire length. Place the fingers of one hand on one side of the groove and the fingers of the other hand on the other side, and press gently. You will thus open the groove and expose a number of odd-shaped structures that, through the lens, appear as cylinders with small disks (suckers). They are hollow and thin-walled and are called tube feet. Each is connected by a short branch to the radial canal and to a rounded muscular sac (ampulla). The ampulla cannot be seen unless the animal is dissected.

Water flowing down the radial canal enters the ampulla, which then contracts by muscular action, forcing the water into the tube foot (the water is prevented from flowing back into the radial canal by a valve). Because of the pressure exerted by the water, the tube foot, being elastic, becomes distended, and the sucker at the end is pressed against the substratum. Once the tube foot has become distended, other muscles are brought into play, forcing the water back into the ampulla. The result of all this action is to draw the animal forward. One tube foot, of course, is not capable of producing much movement, but hundreds of tube feet acting in concert are effective in doing so.

FIGURE 96

*The Starfish
and the Starfish Feeding*

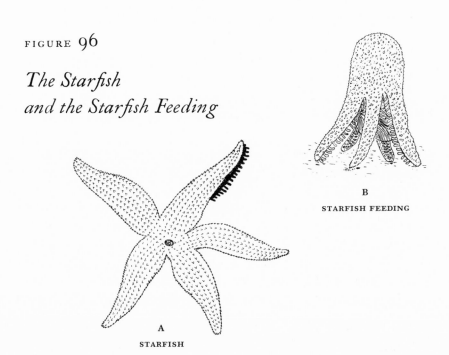

B
STARFISH FEEDING

A
STARFISH

That these tube feet can exert tremendous force is evident by the manner in which a starfish feeds on an oyster or clam. It is not easy to pry open a clam, as you may know; but a starfish does it with relative ease. When feeding on one of these animals, the starfish mounts it in a sort of humped-up position (Fig. 96B), attaches the tube feet to the two valves, and pulls. The clam or oyster reacts by closing its valves tightly. But the starfish continues to pull, and by using its tube feet in relays, it is able to outlast the clam or oyster, which eventually becomes exhausted from the uneven struggle so that its valves open. It is probable that the starfish does not pry the valves open entirely by muscular action but also secretes some kind of a substance that relaxes the clam's muscles. When the valves open, the starfish everts the lower part of its stomach, extends it through its mouth and in between the two valves, and proceeds to feed on the softer parts of its prey.

Before you replace the starfish in the tidal pool, note the movable spines on both sides of the ambulacral groove. They can be brought closely together and thus protect the soft tube-feet in case a starfish is attacked. The starfish, incidentally, is able to move in the manner described only when it is on a hard surface, such as a rock. On soft sand or mud the tube feet function as legs.

Sea anemones (Fig. 94E), though quite flowerlike when their tentacles are fully expanded, and beautiful, too, with their pink, orange, white, salmon, blue-green, and cherry-red colors, are anything but like the flower for which they are named; they resemble chrysanthemums and dahlias far more. We might easily overlook them, as they are very sensitive to environmental changes and are easily alarmed, upon which they contract into a shapeless, unattractive, inconspicuous mass.

Sea anemones are not at all flowerlike in their behavior, for, like the hydra, the tentacles are provided with stinging cells and are a veritable death trap for any small creature that comes within their reach. Of the many species found along the coast, few can equal *Metridium dianthus* in beauty. This sea anemone measures up to 4 inches in height, is usually a dark chocolate-brown, and may have as many as a thousand tentacles. When fully expanded, the animal is truly a marine chrysanthemum. A smaller, more delicate species, *Sagartia leucolena*, is translucent, varying from flesh color to white;

but since it is extremely sensitive to light, we must look for it in the dark corners and crevices of rocks.

Tide pools also have their quota of mollusks. Blue clams or blue mussels (Fig. 91I) seem to be as numerous as barnacles. We find them hanging from the rocks by their byssal threads, or attached to one another, or lying exposed on stones, pebbles, and among seaweed. Periwinkles (Fig. 91J) are there too, colored and striped for camouflage, clinging to the rocks or the wet seaweed, or tracing furrows in the wet sand. These animals seem to do as well out of water as in it, and may be found well above the high-tide limits. It has long been suspected that this marine species is making some progress toward a terrestrial habit. Such a changeover is neither impossible nor improbable, for many land mollusks show evidence of once having led an aquatic or marine existence. Changes in nature are always taking place—slowly perhaps, but inexorably—and a gradual substitution of lungs for gills would not be a startling innovation.

One need only examine any rock to observe the different forms that strive for a place to attach themselves: barnacles, mussels, anemones, and others such as the purple snail or rock purple, an animal so individual that we recognize it at once. Yet its details are difficult to describe, because it is so variable in size, form, color, and markings, that if a series of rock purple shells are collected and examined it almost seems as if they were different species. The purple snail and other members of its family secrete a colored fluid that may be green, crimson, or purple. The Indians used it as a dye. The famous "Tyrian purple" of the ancient Phoenicians was obtained from a related species.

We should also mention the oyster drill (Fig. 91K), another common inhabitant of tide pools. This is the mollusk that does so much damage to the oyster beds, drilling a small, round hole in the shell of the oyster with its sandpaperlike radula and sucking out the body juices. The oyster drill has a small fusiform or spindle-shaped shell with about six convex whorls crossed by numerous vertical, rounded ribs or folds. These in turn are sculptured by a number of raised lines that are quite distinct and most striking when viewed with the lens.

Crabs, of course, are among the more conspicuous animals of

rocks and tide pools, as well as the sea beach, and invite the attention of the most casual observer. As they vary considerably in habits, some creeping, some climbing, and still others swimming or burrowing, there are great differences in shape and structure to accommodate them to their particular mode of life. There is also great diversity in color and markings. Those who have studied crabs have found that they have many amusing habits, are not without a degree of intelligence, and, being somewhat wily, often resort to a clever ruse or stratagem to avoid danger. Many will claim and, with good reason, that they are the most interesting of all the animals found along the seashore. Apart from this, they play an important role in the economy of the marine habitat by being scavengers.

There are several species that are more or less common. The rock crab (Fig. 93c) hides or lies nearly buried in the sand or gravel; the Jonah crab prefers exposure to the action of the waves, and we usually find it on the rocks; the lady crab, in contrast, lives buried in the sand with only its eyes and antennae exposed, alert for both prey and foe. We have all heard of the hermit crab that lives within an empty sea shell. The little green crab (Fig. 93D), the liveliest of them all, ranges from the tide pools to well up on the beach, where it takes refuge under loose stones and debris cast up by the waves.

By the calendar November is no shorter than any other month, but it seems so. Perhaps the holiday has something to do with it, being a sort of hiatus in the succession of routine days. Then, too, we are being constantly reminded that Christmas is just so many days ahead, and our thoughts turn to the multitude of things to be done, so that the days pass quickly. Of course, the days are getting noticeably shorter to remind us that winter is almost here. To be sure, we have a few weeks to go, but most of us accept winter as having arrived when the first snowflakes fly. We may not see them in November, but we are sure to have at least a dusting in December; one of these days we shall awaken to find the windows frosted over and the ground a blanket of white, and as we glance at the calendar we realize that we have slipped into the last month of the year without being conscious of it.

DECEMBER

"But Winter has yet brighter scenes—he boasts
Splendors beyond what gorgeous Summer knows;
Or Autumn with his many fruits, and woods
All flushed with many hues."

WILLIAM CULLEN BRYANT

"A Winter Piece"

DECEMBER

O FFICIALLY WINTER DOESN'T COME UNTIL the twenty-first of the month, and whether we have winter weather before or after that date depends on the meteorological pattern for the year. But whatever the weather December may have in store for us, the plants and animals have made their preparations. And the adjustments that many of them make to survive low temperatures, heavy snows, and biting winds are varied and often quite ingenious.

The grouse, for instance, grows fringes of sharp points on his toes to serve as snowshoes. With these he can run easily over the ground in search of partridge berries and bearberries which, though the bearberries are somewhat dry and flavorless, offer a welcome departure from his normal summer diet of seeds and insects. The varying hare also grows long, stiff hairs along the margins of his feet, permitting him to race with abandon over deep snow in which other animals flounder. Many years ago I saw a fox chase a hare over the snow, but the fox was easily eluded; an owl would have been more successful had it followed the white, bounding figure, practically invisible against the prevailing color of the landscape.

As a special protection against the snow, animals that remain active during the winter have winter underwear in the form of very dense short hairs that sprout among the long true fur. The birds that remain or visit us are protected in a similar manner: their plumage is more dense and more closely interlocked than that which follows a spring molt. Ducks and related birds have a downy undergrowth which helps to keep out the cold. And it is this down, which loosens in the spring, that the females pluck out and use as a bedding for their eggs.

The invertebrates, too, have taken precautions against cold, either by seeking refuge in some protected place or by constructing shelters of various kinds: egg sacs, cocoons, or tubelike structures of leaves and the like, often lined with silk. Plants also have protective devices. The woolly leaves of the mullein, for instance, help the exquisite rosettes, formed by the year-old plants, endure through the winter so that they can send up a flower stalk the second spring. Trees and shrubs protect the tender inner parts of their buds against cold and moisture by covering them with thick scales or coating them with a waxy resinous substance, and they guard them against sudden changes of temperature by lining them with down or wool that serves the same purpose as the underfur of animals.

It may be that berries of certain trees and shrubs remain on the branches well into the winter to prevent the embryos from being killed by the frosts and rains of autumn were they to fall at that time. Beechnuts, acorns, and other nuts and berries that fall in autumn often germinate and send up sprouts, but these usually do not survive the winter.

If snow falls during the night, we need only step outdoors to find that we have had visitors to our garden or backyard. Though none may be in view, their footprints in the snow are evidence that they have been there: a sparrow, perhaps, or a woodpecker, maybe a crow or starling, and doubtless a squirrel or cottontail. In the fields and woods we would find many other prints, and all tell a tale if we knew how to read them.

Learning to recognize the prints of the wild creatures is a study that might engage our attention at any time of the year, for prints may also be found in mud and sand. Along the muddy banks of a stream I have often found the prints of a raccoon, mink, or otter— and those of a heron, too, as the bird stood motionless, perhaps at dawn, waiting for a frog to appear. On the stretches of sandy beaches I have traced the pattern made by hundreds of sandpiper toes, or the prints of gulls and terns. But since it is in winter that prints are more noticeable and more clearly defined, it is then that one is more likely to become interested in them. They are best studied when freshly made in the newly fallen snow, for later when it has become packed or blown about, they are distorted or otherwise mutilated.

That snowflakes have many different patterns is well known, and this you can see for yourself by examining a few of them with your lens as they fall on a piece of glass. You will find them exquisitely simple or intricately complex, forming six-pointed stars, six-sided prismatic columns, or thin geometric plates. A snowflake, of course, is merely a crystal of water or, more accurately, crystals of water, since a snowflake is not just one crystal. It may seem strange to speak of a crystal of water, but when water freezes or becomes solid, it forms a hexagonal or six-sided pyramid. You can see such crystals any time; just scrape some ice from your refrigerator onto a piece of glass and look at it with your lens. Usually, however, the crystals are not well defined, because many of them are grouped together.

By definition, a crystal is any solid having plane surfaces symmetrically arranged. To see a crystal gradually take form and grow is a fascinating experience that few of us have had, yet it is readily available to anyone in his own kitchen with a minimum of effort. That infinitely small particles of a substance can assemble into a certain shape, with plane surfaces set exactly at definite angles to each other, and that the angles are always the same whether the crystal is small or large, seems something of a miracle and quite transcends our understanding. Still, it is no less miraculous than the growth of living things from a unit of protoplasm so small that we must magnify it manyfold to see it at all.

To see a crystal take form, you need only dissolve a little table salt in some water, place a drop of the solution on a piece of glass, and then look at it with your lens. As the water gradually evaporates, small, perfect cubes appear—crystals of salt. If other substances are used in place of salt, they will form crystals of different shapes, but only if such substances are crystalline and not amorphous.

Needless to say, we can find many different kinds of crystals almost anywhere—in the fields and woods, along the wayside; in fact, just outside your doorstep in an ordinary rock or stone. For a rock is made up of one or more minerals, and minerals are compounds of definite composition with a specific crystal structure.

Bare, exposed rocks, where fresh and unweathered surfaces are available, are often productive of minerals, and quarries and places where crushed stone is made are excellent sites. It is a matter of grubbing around and examining various pieces at random. Other likely sites are cavities in cliffs or ledges and places where blasting has occurred in road construction; large boulders should also be explored, for they frequently contain crystals of various minerals. All one needs is a lens and a hammer, preferably a stonemason's hammer, for breaking up the rocks.

FIGURE 97

Bacteria, Diatoms, Desmids, and Yeast

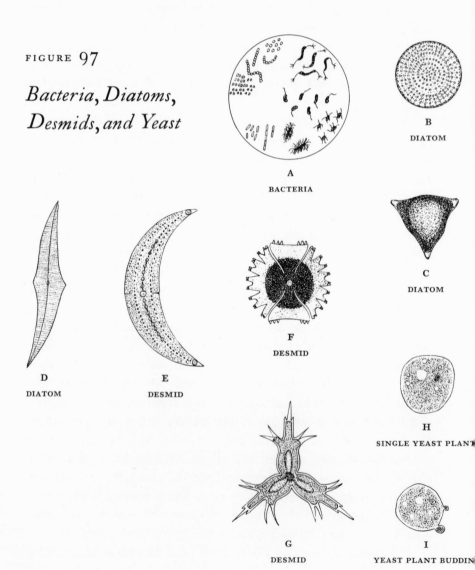

A
BACTERIA

B
DIATOM

C
DIATOM

D
DIATOM

E
DESMID

F
DESMID

G
DESMID

H
SINGLE YEAST PLANT

I
YEAST PLANT BUDDING

Most of us do not realize it, but we live in a sort of triple world: the world we can see with our naked eyes, the world we can see with a hand lens, and the world of animals and plants so small that we need the high magnification of the microscope to see them.

These microscopic plants and animals abound everywhere, even in a stone urn, for a completely dry urn placed outdoors will in time produce a varied assortment of microscopic plants and animals. If you wish, you can duplicate the succession of events that would normally take place in a stone urn within a matter of a few days or weeks. Get a small amount of hay, preferably timothy, steep it in hot water for an hour or so, and strain it into an open dish. Put the dish aside, and after a couple of days add a small amount of pond water. A day or two later, the clear liquid will become cloudy, or it may have already become so. If a drop or two is now examined with the microscope, several forms of bacteria should be seen (Fig. 97A).

From now on, examine the liquid each day. Before long small one-celled animals with one or two whiplike processes will appear and swim about (Fig. 98A). The flagellates, as they are called, feed on the bacteria, making such inroads on the bacterial population that they literally eat themselves out of house and home. With a decrease in bacteria, the flagellates begin to disappear, and their decline is followed by the rise of the ciliated protozoan Colpoda (Fig. 98B). Colpoda is followed by the ciliated hypotrichs (Fig. 98c), and in the course of events, the hypotrichs give way to the paramecia (Fig. 98D). Vorticellae (Fig. 98E) may also appear, and sooner or later some amoebae (Fig. 98F).

Ultimately, green algae will form and multiply, and rotifers and various crustaceans will put in an appearance, bringing to an end the series of events[1] that began with the bacteria. As long as the algae receive enough light to form a food basis for the animal forms that are present—and a few of all that have appeared remain—a relatively stable condition develops (climax stage) that will endure for months or even years, unless the unexpected should occur and bring this microscopic world to an end.

Flagellates, hypotrichs, paramecia, and Vorticellas are all pro-

[1] —Known as protozoan succession.

FIGURE 98

Some Representative
Protozoans

C
STYLONCHIA

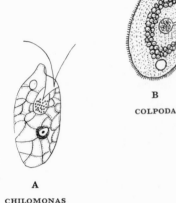

B
COLPODA

D
PARAMECIUM

E
VORTICELLA

A
CHILOMONAS

tozoans, animals that belong to the phylum Protozoa, which means first animals. They are the simplest of animals—merely a unit mass of protoplasm—yet they are capable of carrying on all the activities necessary for existence and survival. They capture their food, digest it, and convert it to protoplasm for growth and repair. They take in oxygen to oxidize the food materials for heat and energy with which to carry on their activities, and they expel the waste products of metabolism as well as undigested food materials. They also move around by various means and are capable of reproducing their kind. They even respond to stimuli, a function essential to survival.

Basically they are like other animals except that one cell performs all the life functions, whereas in the higher animals the cells are differentiated to form tissues and organs that perform these functions. But even among the protozoa, there is a degree of division of labor and a hint of what eventually was to evolve in the course of time. For though most protozoans are solitary and free-living, there are some colonial forms with cells that have become differentiated for reproductive purposes. Some of these colonial

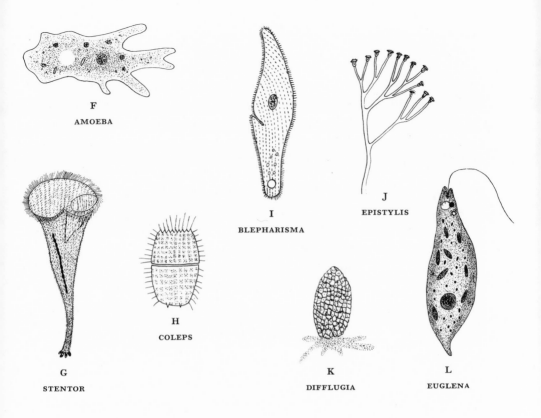

F
AMOEBA

I
BLEPHARISMA

J
EPISTYLIS

H
COLEPS

G
STENTOR

K
DIFFLUGIA

L
EUGLENA

forms serve as a link between the protozoa, the one-celled animals, and the metazoa, the many-celled animals.

In these one-celled animals, the protoplasm, a complex, jelly-like substance containing such vital materials as water, proteins, carbohydrates, fats, and pigments, is capable of performing all the life functions, with the aid of small bodies called mitochondria and other structures (vacuoles) that, in some species, appear and disappear as the need for them arise and that in other species are permanent. And all contain a central body or nucleus with the all-important molecules of deoxyribonucleic acid, DNA, and ribonucleic acid, RNA, of which we have heard so much in recent years. These coordinate all the complex cell activities and bear the coded hereditary information by which each organism reproduces itself.

A drop of water from any natural situation—pool, pond, brook, stream, lake, ocean, mud puddle, rain-filled hollow of a tree or stump, stone urn—will contain any number and variety of protozoans. Almost everyone has heard of the famous amoeba, the shapeless mass of protoplasm that flows erratically on its way, engulfing food materials by extending fingerlike extensions of itself with which it surrounds the materials, and takes them into itself, leaving in its wake what materials it cannot digest. And equally famous is the streamlined, slipper-shaped paramecium, undoubtedly the most studied of all the protozoa, that moves swiftly through the water by the gyrations of thousands of cilia.

There are countless others, of course, so fascinating that we can spend hours just watching them. A little water from a pond, a few dead leaves floating in the water or lying on the bottom, a twig or stick immersed in the water, and a few submerged water plants will provide us with enough of these one-celled animals to keep us engaged for as much time as we care to devote to them. In this world of the miniature you will see animals of all shapes, sizes, and habits of feeding, resting, moving, and reproducing, experiencing dangers and hazards of all kinds, meeting them successfully or succumbing to them; in short, a world as absorbing as the one we live in, with as many trials and tribulations as we ourselves encounter.

You will see the Vorticellas that remind you of a bed of tiny tulips, and as you watch them, you suddenly see the stems shorten and the "blossoms" drop down; than as suddenly you see them rise again. You will see the pear- or trumpet-shaped Stentor (Fig. 98G); the barrel-shaped Coleps (Fig. 98H); the pink Blepharisma (Fig. 98I); the girdle-wearer Gymnodinium; the tumbling Urocentrum; the stalked Epistylis (Fig. 98J); the somersaulting Conium; the walking Stylonchia; the beautiful sun animalcule *Actinophrys sol,* and its larger relative, *Actinosphaerium eichnorni.* You will also see Difflugia (Fig. 98K), that builds a shell of sand grains in which it lives, each grain fitting exactly into its proper place; and Arcella, which also lives within a house, but one constructed of minute hexagons from a substance that it secretes. You can watch such colonial forms as the stationary and branching Carchesium and the free-moving Volvox, that appears as a small globe or sphere spinning gracefully through the water. You will also find Euglena (Fig. 98L), a most unusual animal with both plant and animal characteristics, for it has chlorophyll and thus can make its own food, a typical plant function, but it moves through the water by means of a whiplike process, locomotion being an animal trait. It is mentioned in books on zoology as well as on botany, because for years the scientists did not know how to clarify it. Finally they resolved their dilemma by creating a new kingdom, the Protista, for it and its kind.

Included in this kingdom by some biologists are the diatoms and desmids. Individually diatoms are invisible to the naked eye, but collectively they are often seen in large numbers forming a yellowish-brown film on the surface of the pond, where they may be gathered by skimming them up. Otherwise they can be obtained by scraping up some of the mud on the bottom or by collecting some of the larger water plants, to which they are frequently attached.

Diatoms are usually golden-brown, and have a most peculiar structure. They have often been compared to an old-fashioned pill-box, for as you will see, each diatom is formed of two parts called valves, one of which may be likened to the pillbox proper and the other to the cover as it overlaps or slips over the upright edge of the lower valve.

All diatoms are highly geometric in form—circles, ovals, triangles, with patterns so precise that we might think they had been laid out with drawing instruments. Many of them are decorated with regular lines of partly raised ribs or tubercles and partly sunken pits, slits, or pores, making them some of the most beautiful and delicate things found in the microscopic world. They have been called the jewels of the plant world. Over five thousand species have been described, a few of which I have shown in Figures 97 B to D.

Since the walls of the diatoms are impregnated with silica, the pillbox cases are indestructible. These walls are actually glasslike and transparent, so that the organisms literally live in glass houses. Such a house also may be termed a majestic sarcophagus, for the beautiful shells persist long after the living part has died. As a result, beds of the shells, often immense, may be found in many parts of the world. In many small lakes such beds are composed chiefly of living diatoms on the top layers, their partially decayed remains lower down, and their empty shells at the bottom. These deposits furnish a substance called diatomaceous earth, which has a number of commercial uses.

In some ways similar to the diatoms but in others unlike them, the desmids also occur in fresh water, where they live mostly around the margins and bottom of ponds. Like the diatoms, they too are individually invisible to the naked eye but may be seen collectively —as a Nile green film adhering to the stems of other plants.

Each desmid has a soft, flexible envelope and consists of two similar halves which are related, like an object and its mirror image. The two halves are connected by a (usually) constricted area called the isthmus, in which the nucleus lies. There are about a thousand species of desmids, and as you observe them through the microscope you will be astonished at their great diversity of form and exquisitely sculptured walls. These walls may be finely striated or roughened by minute dots or points, by wartlike elevations, or even by spines of different shapes. Their edges, too, may be notched, prolonged into teeth, or otherwise variously cut and divided. These ornaments, together with their graceful form and beautiful green color, make the desmids objects of admiration and a source of never-ending delight (Figs. 97E,F,G).

A striking feature of many desmids, especially in the crescent-shaped species, is the presence in each extremity of a small colorless, spherical space or vacuole containing minute granules of gypsum crystals that are in constant motion.[2] It has been suggested that these granules function as statoliths—orienting organelles that help the organisms maintain their balance.

Although both the desmids and the diatoms are spread chiefly by water currents, both have the ability to move about from place to place with the aid of projecting threads. Desmids move slowly and in a rather stately fashion in one direction, whereas diatoms may travel quickly halfway across the field of view and then, for no apparent reason, abruptly retrace their course or dart off obliquely on a new one.

There are many one-celled plants that instead of being separate and free are grouped together in large numbers or attached to one another in the form of filaments. The shining green scum which we often see on the surface of ponds and streams, though repulsive-looking, reveals a beauty literally undreamed of when examined with the microscope, for it is composed of countless filaments or threads which appear as beautiful strings of green beads. The beads are individual plant cells; all are alike, and each is capable of living

[2] The Brownian movement.

a separate existence if detached from the others, but through one of nature's whims, they have become fastened together.

Should we focus our attention on one of these individual cells, we find that the cytoplasm, a term given to all the protoplasm exclusive of the nucleus, occupies the largest area and is clearly the working part of the cell—that which transports the materials, builds the wall, produces chemical reactions, and the like. Next in prominence is the nucleus, a rounded body somewhat denser in consistency. This is the control organ of the cell, directing the work of the cytoplasm so that the organism develops along the general lines of its hereditary genes. Scattered throughout the cytoplasm are chloroplasts, small bodies that contain the green pigment chlorophyll. These chloroplasts are the factories, as it were, where food is manufactured for the plant; here the raw food materials are taken in and transformed by the action of sunlight into certain food products that are transported to various parts of the plant.

In young and small cells, the cytoplasm occupies the entire cell, but as they grow older and larger, they develop rifts filled with sap that is composed of various substances in solution. Eventually these rifts enlarge and run together until they form a single great, sap-filled vacuole. Finally there is the cell wall, made of a firm, elastic, water-permeable substance called celluose, whose obvious function is to enclose this microscopic thing of wonder.

One-celled, self-supporting plants are collectively known as algae. One of the most interesting of the many species is Spirogyra (Fig. 99A), which usually gathers in masses on the surface of the water or hangs in long streamers from the leaves of submerged water plants. It gets its name from the spiral green chloroplasts. The species Oscillatoria (Fig. 99B) is equally interesting: under the microscope the filaments twist and writhe, coil and uncoil, or curve to one side and then to another. The green felt which we see on the surface of damp soil in the greenhouse and in similar situations is Vaucheria (Fig. 99C). The dense, pilose, bright-green growths commonly seen on stones in streams and ponds is Ulothrix (Fig. 99D). A single filament, which is attached to the stone at one end by a modified holdfast cell, consists of numerous similar cells, each with a chloroplast that resembles a slightly opened ring.

Stones, flowerpots, and old, shaded tree trunks often have a bright green coating that looks much like patches of green paint. This is composed of numberless spherical green cells of Protococcus (Fig. 99E), grouped together in units of two, four, or six. Since Protococcus usually grows on the north side of trees, the points of the compass may be approximately determined by observing which side of a trunk is green. The little balls of jelly we see in lily ponds, quiet pools, and similar places, if crushed, will contain numerous chains or strands of cells of Nostoc (Fig. 99F) that resemble tiny beads on a necklace.

Like algae, diatoms, and desmids, yeasts are also unicellular

FIGURE 99

Some Representatives of the Algae

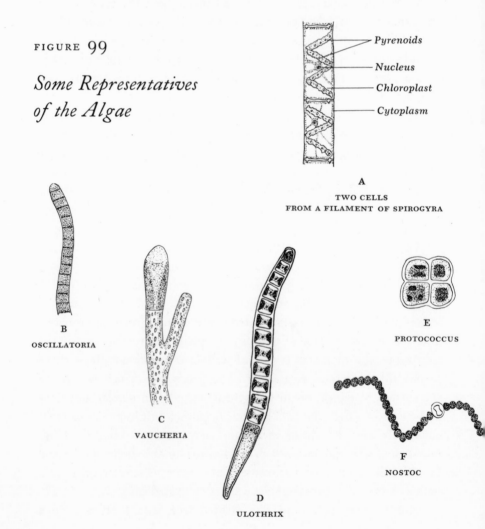

Pyrenoids

Nucleus

Chloroplast

Cytoplasm

A

TWO CELLS
FROM A FILAMENT OF SPIROGYRA

B
OSCILLATORIA

C
VAUCHERIA

D
ULOTHRIX

E
PROTOCOCCUS

F
NOSTOC

plants, but unlike the others, they lack chloroplasts and are incapable of making their own food. They are saprophytic fungi and live in the sweet juices of plants or on the surface of fruits. Most of us are acquainted with these fungi only in the form of yeast we get at the supermarket or neighborhood grocery; few realize that the yeast actually consists of countless numbers of these microscopic plants. Place a little yeast in some water and examine a drop with the microscope. You will see myriads of minute, egg-shaped bodies (Fig. 97H). These are the plant cells, each with a well-defined wall, a nucleus, oil globules, and vacuoles.

Yeast plants convert sugars into alcohol and carbon dioxide by means of an enzyme system called zymase, a process known as fermentation. Hence yeasts are economically important, since they are the principal source of alcoholic beverages and industrial alcohol. They are also important in baking, for the bubbles of carbon dioxide cause the dough to rise. In addition, these lowly plants are used to manufacture vitamins, and one species has become very important in the synthesis of proteins from molasses and ammonia, an activity that has led to the production of protein foods.

Yeast plants reproduce by budding (Fig. 97I). A lateral outgrowth of the plant cell takes place, forming a dilatation or "bud." As this bud continues to grow, the plant cell becomes two unequal parts, and a wall develops between the two. The bud may either become detached from the original or parent cell or remain attached and in turn form another bud; sometimes budding takes place so rapidly that the cells cohere and form a chain of plant cells.

Adventures into the microscopic world are of the indoor variety, and we can spend many an evening at this time of the year peering through the microscope. We are apt to become so engrossed in the many wonders of the new-found world that, except for a hurried trip to the pond to replenish our supply of specimens, we are loath to spend any time outdoors, particularly when there seems so little to see. But December outdoors is not quite the lusterless prospect that many envisage, nor are the fields and woodlands as empty and silent and as deserted as one may think. Meadow mice run along their pathways, and white-footed mice scamper over the snow; shrews hunt for hidden insects, and mink, weasel, otter, fox, and bobcat are on the prowl; to be sure, we may not see any of them unless we venture out at night. During the day the squirrels are always on the move, and wherever we may wander we see chickadees, blue jays, nuthatches, woodpeckers. In the field across the road, a flock of starlings at the moment are walking over the ground, and only this morning I saw pine siskins and redpolls feeding in an alder thicket, and some purple finches feasting on cedar berries.

The landscape is now a canvas on which the distant hills and the rough-hewn edges of rocks are delineated among blue and violet shadows, and trees are etched against the pale-blue sky. To temper the austerity of the winter scene, we find a bit of green here and there: in the crowns of the woodfern that rise above the rocks of a woodland ledge, in the twigs of the sassafras, in the leaves of the pipsissewa, the snowberry, the wintergreen, the laurels, and not in the least in the club mosses, that now really come into their own amidst the snows of the woodland floor.

Look at a club moss and you look forty million years into the past, for it can trace its ancestry back to the Coal Age. Club mosses resemble mosses or small conifers, depending on how you look at them; but they are neither. They creep or trail over the ground and send up more or less erect branches from an inch to as much as a foot high. They differ from the true mosses in that they have roots, stems, and leaves with well-developed vascular tissues, and they differ from the higher flowering plants in that they do not produce seeds. Instead they form spores that develop in small sacs (sporangia), and in this respect resemble the ferns. The sporangia are borne on the leaves at the apices of the stems, and the name club moss is derived from the club-like appearance of the apical cones formed by the spore-bearing leaves. There are, of course, other leaves that function as leaves: small and scalelike when seen with the lens, they overlap like the shingles on a house, covering the stems completely.

One might easily mistake the ground pine for a pine seedling when seen for the first time, because the erect stem produces bushy side branches and seems to rise out of the ground. In a way it does come out of the ground, of course, but actually it arises from a horizontal stem which is often so deep underground that the erect branches appear as separate plants. In the damp woods there may be so many that they create the illusion of a diminutive forest. The ground cedar, also treelike in habit but not quite as tall or erect, is well named too, for its flattened branches with their closely over-lapping leaves look very much like sprays of cedar.

There are other species of club mosses: running pine, perhaps the most beautiful of them all; the stiff-leaved club moss; the foxtail club moss; the shining club moss, which some botanists claim is not a true club moss because it bears its sporangia in the axils of the leaves instead of on "clubs." But if not, then what is it?

A botanist, nature lover, artist, and lumberman all view a tree from a different perspective, as is their right, and each identifies a tree in his own way. The lumberman or forester, for instance, depends on the bark more than any other character to recognize tim-ber trees. This may seem at first a rather uncertain way of identifying a tree; to most of us the bark of one tree seems much like the bark of another, except in extreme instances, such as the distinctive bark of the white birch. But compare the bark of a few trees and differences will become apparent. Bark that peels off in large sheets, as in the shagbark hickory and the sycamore, is readily recognized. The bark of the hop hornbeam also peels off, though not in such large pieces as those of the hickory; yet the two can sometimes be confused by this character. The bark of the paper and yellow birches, too, peels off in thin, papery layers but the texture of the bark is characteristic and is not apt to mislead anyone.

The color of bark is an important mark of distinction, espe-cially in birches, and in some instances both the color and taste of the inner layers are distinctive. Thus the black oak is readily distin-guished from other oaks by the yellow and intensely bitter inner bark; similarly the barks of the black birch, sassafras, and cherries have characteristic flavors. The breathing pores or lenticels, when examined with the lens, also have particular shapes and colors in different species, especially in cherries, and are a clue to their iden-tity. In some trees, such as the beech and American hornbeam, the bark is smooth—steel-gray in the former, blue-gray in the latter. In many others the bark, as the tree grows older, breaks into ridges or scales or otherwise becomes sculptured. In the chestnut the ridges are long, and in the white ash they run together to form more or less perfect diamond-shaped areas. The scales of the white oak and other members of the white oak group are easily rubbed off; those of the black oak and the black oak group come off only with difficulty.

More noticeable at this time of the year are the bracket fungi that we so often see on trees but completely ignore, like so many other things. They are somewhat like the boleti in that they produce spores within tubes, but unlike the fleshy boleti, they are hard, tough, and almost woodlike. All are notorious destroyers of trees, the mycelium spreading beneath the bark of living trees and drawing away their sustenance until they finally succumb to the exhausting and one-sided struggle. No one knows that a tree has become infected with these insidious killers until the "brackets" or "shelves" begin to appear on the sides of the trunk, for long before the mycelium is prepared to send forth these reproductive structures, it has been worming its way throughout the tree. Then when the brackets do appear, more and more of them may take form until the trunk is covered with them. They come in cream-white or green or red or brown, in many shapes and sizes; one of them resembles a horse's hoof and is the main ingredient in the manufacture of the sticks commonly known as punk, used to set off fireworks.

With the trees now divested of their leaves and vegetation in general less dense than it has been, many bird's nests become conspicuous on the branches and among the tangled undergrowth of roadside thickets and woodland borders. These now-deserted and empty cradles can become a facet of nature study, and all may be examined with the lens for insects and other invertebrates that may have found them desirable winter quarters.

With all its discomforts, winter has its compensations: what better panacea for the spirit and body than the exhilarating glow that comes from walking on the crusty snow, with the cold stinging one's face, to stop and watch the goldfinches feed in a windswept field or the chickadees as they search the red plumes of the sumac for hidden insects, or to hear the crows and blue jays break the frozen silence with their cries as they streak across the sky. Then to sit by the fire in the evening and watch the burning logs cast dancing shadows on the walls and send out flashing sparks in complete abandon as the wind sighs beneath the eaves and an owl calls far in the distance.

A LIST OF SUPPLY HOUSES

GLOSSARY

SELECTED BIBLIOGRAPHY

INDEX

A List of Supply Houses

General Biological Supply House, Inc.
8200 South Hoyne Avenue
Chicago, Illinois 60620

Ward's Natural Science Establishment, Inc.
3000 Ridge Road East
Rochester 9, New York 14622

Glossary

Abdomen	In mammals, that part of the body that extends from the diaphram to the pelvis; in insects, the hindermost of the three body regions.
Acute	Terminating with a sharp or well-defined angle.
Adult	A fully developed animal, especially an insect.
Adventitious Bud	A bud which develops in some other place than the axils of leaves or stem apices.
Aeciospore	A rust spore produced in an aecium.
Aecium	A cup-shaped structure within which one type of rust spore is produced.
Aftershaft	An accessory plume arising from the posterior side of the stem of the feathers of many birds.
Aggregate Fruit	A cluster of fruits developed from the ovaries of a single flower.
Akene	A small, dry, hard one-celled, one-seeded indehiscent fruit.
Albinism	A condition in which the normal pigment is missing.
Ambulacral Groove	A groove in the arm of a starfish containing rows of openings through which the tube feet are extended.
Ament	A catkin or scaly spike.
Ampulla	A small, bladder-shaped enlargement or water sac attached to the tube foot of a starfish.
Androconium	A modified scale that produces an odor. Found on the fore wings of males of certain butterflies.
Annulus	A row of specialized cells in the wall of a fern sporangium; the ring on the stipe of a mushroom.
Antenna	A movable sense organ on the heads of insects, myriopods, and crustaceans.
Anther	The pollen-bearing part of a stamen.
Antheridium	The organ in plants that produces sperms.
Anthocyanin	A blue, red, or purple pigment of plants.
Anus	The posterior opening of the digestive tract.
Apical	The end or outermost part; forming the apex.
Appendage	A portion of the body that projects and has a free end, as a limb.
Appressed	Lying close and flat against.
Archegonium	The organ in plants that produces eggs.

Arthropod	An invertebrate animal with jointed appendages.
Ascospore	A spore produced by sac fungi in asci.
Ascus	A saclike structure within which ascospores are formed.
Auricle	An angular or earlike lobe or process.
Awn	A bristle-shaped appendage.
Axil	The angle formed by a leaf or branch with the stem.
Axillary Bud	A bud borne in the axil of a leaf.
Banner	The upper petal of a papilionaceous flower.
Barb	One of the side branches of the shaft of a feather.
Barbicel	One of the small processes on the barbule of a feather.
Barbule	One of the processes along the edges of the barbs of a feather.
Basidiospore	A spore produced in a basidium.
Basidium	A club-shaped spore-producing structure.
Berry	A thin-skinned fleshy fruit with numerous scattered seeds.
Bladder	Any thin-walled sac enclosing a fluid.
Blade	The expanded portion of a leaf.
Blastostyle	In certain hydroids, a process that may be regarded as a zooid without mouth or tentacles whose function is to produce medusoid buds.
Book Lung	A lunglike saccular breathing organ containing numerous thin folds of membrane arranged like the leaves of a book.
Bract	A modified leaf associated with a flower or inflorescence.
Bractlet	A secondary bract, as one upon the pedicel of a flower.
Branchial	Pertaining to the gills.
Bud	A terminal or axillary structure on a stem consisting of a small mass of meristematic tissue covered, wholly or in part, by overlapping leaves; an expanded flower. Also a developing lateral branch of an organism.
Budding	The production of offspring by the development of a lateral branch from a part of the body.
Bud Scale	A specialized protective leaf of a bud.
Bundle Scar	Scar left in a leaf scar at the time of leaf fall by the breaking of the vascular bundles passing from a stem into a petiole.
Button	An immature mushroom.
Byssus	A tuft of long, tough filaments which are secreted by a gland in the groove of the foot and issue between the valves of certain mollusks.
Calamus	The quill of a feather.
Calyptra	A thin hoodlike covering of the capsule of a moss.
Calyx	A collective term for the sepals of a flower.
Capillitium	A delicate, branched network in the sporangia of slime fungi.
Capitate	Shaped like a head or with a distinct knob at the end.
Capsule	A dry, dehiscent fruit composed of more than one carpel.
Carapace	The dorsal shell or shield of certain animals.

Carbohydrate	An organic compound composed of carbon, hydrogen, and oxygen, with the hydrogen and oxygen in the ratio of 2 to 1.
Carnivorous	Eating or living on other animals.
Carotene	A reddish-orange pigment found in plants.
Carpel	A floral organ which bears and encloses ovules; a simple pistil or one member of a compound pistil.
Caruncle	The horny tubercle on the snouts of embryonic turtles used in slitting the eggshells at hatching.
Caryopsis	A dry, one-seeded, indehiscent fruit with the seed coat and pericarp completely united.
Caterpillar	The elongated wormlike larva of certain insects, such as butterflies and moths.
Catkin	A spike inflorescence bearing staminate or pistillate apetalous (no petals) flower.
Caudal	Pertaining to the tail or posterior part of the body.
Caudals	The scales on the lower surface of the tail of a snake.
Cell	A small mass or unit of protoplasm surrounded by a cell membrane containing one or more nuclei; in plants the protoplasm is also surrounded by a wall.
Cephalothorax	The body division formed by the fusion of the head and thorax in some arthropods.
Cercus	A pair of short appendages at the end of the abdomen.
Chelicera	The anterior pair of appendages in spiders.
Chin Shields	The scales bordering the mental groove in snakes.
Chitin	A complex organic substance occurring in the exoskeleton of arthropods and some other animals.
Chloroplast	A body in the cytoplasm of a plant cell containing chlorophyll.
Chlorophyll	A green pigment in plants involved in photosynthesis.
Chordotontal	Pertaining to or designating certain organs of insects found in various parts of the body and believed to be auditory.
Chrysalis	A hard case containing the pupa stage of a butterfly.
Cilia	Microscopic hairlike, protoplasmic processes projecting from the free surface of certain cells and capable of vibration.
Cirri	Small, slender projections or appendages.
Clavate	Club-shaped or enlarged at the tip.
Clavola	A term applied to all parts of an antenna except the first and second segments.
Cleistogamous	Fertilized in the bud, without the opening of the flower.
Cleistothecium	A small closed ascocarp.
Climax	A relatively permanent biome which maintains itself with little change in a given region so long as there are no major changes in environmental conditions.
Cloaca	The common passageway at the posterior end of the body into which the intestine, kidneys, and genital organs discharge their products.
Club	The more or less enlarged distal segments of the antennae of certain insects.

Cocoon	A silken protective case about a mass of eggs, a larva, or a pupa of an insect.
Coenosarc	The inner, cellular part of a hydroid as distinguished from the outer surrounding perisarc.
Collateral Bud	An accessory bud at the side of the axillary bud.
Colorational Antigeny	Secondary sexual color differences.
Columella	A dome-shaped structure in the sporangia of bread mold and related fungi; a central mass of sterile tissue in the sporophytes of mosses and liverworts; also, the axis of a spiral snail shell.
Column	The united stamens and pistils in orchids.
Complete Flower	A flower that bears sepals, petals, stamens, and pistils.
Conical	Cone-shaped, round, and tapering to a point.
Conidiophore	A hypha that produces conidia.
Conidium	An asexual reproductive structure; usually produced in chains by the terminal portion of a hypha.
Contour Feather	One of the feathers that forms the general covering of a bird.
Convex	Curved outwardly.
Corbicula	The pollen basket of a bee.
Corm	A short, often globose, upright, underground stem that stores food.
Cornea	The outer, transparent layer of the eye.
Corolla	Collectively, the petals of a flower.
Corona	The ciliated disk of rotifers.
Corymb	A simple inflorescense in which the pedicels, growing along the peduncle, are of unequal length, those of the lowest flowers being longest, those of the upper flowers shortest.
Costa	The front margin of an insect's wing, considered the first vein.
Cotyledon	A food-digesting and food-storing part of an embryo.
Coxa	The basal joint of the leg of an insect.
Cremaster	A variously shaped process of the posterior end of many insect pupae, especially of the Lepidoptera.
Crochets	The hooks with which the prolegs of caterpillars are armed.
Crop	An enlarged portion of the anterior part of the digestive tract specialized for storage.
Cross-Pollination	The transfer of pollen from the stamen of a flower of one plant to the stigma of a flower of another plant.
Crown	The head of foliage in a tree or shrub.
Crozier	The young frond of a fern which is coiled in vernation.
Ctenoid Scale	A fish scale having a comblike margin.
Cubitus	The fifth of the main veins of an insect's wing.
Cuticle	A thin, noncellular outermost covering of an organism.
Cycloid Scale	A fish scale which is thin and not enameled and shows concentric lines of growth, without serrations on the margin.

Cyme	A flower cluster in which the growing apex ceases growth early, all its meristematic tissues being used up in the formation of an apical flower, with other flowers developing farther down on the axis, the youngest flower appearing farthest from the apex.
Cytoplasm	The protoplasm of a cell exclusive of the nucleus or of nuclear material.
Deciduous	Refers to plants that regularly lose their leaves each year, as opposed to those that retain their leaves for more than one year.
Dehiscent	Splitting or opening along definite seams when mature.
Dermis	The inner layer of the skin below the epidermis.
Dioecious	Having the male (staminate) and female (pistillate) flowers on separate plants.
Discal Cell	In Lepidoptera, a large cell near the base of the wing.
Dorsal	Pertaining or relating to the back of an organism or structure.
Down Feather	A soft feather without a shaft.
Downy	Covered with down or with pubescence or soft hairs.
Drupe	A simple fleshy fruit in which the inner wall of the ovary becomes hard and stony and encloses one or two seeds.
Drupelet	A diminutive drupe.
Egg	The female germ cell of a plant or animal.
Egg Cell	The female germ cell, or egg proper, exclusive of any envelopes derived from or consisting of other cells.
Elliptical	Oblong with rounded ends.
Elytra	The horny upper wings, or wing covers, of beetles.
Embryo	The stage in the development of an animal still contained in the egg or in the body of the mother; also, the rudimentary plant within the seed.
Endosperm	The food storage tissue in seeds.
Enzyme	An organic substance, produced by living protoplasm, that brings about chemical changes but which does not itself undergo any significant change.
Epicotyl	The upper part of the stem of an embryo above the cotyledons.
Epidermis	The outer cellular layer or layers covering the external surface of a metazoan; also, the surface layer of cells of leaves and other soft plant parts.
Epipharynx	A median lobe sometimes present on the under surface of the labrum.
Epiphragm	A membrane closing the aperture of the capsule in certain mosses; also, a membrane with which many mollusks close the shell opening.
Esophagus	The part of the digestive tract extending from the pharynx to the stomach.
Evergreen	Having green leaves all the year.
Facet	The external surface of a single ommatidium.

Fat	An organic compound, insoluble in water, composed of carbon, hydrogen, and oxygen, with proportionately less oxygen than in carbohydrates.
Fecula	Waste material or excrement voided by insects.
Femur	The third segment of an insect's leg; it is articulated to the trochanter and sometimes fused with it.
Fertilization	The union of a mature sperm cell with a mature egg cell.
Fibrovascular Bundle	A strand containing the conducting tissues or vessels in plants.
Fibula	A structure at the base of the front wing that extends back above the base of the hind wing, and is clasped over an elevated part of the hind wing.
Filament	The part of a stamen which supports the anther; also a threadlike row of cells.
Filamentous	Composed of threads.
Filiform	Threadlike.
Filoplume	A minute hairlike feather with a slender barbless shaft and an inconspicuous tuft of weak barbs and barbules at tip.
Flagellum	A long whiplike cytoplasmic process of a cell or single-celled animal, capable of vibration; similar to a cilium but longer.
Flame Cell	The terminal cell of an excretory tubule in the kidney of the flatworms.
Flower	The characteristic reproductive structure of angiosperms or flowering plants.
Follicle	A simple, dry, dehiscent fruit, producing several to numerous seeds and composed of one carpel which splits along one seam.
Foot	In certain plants, as the ferns, a portion of a young sporophyte which attaches the sporophyte to the gametophyte and often absorbs food from the latter.
Formicary	An ant's nest; the dwelling of a community of ants.
Frass	The refuse or excrement voided by insect larvae, especially leaf miners.
Frenulum	A single spine (males) or several (females) arising from the humeral angle of the hind wings and projecting over the elevated part of the hind wing.
Frond	The leaf of a fern.
Fruit	A matured ovary or cluster of matured ovaries; the seed-bearing product of a plant.
Funicle	The part of the clavola between the club and the ring joints.
Fusiform	Spindle-shaped; swollen in the middle, tapering toward each end.
Ganoid	A fish scale composed of an inner layer of bone and an outer layer of shining enamel; articulates to another scale edge-to-edge by a peglike process of one scale fitting into a recess or socket of another.
Gastral Cavity	A central chamber in sponges into which water from the excurrent canals enters.

Gemma	An asexual or vegetative outgrowth capable of growing into a new individual.
Gemma Cup	A structure that produces gemmae.
Gemmule	A capsule of sponge cells which live through the winter and from which a new colony develops in the spring.
Gene	A unit of heredity.
Geniculate	Abruptly bent; elbowed.
Gill	An organ for breathing the air in water.
Gizzard	A muscular part of the digestive tract for grinding ingested food.
Glume	A chafflike bract, specifically, one found at the base of a grass spikelet.
Grub	The larva of certain insects, such as beetles.
Hair	A slender, threadlike outgrowth of the epidermis.
Halter	A small knobbed appendage on each side of the thorax replacing the hind wing.
Hamulus	A hook or hooklike process.
Haustorium	A suckerlike structure which penetrates host tissues and cells and absorbs nourishment from them.
Head	The anterior body division of an animal; a dense cluster of sessile or nearly sessile flowers on a short axis or receptacle.
Head Shield	A scale on the head of a snake.
Hemelytron	A partially thickened fore wing of certain insects.
Hemoglobin	The red pigment in blood capable of carrying oxygen.
Herbaceous	Referring to an herb, a plant that does not develop much woody tissue.
Herbivorous	Eating or living on plants.
Hesperidium	A type of berry in which the outer fruit wall becomes leathery.
Hibernaculum	A case or covering for protection during the winter.
Hilum	Scar on a seed coat marking the point of attachment of the seed stalk to the seed.
Holdfast	The basal portion of a thallus that anchors it to a solid object in water.
Honeydew	A sweet honeylike secretion produced by certain insects.
Humeral	Pertaining to the anterior basal angle of an insect's wing.
Hydranth	An individual feeding polyp in a hydroid colony.
Hydroid	The sessile asexual generation of certain coelenterates.
Hymenium	The spore-bearing surface of certain fungi.
Hypha	One of the threadlike strands of the mycelium of a fungus.
Hypocotyl	The portion of an embryo stem below the cotyledons.
Hypodermal	Pertaining to the hypodermis, the cellular layer which lies beneath the cuticle and which secretes the chitinous cuticle of arthropods.
Hypopharynx	An appendage or membranous fold on the floor of the mouth of many insects.
Hyporhachis	The stem of the aftershaft of a feather.

Imperfect Flower	A flower which bears only stamens or only pistils.
Incomplete Flower	A flower which lacks one or more of the four kinds of floral organs.
Indehiscent	A fruit which does not split open along regular seams.
Inferior Umbilicus	An opening in the quill of a feather through which nutrients and pigments are supplied to the developing feather.
Inflorescence	A flower cluster.
Inquiline	An animal, especially an insect, that lives habitually in the nest or abode of another; a guest.
Instar	Any of the several stages in the life of an insect marked by molts.
Integument	The outer covering or an outer enveloping cell layer.
Intestine	Part of the digestive tract posterior to the stomach in which absorption occurs.
Involucre	The circle or collection of bracts surrounding a flower cluster or a single flower.
Isthmus	The zone connecting the two halves of a desmid.
Jugum	A lobelike process at the base of the fore wings overlapping the hind wings.
Keel	A ridge like the keel of a boat.
Kidney	An organ for the excretion of liquid nitrogenous wastes in vertebrates, loosely applied to analogous organs in other animals.
Labellum	The expanded sensitive tip of the proboscis.
Labial Palpus	The appendage attached to the lower lip of an insect.
Labium	The lower lip of an insect.
Labrum	The upper lip of an insect.
Lamellate	Composed of thin plates or lamellae.
Larva	The earlier stage in the life history of certain insects after hatching from the egg and before the pupal period.
Lateral Bud	An axillary bud.
Leaf	A lateral green expanded outgrowth of a stem.
Leaf Axil	The upper angle between a leaf petiole and the stem from which it grows.
Leaf Scar	The scar left on the twig after the fall of a leaf.
Legume	A simple, dry, dehiscent fruit formed of a single pistil and splitting along two sutures.
Lemma	The lower of the two bracts enclosing the flower in the grasses.
Lenticel	In plants, a pore through which exchange of gases takes place.
Lip	A fleshy lobe at the anterior end of the earthworm; also, each of the upper and lower divisions of a bilabiate corolla or calyx.
Lodicule	A minute, thin, translucent scale at the base of a grass flower.
Lophophore	A circular or horseshoe-shaped ridge bearing tentacles.
Loreal	Pertaining to the sides of the snout in snakes.

Madreporite	The strainerlike cover of the opening to the water-vascular system in the starfish.
Maggot	The wormlike legless larva of the fly.
Mandible	A strong, cutting mouthpart of arthropods.
Mastax	The muscular pharynx of a rotifer.
Maxilla	One of a pair of mouthparts of an insect.
Maxilliped	
Maxillary Palpus	A fingerlike jointed appendage on each maxilla. One of the first three pairs of thoracic appendages in crustaceans.
Media	The fourth of the principal veins in an insect's wing.
Medusa	The free-swimming sexual stage in the life history of a coelenterate.
Melanin	A dark brown or black pigment.
Melanism	An unusual development of a black or nearly black color in the skin, plumage, or pelage.
Membrane	A thin, pliable sheet or layer of cells or material secreted by cells which covers a surface or divides a space.
Membranous	Pertaining to or resembling a membrane.
Meniscus	The surface of a liquid column.
Meristematic Tissue	A tissue whose cells are capable of frequent division and thus are responsible for growth.
Mesothorax	The middle of the three thoracic divisions.
Metabolism	The sum of the chemical processes involved in the building up and breaking down of protoplasm.
Metatarsus	The first joint of the tarsus, next to the tibia.
Metathorax	The third division of the thorax.
Metazoa	All multicellular animals; all animals exclusive of the protozoa.
Micropyle	The small opening in the integument of an ovule or seed through which the pollen tube enters.
Midrib	The central or main rib of a leaf.
Mitochondria	Small, spherical or rodlike cytoplasmic structures in a cell which are associated with important metabolic reactions.
Moniliform	Resembling a string of beads.
Monoecious	Having staminate and pistillate flowers on the same plant.
Molting	The casting off or shedding of the outer layer of the exoskeleton of arthropods, or scale layer of reptiles, or plumage of birds.
Multiple Fruit	A cluster of matured ovaries produced by several flowers.
Mycelium	The mass of hyphae forming the vegetative body of fungi.
Naiad	The young of an aquatic insect with incomplete metamorphosis.
Nauplius	A young larval form of many crustaceans.
Nectar	A sweet liquid produced by flowers.
Nectary	A floral gland which secretes nectar.
Nematocyst	A stinging capsule found in coelenterates.
Nephridiopore	The external opening of a nephridium.

Nephridium	A tubular excretory structure found in certain invertebrates.
Neuration	The arrangement of the veins in an insect wing.
Nucleus	A spherical or ovoid protoplasmic body found in most cells and considered to control the activities of the cell.
Nut	An indehiscent, dry, one-seeded, hard-walled fruit, produced from a compound ovary.
Nymph	The young of an insect having direct metamorphosis.
Obtuse	Blunt or round at the end.
Ocellus	A simple type of eye found in many invertebrates, especially insects.
Olfactory	Pertaining to the sense of smell.
Ommatidium	An elongated, rodlike unit of the compound eye of an arthropod.
Omnivorous	Eating all kinds of food, both plant and animal.
Operculum	A structure that covers and protects the gills of fish or that serves to cover the opening of snail shells.
Organ	An association of tissues for the performance of a particular function.
Organelle	A specialized part within a single-celled animal differentiated to perform a certain function.
Osculum	The external opening of the central cavity through which water leaves a sponge.
Osmeterium	A protrusile forked process emitting a disagreeable odor, borne on the first thoracic segment of the larva of certain butterflies.
Ostiole	A small inhalant opening in a sponge.
Oval	A body or figure in the shape of the longitudinal section of an egg.
Ovate	Egg-shaped, or having an outline like that of an egg.
Ovary	In animals, the female reproductive organ in which the egg cells are formed; in plants, the basal part of the pistil within which seeds develop.
Oviduct	A duct through which the egg passes from the ovary.
Ovipositor	An egg-laying organ in insects.
Ovoid	Egg-shaped.
Ovule	A structure in the ovary of a flower which, when fertilized, can become a seed.
Palea	The upper bract with which the lemma encloses the flower.
Palpus	A fingerlike appendage of the mouthpart of an arthropod.
Panicle	A compound inflorescence which has several main branches with pedicellate flowers arranged along its axis.
Papilla	A small nipple-shaped projection or elevation.
Papilionaceous	Having a standard, wings, and keel, as in the peculiar corolla of the sweet pea.
Pappus	Any appendage or tuft of appendages forming a crown of various characters at the summit of the akene.

Paraphyses	Sterile, hairlike structures frequently associated with sex structures in certain plants, such as the fungi.
Parapodium	A flattened movable appendage of annelids.
Parasite	An animal or plant which lives on or within another animal or plant and from which it derives nourishment.
Parenchyma	In plants a tissue composed of thin-walled cells which often store food.
Pectinate	Comblike.
Pedicel	The second segment of an insect's antenna; the stalk of an individual flower of an inflorescence.
Peduncle	The stalk of a solitary flower, or the main stalk of an inflorescence.
Pepo	An indehiscent, fleshy, many-seeded fruit with a hard rind.
Perfect Flower	A flower which bears both stamens and pistils.
Perianth	The floral envelope: the calyx and corolla taken together.
Perisarc	The cuticular outer covering of a hydroid.
Peristome	The region around the mouth of a radically symmetrical animal; the outer lip of the aperture of a snail shell. In mosses the fringe of teeth that surrounds the opening of the capsule.
Petal	A floral leaf in the whorl between the stamens and sepals.
Petiole	A leaf stalk.
Pharynx	The anterior portion of the digestive tract extending from the mouth to the esophagus.
Pileus	The cap or umbrella-shaped part of a mushroom.
Pilose	Hairy—especially with soft hairs.
Pinna	One of the primary leaflets of a compound frond.
Pinnule	One of the leaflets of a divided pinna.
Pistil	The ovule-producing part of a flower, consisting of ovary, style, and stigma.
Pistillate	With pistils only; applied to a flower with a pistil or pistils but no stamens.
Planula	A ciliated, free-swimming larval form of coelenterates.
Plasmodium	The naked, protoplasmic mass of slime fungi.
Pleopod	An abdominal appendage in crustaceans.
Plumule	The part of a plant embryo from which the shoot develops.
Pod	A dry, dehiscent fruit splitting along two sutures.
Pollen Comb	Stiff hairs on the leg of a honeybee with which the insect gathers the pollen grains from its body.
Pollen Grain	The male reproductive tissue of flowering plants.
Pollinia	Coherent masses of pollen grains, often with a stalk bearing an adhesive disk.
Polyp	An invertebrate having typically a hollow, cylindrical body closed at one end, with a mouth and tentacles at the free oral end.
Polypide	An individual zooid of a polyzoan colony.
Pome	A fleshy fruit consisting of a ripened receptacle surrounding the ovary.

Powderdown Feather	A modified down feather that grows continuously and disintegrates at the end, producing a sort of powder.
Predatory	Preying on other animals.
Proboscis	The extended trunklike or beaklike mouthparts of an insect.
Proleg	A fleshy leg found on the abdominal segments of the larva of certain insects.
Prostomium	The anterior portion of the first segment of the annelids, overhanging the mouth.
Protein	A complex organic compound containing nitrogen, carbon, hydrogen, and oxygen, and often other elements. Made up of amino acids and constituting an essential part of protoplasm.
Prothallus	A tiny, delicate, heart-shaped structure which develops from the spore of a fern.
Prothorax	The first division of the thorax.
Protonema	A branching filamentous structure forming an early stage in the life history of a moss.
Protoplasm	The living substance in the cells of all plants and animals.
Protozoa	The group of one-celled animals.
Pubescence	An epidermal covering of soft, short hairs or down.
Pulvillus	A pad often covered with short hairs, or a cushionlike or suckerlike organ or process between the tarsal claws of an insect's leg.
Pupa	The resting stage between the larva and adult in the life of an insect.
Pycniospore	One of the conidia produced within a pycnium.
Pycnium	A special cavity or receptacle of varying form bearing conidia on its inner walls.
Quill	The part of the shaft of a feather which is webless.
Raceme	A simple inflorescence in which the flowers, each with its own pedicel, are spaced along a common, more or less elongated, axis.
Rachilla	The axis of a spikelet in grasses, specifically the floral axis as opposed to that of the spikelet.
Rachis	The part of a feather which bears the web; the elongated axis of a spike or raceme; and in a pinnately compound leaf, the extension of the petiole bearing the leaflets.
Radicle	The primary root of an embryo plant or seedling.
Radius	The third of the principal veins of an insect's wing.
Radula	The band of calcareous teeth in the pharynx of most mollusks.
Raphe	A ridge along the surface of a seed.
Raptorial	Adapted to seize prey.
Receptacle	The enlarged end of the stem to which the parts of a flower are attached.
Regeneration	The regrowth of a part of an organism lost or destroyed.
Rhizoid	In certain plants a hairlike appendage which penetrates the

	soil or other substratum, anchoring the plant and absorbing water and other substances.
Rhizome	A horizontal underground stem.
Ring Joints	In the antennae of certain insects, the proximal segment or segments of the clavola, being much shorter than the succeeding segments.
Rostellum	In orchids the apex of the column.
Rostrum	A beak or snout.
Samara	A dry, indehiscent one-seeded winged fruit.
Saprophyte	A plant which lives on dead or decaying organic matter.
Sapwood	Active tissue in the outer area of wood in a stem, consisting of the usually light-colored outermost annual rings.
Scale	In animals, a more or less flattened, rigid, and definitely circumscribed plate forming part of the external body covering; a rudimentary leaf serving to protect a bud before expansion, such as a bud scale.
Scape	The first or proximal segment in the antenna of an insect; a peduncle rising from the ground, naked or without proper foliage.
Schizocarp	A simple, dry, indehiscent fruit composed of two fused carpels which split apart at maturity, each part usually with one seed.
Scleroblast	One of the cells of a sponge, by which a spicule is formed.
Seed	A complete embryo plant protected by one or more seed coats.
Segment	One part of a metameric animal; one of the parts of a leaf or other like organ that is cleft or divided.
Self-fertiliza- *tion*	The transfer of pollen from the stamen to the stigma of the same flower or of another flower on the same plant.
Self-pollina- *tion*	See self-fertilization.
Seminal *Receptacle*	A saclike organ which receives and stores sperms after their release.
Semiplume	A down feather with a short shaft but lacking the interlocking mechanism of a contour feather.
Sepal	The outermost floral organ, usually green, which encloses the other parts of the flower in the bud.
Serrate	Having sharp teeth pointing forward; saw-toothed.
Sessile	Lacking a stalk; attached by the base without a stalk or stem.
Seta	In mosses and liverworts, the stalk that supports the capsule of the sporophyte; a bristle.
Setaceous	Bristlelike.
Shaft	The stem or midrib of a feather.
Silicle	A silique broader than it is long.
Silique	A simple, dry, dehiscent fruit developed from two fused carpels which separate at maturity, leaving a persistent partition between.
Sorus	A cluster of sporangia on a fern.

Spadix	A spike with a fleshy axis.
Spathe	A large bract or pair of bracts enclosing an inflorescence.
Sperm Cell	The male reproductive unit.
Sperm Duct	The duct carrying sperms from the testis.
Spicule	One of the small, solid structures that forms the skeleton of a sponge.
Spiderling	A tiny, immature spider, usually the form just emerged from the egg sac.
Spike	An inflorescence in which the sessile flowers are arranged on a more or less elongated common axis.
Spikelet	A unit of a grass inflorescence with one or more flowers and their bracts.
Spiracle	The external opening of the respiratory system in insects.
Sporangiophore	A structure which bears sporangia or spore cases.
Sporangium	A saclike structure that produces spores.
Sporophyte	The spore-producing phase in plants having an alternation of generations.
Spur	A hollow saclike or tubular extension of some part of a blossom, usually nectariferous; a structure on the leg of a honeybee for removing the pollen from the pollen basket.
Stadium	The period between molts in an insect.
Stamen	The pollen-producing organ of a flower.
Staminate	Having stamens only; applied to a flower with stamens but no pistils.
Standard	The upper dilated petal of a papilionaceous corolla.
Statoblast	A disclike bud in freshwater Bryozoa which lives through the winter.
Statocyst	An organ of equilibrium.
Statolith	A solid body within a statocyst.
Stem	The main ascending axis of a plant.
Sterigma	A stalk on which a basidiospore is borne.
Stigma	The part of the pistil, usually the apex, which receives pollen and on which pollen gains germinate.
Stipe	The stalk of a mushroom.
Stomach	A division of the digestive tract in which digestion occurs.
Stone Canal	A tube joining the madreporite with the ring canal in the starfish.
Stridulate	To make a chirping or creaking sound.
Strobilus	A conelike aggregation of sporophylls (spore-bearing leaves) on a stem axis.
Style	The usually attenuated portion of a pistil connecting the ovary and stigma.
Subcosta	The second of the principal veins of an insect's wings.
Substratum	The substance or base on which a plant grows.
Superior Umbilicus	A pit at the outer end of the calamus, at a point about where the web of the feather begins.

Superposed Bud	An accessory bud above the axillary bud.
Symbiosis	The living together of two different species or organisms to their mutual advantage.
Tactile	Pertaining to the sense of touch.
Tarsus	The foot; the jointed portion of an insect's leg beyond the tibia.
Tegmina	The toughened upper wings of grasshoppers, etc.
Teliospore	The winter spore of a rust.
Tentacle	A flexible, armlike extension from the body of many animals; used for grasping and movement.
Terminal Bud	The bud formed at the tip of a twig.
Thallus	A simple plant body not differentiated into roots, stems, and leaves.
Thigmotropism	The behavior response of an organism to contact.
Thorax	The second of the main divisions of an insect's body between the head and abdomen.
Thyrse	A form of mixed inflorescence in which the main axis is racemose and the secondary or later axes are cymose.
Tibia	The part of an insect's leg between the femur and tarsus.
Tissue	A group of cells of similar structure for the performance of a specialized function.
Trachea	A respiratory tube in air-breathing arthropods.
Tracheal Gills	External filaments connected with the tracheae of the body which form part of the respiratory system of some aquatic insect larvae.
Tracheole	A small respiratory tube in insects.
Trimorphous	Occurring under three forms.
Trochanter	The small joint of an insect's leg between the coxa and the femur.
Truncate	Ending abruptly, as if cut off transversely.
Tube Feet	Tubular organs of locomotion found in the starfish.
Tubercle	A small knoblike prominence or excrescence on some part of a plant or animal.
Twig	A young shoot or branch; a portion of a stem.
Tympanum	A thin, tense membrane covering an organ of hearing in the leg or other part of an insect.
Umbel	An inflorescence in which the stems of the flowers are of approximately the same length and grow from the same point.
Vacuole	A cavity in the cytoplasm of a cell containing a liquid and/or other products.
Valvate	Opening by valves.
Vane	The flat expanded part of a feather.

Velum An appendage on the leg of the honeybee that forms part of a circular comb; also the circular muscular locomotory membrane of certain coelenterates.

Venation The arrangement of the veins in a leaf or in an insect's wing.

Ventral Pertaining to the underside of the body.

Vitta An oil tube.

Volva A membranous sac enclosing the young sporophore of certain mushrooms. It is ruptured and usually remains as a cup at the base of the stipe.

Web See vane.

Wing An organ of aerial flight; one of the movable paired appendages by means of which certain animals are able to fly.

Xanthophyll A yellow pigment found in chloroplasts.

Zooecium The outer cuticular covering of a bryozoan.

Zooid One of the members of a hydroid or polyzoan colony.

Zoospore A spore able to swim by movements of cilia or flagella.

Selected Bibliography

Algae, Desmids, and Diatoms

HYLANDER, C. J. *The World of Plant Life.* New York: Macmillan Company; 1956.

MORGAN, A. H. *Field Book of Ponds and Streams.* New York: G. P. Putnam's Sons; 1930.

NEEDHAM, J. G., and P. R. NEEDHAM *A Guide to the Study of Freshwater Biology.* San Francisco: Holden-Day; 1962.

PRESCOTT, G. W. *How to Know the Freshwater Algae.* Dubuque, Iowa: William C. Brown; 1954.

SMITH, G. M. *The Freshwater Algae of the United States.* New York: McGraw-Hill; 1950.

WARD, H. B., and G. C. WHIPPLE *Freshwater Biology.* New York: John Wiley & Sons; 1959.

Animal Camouflage

PORTMAN, A. *Animal Camouflage.* Ann Arbor: University of Michigan Press; 1959.

Animal Navigation

CARTHY, J. D. *Animal Navigation.* New York: Charles Scribner's Sons; 1956.

LINCOLN, F. C. *Migration of Birds.* Washington, D.C.: GPO; 1950.

Animal Photoperiodism

BECK, S. D. *Animal Photoperiodism.* New York: Holt, Rinehart and Winston; 1963.

Animal Tracks

MURIE, O. J. *A Field Guide to Animal Tracks.* Boston: Houghton Mifflin; 1954.

Birds

ALLEN, A. A. *The Book of Bird Life.* Princeton, N.J.: Van Nostrand; 1961.

CHAPMAN, F. M. *Handbook of Birds of Eastern North America.* New York: Dover Publications; 1931.

FORBUSH, E. H., and J. B. MAY *Natural History of the Birds of Eastern and Central North America.* Boston: Houghton Mifflin; 1939.

HEADSTROM, B. R. *Birds' Nests.* New York: Ives Washburn; 1961.

HEADSTROM, B. R. *Birds' Nests of the West.* New York: Ives Washburn; 1951.

MORGAN, A. H. *Field Book of Animals in Winter.* New York: G. P. Putnam's Sons; 1939.

PETERSON, R. T. *A Field Guide to the Birds.* Boston: Houghton Mifflin; 1947.

WALLACE, G. J. *An Introduction to Ornithology.* New York: Macmillan; 1963.

Bryozoans

BUCHSBAUM, R. *Animals without Backbones.* Chicago: University of Chicago Press; 1948.

HEADSTROM, B. R. *Adventures with Freshwater Animals.* Philadelphia: Lippincott; 1964.

MINER, R. W. *Field Book of Seashore Life.* New York: G. P. Putnam's Sons; 1950.

MORGAN, A. H. *Field Book of Animals in Winter.* New York: G. P. Putnam's Sons; 1939.

MORGAN, A. H. *Field Book of Ponds and Streams.* New York: G. P. Putnam's Sons; 1930.

NEEDHAM, J. G., and P. R. NEEDHAM *A Guide to the Study of Freshwater Biology.* San Francisco: Holden-Day; 1962.

PRATT, H. S. *A Manual of the Common Invertebrate Animals.* Philadelphia: Blakiston; 1935.

WARD, H. B., and G. C. WHIPPLE *Freshwater Biology.* New York: John Wiley & Sons; 1959.

Butterflies

COMSTOCK, J. H. *An Introduction to Entomology.* Ithaca, N.Y.: Comstock Publishing Associates; 1940.

LUTZ, F. E.   *Field Book of Insects.* New York: G. P. Putnam's Sons; 1948.

KLOTS, A. B.   *A Field Guide to the Butterflies.* Boston: Houghton Mifflin; 1951.

Coelenterates
(Hydra, Sea Anemone, Obelia)

BUCHSBAUM, R.   *Animals without Backbones.* Chicago: University of Chicago Press; 1948.

HEADSTROM, B. R.   *Adventures with Freshwater Animals.* Philadelphia: Lippincott; 1964.

MINER, R. W.   *Field Book of Seashore Life.* New York: G. P. Putnam's Sons; 1950.

MORGAN, A. H.   *Field Book of Animals in Winter.* New York: G. P. Putnam's Sons; 1939.

MORGAN, A. H.   *Field Book of Ponds and Streams.* New York: G. P. Putnam's Sons; 1930.

PRATT, H. S.   *A Manual of the Common Invertebrate Animals.* Philadelphia: Blakiston; 1935.

WARD, H. B., and
G. C. WHIPPLE   *Freshwater Biology.* New York: John Wiley & Sons; 1959.

Crustaceans

BUCHSBAUM, R.   *Animals without Backbones.* Chicago: University of Chicago Press; 1948.

HEADSTROM, B. R.   *Adventures with Freshwater Animals.* Philadelphia: Lippincott; 1964.

MINER, R. W.   *Field Book of Seashore Life.* New York: G. P. Putnam's Sons; 1950.

MORGAN, A. H.   *Field Book of Animals in Winter.* New York: G. P. Putnam's Sons; 1939.

MORGAN, A. H.   *Field Book of Ponds and Streams.* New York: G. P. Putnam's Sons; 1930.

NEEDHAM, J. G.,
and
P. R. NEEDHAM   *A Guide to the Study of Freshwater Biology.* San Francisco: Holden-Day; 1962.

PRATT, H. S.   *A Manual of the Common Invertebrate Animals.* Philadelphia; Blakiston; 1935.

WARD, H. B., and
G. C. WHIPPLE   *Freshwater Biology.* New York: John Wiley & Sons; 1959.

Ecology

BUCHSBAUM, R., and
   M. BUCHSBAUM   *Basic Ecology.* Pittsburgh: Boxwood Press; 1957.

STORER, J. H.   *The Web of Life.* New York: New American Library; 1961.

Ferns and Club Mosses

COBB, B.   *A Field Guide to the Ferns.* Boston: Houghton Mifflin; 1956.

DURAND, H.   *Field Book of Common Ferns.* New York: G. P. Putnam's Sons; 1949.

HYLANDER, C. J.   *The World of Plant Life.* New York: Macmillan; 1956.

Fishes

CURTIS, B.   *The Life Story of the Fish.* New York: Dover Publications; 1961.

HEADSTROM, R.   *Adventures with Freshwater Animals.* Philadelphia: Lippincott; 1964.

HUBBS, C. S., and
   K. F. LAGLER   *Fishes of the Great Lakes Region.* Bloomfield Hills, Mich.: Cranbrook Institute of Science; 1964.

LAGLER, K. F.   *Freshwater Fishery Biology.* Dubuque, Iowa: William C. Brown; 1956.

MORGAN, A. H.   *Field Book of Animals in Winter.* New York: G. P. Putnam's Sons; 1939.

MORGAN, A. H.   *Field Book of Ponds and Streams.* New York: G. P. Putnam's Sons; 1930.

NEEDHAM, J. G.,
   and
P. R. NEEDHAM   *A Guide to the Study of Freshwater Biology.* San Francisco: Holden-Day; 1962.

SCHRENKEISEN, R.   *Field Book of Freshwater Fishes of North America.* New York: G. P. Putnam's Sons; 1962.

WARD, H. B., and
   G. C. WHIPPLE   *Freshwater Biology.* New York: John Wiley & Sons, 1959.

Frogs, Toads, and Salamanders

CONANT, R.   *A Field Guide to Reptiles and Amphibians.* Boston: Houghton Mifflin; 1958.

HEADSTROM, B. R.   *Adventures with Freshwater Animals.* Philadelphia: Lippincott; 1964.

MORGAN, A. H.   *Field Book of Animals in Winter.* New York: G. P. Putnam's Sons; 1939.

MORGAN, A. H.   *Field Book of Ponds and Streams.* New York: G. P. Putnam's Sons; 1930.

NOBLE, G. K.   *The Biology of the Amphibia.* New York: Dover Publications; 1931.

OLIVER, J. A.   *The Natural History of North American Amphibians and Reptiles.* Princeton, N.J.: Van Nostrand; 1955.

WRIGHT, A. A., and A. H. WRIGHT   *Handbook of Frogs and Toads.* Ithaca, N.Y.: Comstock Publishing Associates; 1949.

Grasses

HYLANDER, C. J.   *The World of Plant Life.* New York: Macmillan; 1956.

POHL, R. W.   *How to Know the Grasses.* Dubuque, Iowa: William C. Brown; 1954.

Insects

BUCHSBAUM, R.   *Animals without Backbones.* Chicago: University of Chicago Press; 1948.

COMSTOCK, J. H.   *An Introduction to Entomology.* Ithaca, N.Y.: Comstock Publishing Associates; 1940.

FROST, S. W.   *Insect Life and Insect Natural History.* New York: Dover Publications; 1959.

HEADSTROM, B. R.   *Adventures with Insects.* Philadelphia: Lippincott; 1963.

LUTZ, F. E.   *Field Book of Insects.* New York: G. P. Putnam's Sons; 1948.

MORGAN, A. H.   *Field Book of Animals in Winter.* New York: G. P. Putnam's Sons; 1939.

MORGAN, A. H.   *Field Book of Ponds and Streams.* New York: G. P. Putnam's Sons; 1930.

NEEDHAM, J. G., and P. R. NEEDHAM   *A Guide to the Study of Freshwater Biology.* San Francisco: Holden-Day; 1962.

WARD, H. B., and G. C. WHIPPLE   *Freshwater Biology.* New York: John Wiley & Sons; 1959.

Lichens

NEARING, G. G.   *The Lichen Book.* (Published by the author, Ridgewood, N.J.) 1962.

Lizards

CONANT, R.   *A Field Guide to Reptiles and Amphibians.* Boston: Houghton Mifflin; 1958.

DITMARS, R. L.   *The Reptiles of North America.* New York: Doubleday; 1936.

OLIVER, J. A.   *The Natural History of North American Amphibians and Reptiles.* Princeton, N.J.: Van Nostrand; 1955.

SMITH, H. M.   *Handbook of Lizards.* Ithaca, N.Y.: Comstock Publishing Associates; 1946.

Mammals

ANTHONY, H. E.   *Field Book of North American Mammals.* New York: G. P. Putnam's Sons; 1928.

BURT, W. H., and R. P. GROSSENHEIDER   *Field Guide to the Mammals.* Boston: Houghton Mifflin; 1964.

CAHALANE, V. H.   *Mammals of North America.* New York: Macmillan; 1961.

HAMILTON, W. J.   *American Mammals.* New York: McGraw-Hill; 1939.

HAMILTON, W. J.   *The Mammals of Eastern United States.* Ithaca, N.Y.: Comstock Publishing Associates; 1943.

MORGAN, A. H.   *Field Book of Animals in Winter.* New York: G. P. Putnam's Sons; 1939.

Microscope

CORRINGTON, J. D.   *Exploring with Your Microscope.* New York: McGraw-Hill; 1957.

HEADSTROM, B. R.   *Adventures with a Microscope.* Philadelphia: Lippincott; 1941.

Mildews, Molds and Slime Fungi

HYLANDER, C. J.   *The World of Plant Life.* New York: Macmillan; 1956.

SMITH, G. M.   *Cryptogamic Botany.* New York: McGraw-Hill; 1938.

Minerals

LOOMIS, F. B.   *Field Book of Common Rocks and Minerals.* New York: G. P. Putnam's Sons; 1948.

POUGH, F. H.   *A Field Guide to the Rocks and Minerals.* Boston: Houghton Mifflin; 1953.

Mosses and Liverworts

CONRAD, H. S.   *How to Know the Mosses and Liverworts.* Dubuque, Iowa: William C. Brown; 1959.

HYLANDER, C. J.   *The World of Plant Life.* New York: Macmillan; 1956.

Mushrooms

SMITH, A. H.   *The Mushroom Hunter's Field Guide.* Ann Arbor: University of Michigan Press; 1964.

THOMAS, W. S.   *The Field Book of Common Gilled Mushrooms.* New York; G. P. Putnam's Sons; 1948.

Protozoans

BUCHSBAUM, R.   *Animals without Backbones.* Chicago: University of Chicago Press; 1948.

HEADSTROM, B. R.   *Adventures with Freshwater Animals.* Philadelphia: Lippincott; 1964.

JAHN, T. L.   *How to Know the Protozoa.* Dubuque, Iowa: William C. Brown Company; 1949.

MORGAN, A. H.   *Field Book of Ponds and Streams.* New York. G. P. Putnam's Sons; 1930.

NEEDHAM, J. G., and P. R. NEEDHAM   *A Guide to the Study of Freshwater Biology.* San Francisco: Holden-Day; 1962.

PRATT, H. S.   *A Manual of the Common Invertebrate Animals.* Philadelphia: Blakiston; 1935.

WARD, H. B., and G. C. WHIPPLE   *Freshwater Biology.* New York: John Wiley & Sons; 1959.

Rotifers

BUCHSBAUM, R.   *Animals without Backbones.* Chicago: University of Chicago Press; 1948.

HEADSTROM, B. R.   *Adventures with Freshwater Animals.* Philadelphia: Lippincott; 1964.

MINER, R. W.   *Field Book of Seashore Life.* New York: G. P. Putnam's Sons; 1950.

MORGAN, A. H.   *Field Book of Animals in Winter.* New York: G. P. Putnam's Sons; 1939.

MORGAN, A. H.   *Field Book of Ponds and Streams.* New York: G. P. Putnam's Sons; 1930.

NEEDHAM, J. G., and P. R. NEEDHAM   *A Guide to the Study of Freshwater Biology.* San Francisco: Holden-Day; 1962.

PRATT, H. S.   *A Manual of the Common Invertebrate Animals.* Philadelphia: Blakiston; 1935.

WARD, H. B., and G. C. WHIPPLE   *Freshwater Biology.* New York: John Wiley & Sons; 1959.

Shells

MORRIS, F. A. *A Field Guide to the Shells.* Boston: Houghton Mifflin; 1951.

Snails and Mussels (Mollusks)

BUCHSBAUM, R. *Animals without Backbones.* Chicago: University of Chicago Press; 1948.

HEADSTROM, B. R. *Adventures with Freshwater Animals.* Philadelphia: Lippincott; 1964.

MINER, R. W. *Field Book of Seashore Life.* New York: G. P. Putnam's Sons; 1950.

MORGAN, A. H. *Field Book of Animals in Winter.* New York: G. P. Putnam's Sons; 1939.

NEEDHAM, J. G.,
and
P. R. NEEDHAM *A Guide to the Study of Freshwater Biology.* San Francisco: Holden-Day; 1962.

PRATT, H. S. *A Manual of the Common Invertebrate Animals.* Philadelphia: Blakiston; 1935.

WARD, H. B., and
G. C. WHIPPLE *Freshwater Biology.* New York: John Wiley & Sons; 1959.

Snakes

CONANT, R. *A Field Guide to Reptiles and Amphibians.* Boston: Houghton Mifflin; 1958.

DITMARS, R. L. *The Reptiles of North America.* New York: Doubleday; 1936.

MORGAN, A. H. *Field Book of Animals in Winter.* New York: G. P. Putnam's Sons; 1939.

OLIVER, J. A. *The Natural History of North American Amphibians and Reptiles.* Princeton, N.J.: Van Nostrand; 1955.

SCHMIDT, K. D.,
and D. D. DAVIS *Field Book of Snakes.* New York: G. P. Putnam's Sons; 1941.

Spiders and Mites

COMSTOCK, J. H. *The Spider Book.* (Rev. ed. by W. J. Gertsch.) Ithaca, N.Y.: Comstock Publishing Associates; 1940.

EMERTON, J. H. *The Common Spiders of the United States.* New York: Dover Publications; 1961.

GERTSCH, W. J. *American Spiders.* Princeton, N.J.: Van Nostrand; 1949.

MORGAN, A. H. *Field Book of Animals in Winter.* New York: G. P. Putnam's Sons; 1939.

PRATT, H. S.   *A Manual of the Common Invertebrate Animals.* Philadel-
                       phia: Blakiston; 1935.
WARD, H. B., and
  G. C. WHIPPLE   *Freshwater Biology.* New York: John Wiley & Sons; 1959.

Sponges

BUCHSBAUM, R.   *Animals without Backbones.* Chicago: University of Chi-
                       cago Press; 1948.
HEADSTROM, B. R.   *Adventures with Freshwater Animals.* Philadelphia: Lip-
                       pincott; 1964.
MINER, R. W.   *Field Book of Seashore Life.* New York: G. P. Putnam's
                       Sons; 1950.
MORGAN, A. H.   *Field Book of Animals in Winter.* New York: G. P. Put-
                       nam's Sons; 1939.
MORGAN, A. H.   *Field Book of Ponds and Streams.* New York: G. P. Put-
                       nam's Sons; 1930.
PRATT, H. S.   *A Manual of the Common Invertebrate Animals.* Philadel-
                       phia: Blakiston; 1935.
WARD, H. B., and
  G. C. WHIPPLE   *Freshwater Biology.* New York: John Wiley & Sons; 1959.

Starfishes

BUCHSBAUM, R.   *Animals without Backbones.* Chicago: University of Chi-
                       cago Press; 1948.
MINER, R. W   *Field Book of Seashore Life.* New York: G. P. Putnam's
                       Sons; 1950.

Trees and Shrubs

HARLOW, W. M.   *Fruit Key and Twig Key to Trees and Shrubs.* New York:
                       Dover Publications; 1959.
HARRAR, E. S., and   *Guide to Southern Trees.* New York: Dover Publications;
  J. G. HARRAR   1962.
MATHEWS, F. S.   *Field Book of American Trees and Shrubs.* New York:
                       G. P. Putnam's Sons; 1915.
PETRIDES, G. A.   *A Field Guide to the Trees and Shrubs.* Boston: Houghton
                       Mifflin; 1958.
SARGENT, C. S.   *Manual of the Trees of North America.* New York: Dover
                       Publications; 1961.

Turtles

CONANT, R. — *A Field Guide to Reptiles and Amphibians.* Boston: Houghton Mifflin; 1958.

DITMARS, R. L. — *The Reptiles of North America.* New York: Doubleday; 1936.

MORGAN, A. H. — *Field Book of Animals in Winter.* New York. G. P. Putnam's Sons; 1939.

MORGAN, A. H. — *Field Book of Ponds and Streams.* New York: G. P. Putnam's Sons; 1930.

OLIVER, J. A. — *The Natural History of North American Amphibians and Reptiles.* Princeton, N.J.: Van Nostrand; 1955.

POPE, C. H. — *Turtles of the United States and Canada.* New York: Alfred A. Knopf; 1939.

Wild Flowers

GREENE, W. F., and H. L. BLOMQUIST — *Flowers of the South.* Chapel Hill, N.C.: University of North Carolina Press; 1953.

HYLANDER, C. J., and E. F. JOHNSTON — *The Macmillan Wildflower Book.* New York: Macmillan; 1954.

LEMMON, R. S., and C. C. JOHNSON — *Wildflowers of North America.* Garden City, N.Y.: Hanover House; 1961.

MATHEWS, F. S. — *Field Book of American Wildflowers.* (Rev. ed. by Norman Taylor.) New York: G. P. Putnam's Sons; 1955.

Worms

BUCHSBAUM, R. — *Animals without Backbones.* Chicago: University of Chicago Press; 1948.

HEADSTROM, B. R. — *Adventures with Freshwater Animals.* Philadelphia: Lippincott; 1964.

MINER, R. W. — *Field Book of Seashore Life.* New York: G. P. Putnam's Sons; 1950.

MORGAN, A. H. — *Field Book of Animals in Winter.* New York: G. P. Putnam's Sons; 1939.

MORGAN, A. H. — *Field Book of Ponds and Streams.* New York: G. P. Putnam's Sons; 1930.

NEEDHAM, J. G., and P. R. NEEDHAM — *A Guide to the Study of Freshwater Biology.* San Francisco: Holden-Day; 1962.

PRATT, H. S. — *A Manual of the Common Invertebrate Animals.* Philadelphia: Blakiston; 1935.

WARD, H. B., and G. C. WHIPPLE — *Freshwater Biology.* New York: John Wiley & Sons; 1959.

Index

A NOTE ON THE TYPE

The text of this book was set on the linotype in Caslon Old Face, a modern adaptation of a type designed by the first William Caslon (1692-1766) of the famous English family of type designers and founders. Its characteristics are remarkable regularity and symmetry, as well as beauty in the shape and proportion of the letters; its general effect is clear and open, but not weak or delicate. For uniformity, clearness, and readability it has perhaps never been surpassed.

The Caslon face has had two centuries of ever-increasing popularity in the United States—it is of interest to note that the first copies of the Declaration of Independence and the first paper currency distributed to the citizens of the new-born nation were printed in this type face.

This book was composed by The Haddon Craftsmen, Inc., Scranton, Pennsylvania; printed by Halliday Lithograph Corporation, West Hanover, Massachusetts; and bound by The Book Press Incorporated, Brattleboro, Vermont. Typography and binding design by Betty Anderson.